Biomedical Polymers and
Polymer Therapeutics

Biomedical Polymers and Polymer Therapeutics

Edited by

Emo Chiellini

University of Pisa
Pisa, Italy

Junzo Sunamoto

Niihama National College of Technology
Niihama, Japan

Claudio Migliaresi

University of Trento
Trento, Italy

Raphael M. Ottenbrite

Virginia Commonwealth University
Richmond, Virginia

and

Daniel Cohn

The Hebrew University of Jerusalem
Jerusalem, Israel

Kluwer Academic / Plenum Publishers
New York, Boston, Dordrecht, London, Moscow

Library of Congress Cataloging-in-Publication Data

Biomedical polymers and polymer therapeutics/edited by Emo Chiellini.
    p.   cm.
    "Proceedings of the Third International Symposium on Frontiers in Biomedical
Polymers including Polymer Therapeutics: From Laboratory to Clinical Practice, held
May 23–27, 1999, in Shiga, Japan"—T.p. verso.
    Includes bibliographical references and index.
    ISBN 0-306-46472-1
    1. Polymers in medicine—Congresses. 2. Polymeric drug delivery systems—Congresses.
I. Chiellini, Emo. II. International Symposium on Frontiers in Biomedical Polymers
including Polymer Therapeutics: From Laboratory to Clinical Practice (3rd: 1999:
Shito, Japan)

    R857 .P6 B564 2001
    610′.28—dc21

                                                           00-052736

Proceedings of the Third International Symposium on Frontiers in Biomedical Polymers including Polymer
Therapeutics: From Laboratory to Clinical Practice, held May 23–27, 1999, in Shiga, Japan

ISBN 0-306-46472-1

©2001 Kluwer Academic / Plenum Publishers, New York
233 Spring Street, New York, N.Y. 10013

http://www.wkap.nl/

10 9 8 7 6 5 4 3 2 1

A C.I.P. record for this book is available from the Library of Congress

Preface

This book is the third of a series that focuses on the progress and unique discoveries in an interdisciplinary scientific and technological area of biomedical application of polymers. The topics include polymeric materials for biomedical and pharmaceutical applications, as well as polymeric materials in therapeutics. The chapters in this book are based on the presentations made at the *Third International Symposium on Frontiers in Biomedical Applications* in conjunction with *the Polymer Therapeutics Symposium* that was held in Shiga (Lake Biwa) Japan on May 23–27, 1999.

The goal of the Joint Symposium was to gather scientists and practitioners with different backgrounds and expertise in a unique forum. The broad scope of presentations and debate of the multiform aspects connected to the design and formulation of new polymeric materials were aimed at implementing novel developments in Life Science and Medicine. The achievement of tangible results capable of meeting the expectations and the needs to prevent and provide adequate remedies for various human diseases, constituted the ultimate goal of the studies presented.

There are actually two related fields using polymers for bioapplications. One is comprised of implant materials, which may be permanent prostheses, such as hip joints and tendons, or temporary implants made of bioabsorbable materials for sutures and splints for bone fractures. In this area, tissue engineering constitutes an emerging branch in which the polymeric materials play a primary scaffolding role for the sustainable cell growth and their designed superstructural assembly. The other major field is polymeric drugs and drug delivery systems, including gene therapy. Polymers for this application have to be not only biocompatible but provide optimal drug bioavailability.

The prospects for both applications have been profound. The replacement of hip and knee joints are now common procedures, with a number of patients enjoying freedom from pain, as well as freedom of mobility. In the area of drug delivery, new technologies are being developed, such as mechanisms for systemic drug transports as well as transport across compartmental membranes; protein and macromolecular delivery, and pharmacokinetics and pharmacodynamics of controlled drug delivery, as well as targeting and sitespecific delivery. Now new developments in cellular uptake of macromolecular species and intercellular enzymatic interactions on macromolecular drugs and drug carriers have progressed to the point that significant clinical results are being realized.

The Editors wish to thank all the contributors for their outstanding contributions and for their effort and patience in helping to compile this publication that was more time and energy consuming than expected. A specific personal appreciation is due to the members of the International Scientific Advisory Board of the 3rd International Symposium on Frontiers in Biomedical Polymers Including Polymer Therapeutics, R. Duncan, Y. Maeda and H. Ringsdorf who kindly supported this Joint Symposium. Finally a most grateful thank you and recognition of the following institutions and industries who, with their financial contributions, helped to make the Symposium pleasant and scientifically rewarding: University of London, School of Pharmacy; Johanes-Gutenberg University, Biodynamics Research Foundation; The Kao Foundation for Arts and Sciences; Nagase Science and Technology Foundation; The Naito Foundation; Suntory Institute for Biorganic Research; Terumo Life Science Foundation; The Pharmaceutical Manufacturers' Association of Tokyo; Osaka Pharmaceutical Manufacturers' Association; Daikin Industries Ltd; Daito Chemix Corporation.

Emo Chiellini, Daniel Cohn, Claudio Migliaresi,
Raphael Ottenbrite, and Junzo Sunamoto

Acknowledgments

Emo Chiellini, on behalf of all the other Editors, wishes to acknowledge the continuous and tireless dedication provided by Ms. Maria Viola in handling all the correspondence with the contributors to the book and also the publisher's contact person, Ms. Joanna Lawrence.

Maria Viola spent a lot of effort in setting up the format of the various manuscripts and without her irreplaceable help, it would have been impossible to produce the final manuscript for this book.

Contents

Chapter 3

**APPLICATION TO CANCER CHEMOTHERAPY OF
SUPRAMOLECULAR SYSTEM**
*Katsuro Ichinose, Masayuki Yamamoto, Ikuo Taniguchi, Kazunari Akiyoshi,
Junzo Sunamoto, and Takashi Kamematsu*

Chapter 4

DDS IN CANCER CHEMOTHERAPY
Yasuhiro Matsumura

Chapter 5

CELLULOSE CAPSULES – AN ALTERNATIVE TO GELATIN
Shunji Nagata

Chapter 6

POLYMERIC HYDROGELS IN DRUG RELEASE
Federica Chiellini, Federica Petrucci, Elisabetta Ranucci, and Roberto Solaro

Chapter 7

BIODEGRADABLE POLYROTAXANES AIMING AT BIOMEDICAL AND PHARMACEUTICAL APPLICATIONS
Tooru Ooya and Nobuhiko Yui

Chapter 8

LIPOSOMES LINKED WITH TIME-RELEASE SURFACE PEG
Yuji Kasuya, Keiko Takeshita, and Jun'ichi Okada

Chapter 9

CARRIER DESIGN: INFLUENCE OF CHARGE ON INTERACTION OF BRANCHED POLYMERIC POLYPEPTIDES WITH PHOSPHOLIPID MODEL MEMBRANES
Ferenc Hudecz, Ildikó B. Nagy, György Kóczàn, Maria A. Alsina, and Francesca Reig

Chapter 10

**BLOCK COPOLYMER-BASED FORMULATIONS OF
DOXORUBICIN EFFECTIVE AGAINST DRUG RESISTANT
TUMOURS**
Valery Alakhov and Alexander Kabanov

Chapter 11

**PREPARATION OF POLY(LACTIC ACID)-GRAFTED
POLYSACCHARIDES AS BIODEGRADABLE AMPHIPHILIC
MATERIALS**
Yuichi Ohya, Shotaro maruhashi, Tetsuya Hirano, and Tatsuro Ouchi

Chapter 12

**PHOSPHOLIPID POLYMERS FOR REGULATION OF
BIOREACTIONS ON MEDICAL DEVICES**
Kazuhiko Ishihara

Chapter 13

RELATIVE POLYCATION INTERACTIONS WITH WHOLE
BLOOD AND MODEL MEDIA
*Dominique Domurado, Élisabeth Moreau, Roxana Chotard-Ghodsnia,
Isabelle Ferrari, Pascal Chapon, and Michel Vert*

Chapter 14

INFLUENCE OF PLASMA PROTEIN- AND PLATELET ADHESION
ON COVALENTLY COATED BIOMATERIALS WITH MAJOR
SEQUENCES OF TWENTY REGIOSELECTIVE MODIFIED N- OR
O-DESULPHATED HEPARIN/HEPARAN DERIVATIVES
H. Baumann and M. Linßen

Chapter 15

PREPARATION AND SURFACE ANALYSIS OF HAEMO-
COMPATIBLE NANOCOATINGS WITH AMONO- AND
CARBOXYL GROUP CONTAINING POLYSACCHARIDES
*H. Baumann, V. Faust, M. Hoffmann, A. Kokott, M. Erdtmann, R. Keller,
H.-H. Frey, M. Linßen, and G. Muckel*

Chapter 16

Chapter 17

Chapter 18

Chapter 19

Chapter 20

TEMPERATURE-CONTROL OF INTERACTIONS OF THERMO-SENSITIVE POLYMER MODIFIED LIPOSOMES WITH MODEL MEMBRANES AND CELLS

Kenji Kono, Akiko Henmi, Ryoichi Nakai, and Toru Takagishi

PART 2. POLYMERS IN DIAGNOSIS AND VACCINATION

Chapter 21

THE USE OF POLYCHELATING AND AMPHIPHILIC POLYMERS IN GAMMA, MR AND CT IMAGING

Vladimir P. Torchilin

Chapter 22

CONJUGATION CHEMISTRIES TO REDUCE RENAL RADIOACTIVITY LEVELS OF ANTIBODY FRAGMENTS

Yasushi Arano

Chapter 23

SYNTHETIC PEPTIDE AND THEIR POLYMERS AS VACCINES
*D. Jackson, E. Brandt, L. Brown, G. Deliyannis, C. Fitzmaurice, S. Ghosh,
M. Good, L. Harling-McNabb, D. Dadley-Moore, J. Pagnon, K. Sadler, D.
Salvatore, and W. Zeng*

Chapter 24

**DEVELOPMENT OF FUSOGENIC LIPOSOMES AND ITS
APPLICATION FOR VACCINE**
Akira Hayashi and Tadanori Mayumi

Chapter 25

**A NOVEL HYDROPHOBIZED POLYSACCHARIDE/
ONCOPROTEIN COMPLEX VACCINE FOR HER2 GENE
EXPRESSING CANCER**
Hiroshi Shiku, Lijie Wang, Kazunari Akiyoshi, and Junzo Sunamoto

PART 3. POLYMERS IN GENE THERAPY

Chapter 26

PHARMACEUTICAL ASPECTS OF GENE THERAPY

Chapter 27

HIGH-MOLECULAR WEIGHT POLYETHYLENE GLYCOLS CONJUGATED TO ANTISENSE OLIGONUCLEOTIDES
Gian Maria Bonora

Chapter 28

PHOTODYNAMIC ANTISENSE REGULATION OF HUMAN CERVICAL CARCINOMA CELL
Akira Murakami, Asako Yamayoshi, Reiko Iwase, Jun-ichi Nishida, Tetsuji Yamaoka, and Norio Wake

Chapter 29

IN VITRO GENE DELIVERY BY USING SUPRAMOLECULAR SYSTEMS
Toshihori Sato, Hiroyuku Akino, Hirotake Nishi, Tetsuya Ishizuka, Tsuyoshi Ishii, and Yoshio Okahata

MOLECULAR DESIGN OF BIODEGRADABLE DEXTRAN HYDROGELS FOR THE CONTROLLED RELEASE OF PROTEINS

W.E. Hennink, J.A. Cadée, S.J. de Jong, O. Franssen, R.J.H. Stenekes, H. Talsma, and W.N.E. van Dijk-Wolthuis
Faculty of Pharmacy, Department of Pharmaceutics, Utrecht Institute for Pharmaceutical Sciences (UIPS), University of Utrecht, P.O. Box 80 082, 3508 TB Utrecht, The Netherlands

1.   INTRODUCTION

Recently, biotechnology has made it possible to produce large quantities of pharmaceutically active proteins. These proteins can be used for prophylactic (e.g. vaccines) as well as for therapeutic purposes (e.g. treatment of cancer or vascular diseases)[1]. However, there is a number of problems associated with the use of these protein-based drugs. Frist, the stability of the drug has to be ensured during manufacturing and storage of the pharmaceutical formulation[1-4]. Second, oral administration is not possible due to the low pH and proteolytic activity in the gastrointestinal tract[5]. Third, in general, proteins have a short half-life after parenteral administration (e.g. intravenous injection), which makes repeated injections or continuous infusion of the protein necessary to obtain a therapeutic effect1. To overcome one or more of these problems, a large number of delivery systems have been designed and evaluated for the release of pharmaceutically active proteins during the last 15 years[1,6-13]. Among these systems, those based on polymers are the most successful so far[13].

Until now, the most popular polymers for the preparation of drug delivery systems are poly(lactic acid) (PLA) and its co-polymers with glycolic acid, poly(lactic-co-glycolic acid) (PLGA)[11,12,15]. The PL(G)A system is not especially designed for the controlled release of proteins and has some inherent drawbacks for this purpose. First, organic solvents have to be used to encapsulate proteins in PL(G)A microspheres[12,16-19]. Second, acidic products are formed during degradation, which might result in a low pH inside the device[20-23]. Both a low pH and organic solvents are known to affect protein stability[3,24,25]. Furthermore, it appears to be difficult to control the protein release from these matrices, and a burst release is hard to avoid[17].

· Hydrogels are based on hydrophilic polymers, which are crosslinked to prevent dissolution in water. The crosslink density is an important factor that determines the water content of the gels at equilibrium swelling. Because hydrogels can contain a large amount of water, they are interesting devices for the delivery of proteins[26-28]. First, the hydrated matrix results in good compatibility with proteins[29,30] as well as living cells and body fluids[31]. Second, many parameters, such as the water content, the amount of crosslinks and possible protein-matrix interactions, can be used to control the release of proteins from hydrogels,. These parameters can be made time-dependent by swelling and/or degradation of the gel.

A variety of water-soluble polymers has been used for the preparation of hydrogels. This includes polymers of (semi) natural origin, such as proteins[26] and polysaccharides[28], but also synthetic polymers, like block copolymers of poly(ethylene glycol) and PLA[32], poly(vinyl alcohol)[33] or poly(2-hydroxy-ethyl methacrylate-co-glycoldimethacrylate)[28,34].

Hydrogels are prepared from these polymers by both chemical (covalent) and physical (non-covalent) crosslinking[26-28]. In our Department, biodegradable dextran hydrogels have been designed and evaluated as systems for the controlled release of proteins. This contribution gives an overview of the synthesis and characterization of degradable dextran hydrogels, and their application as protein releasing matrices.

2.   PROTEIN RELEASE FROM NON-DEGRADING DEXTRAN HYDROGELS

Dextran is a bacterial exo-polysaccharide which consists essentially of α-1,6-linked D-glucopyranose residues with a low percentage of α-1,2 and 1,3 side chain[35] (Fig.1).

The low molecular weight fractions of dextran, between 40 and 110 kDa, have been used as plasma expander for several decades. Dextran hydrogels can be obtained by several ways of crosslinking. Single-step chemically crosslinked dextran hydrogels can be obtained by reaction with epichlorohydrin or diisocyanates[36]. Since crosslinking is performed with highly reactive crosslinkers and at a high pH, the gels have to be loaded after preparation. This leads to a low entrapment of the protein and a rapid release. Two-step crosslinking has been investigated with (meth)acrylated dextrans. (Meth)acrylation has been achieved by coupling glycidyl-(meth)acrylate to dextran in water[29,37]. However, the degree of substitution was low and additional crosslinkers were required to obtain gels with sufficient mechanical strength[29].

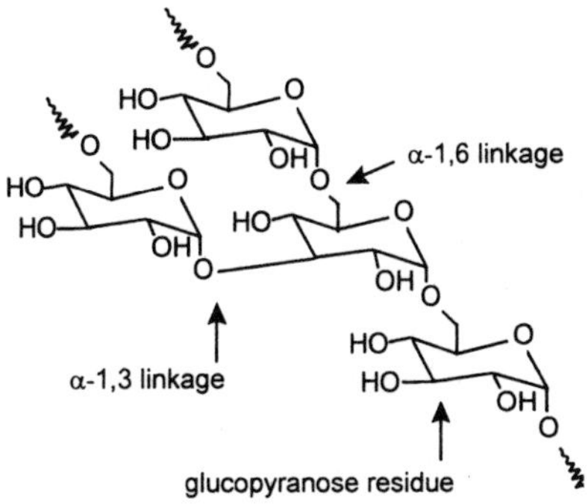

Figure 1. The chemical structure of dextran.

In our procedure, methacrylated dextran (dex-MA) is obtained by reacting glycidyl methacrylate (GMA) with dextran in a suitable aprotic solvent (dimethylsulfoxide, (DMSO)), and 4-(N,N-dimethylamino)pyridine (DMAP) as a catalyst[38,39] (Fig.2).

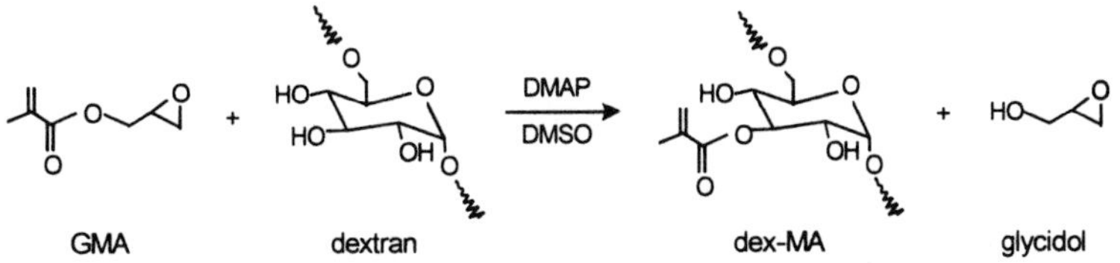

Figure 2. Reaction scheme for the synthesis of dex-MA

The methacrylate groups are directly coupled to the dextran chain, in a ratio of 1:1 on the 2 and 3 position[39]. This route yields dex-MA of which the degree of substitution (DS, the number of methacrylates per 100 gluco-pyranose residues) can be fully controlled.

After dissolution of dex-MA in water, a dextran hydrogel is obtained after the addition of an initiator system consisting of potassium peroxodisulfate (KPS) and N,N,N',N'-tetramethylethylenediamine (TEMED) (Fig.3).

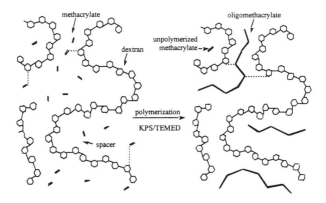

Figure 3. Schematic representation of the formation of dextran hydrogels.

The release of three model proteins with different molecular weights (lysozyme (M=14 kDa), albumin (M=66 kDa), immunoglobulin G (IgG; M=155 kDa)) from dex-MA gels with a varying initial water content and DS was studied[40]. Protein loaded dextran hydrogels were prepared by polymerization of an aqueous solution of dex-MA (varying concentration) and the protein. The reaction mixture was polymerized in a syringe (radius 0.23 cm, length 1 cm) yielding a cylinder shaped gel. It appeared that the cumulative release was proportional to the square root of time up to 60 % protein release. This indicates that the release is controlled by a Fickian diffusion process. Some representative release profiles are given in Figures 4 and 5.

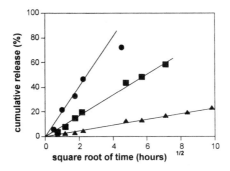

Figure 4. Cumulative release of lysozyme (●), BSA (■) and IgG (▲) from a dextran hydrogel (water content 80 % (w/w); DS 4).

Figure 4 shows that, as expected, the release rate decreased with an increasing protein size. For a given protein, the initial water content of the dextran hydrogel has a large effect on the release rate (Fig.5).Further, a decreasing water content of the gel resulted in a decreasing release rate. For gels with a high equilibrium water content (Fig.4 and 5) the diffusion of proteins in these highly hydrated gels could be effectively described by the free volume theory. On the other hand, the release of the proteins from hydrogels with a low hydration level was marginal and did not follow the free volume theory, indicating that in these gels screening occurred[40].

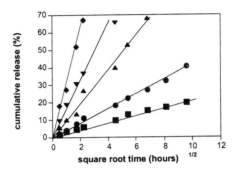

Figure 5.   Cumulative release (%) of lysozyme from dextran hydrogels (DS 9) varying in water content: 90% w/w (◆), 80% w/w (▼), 70% w/w (▲), 60% w/w (●), 50% w/w (■).

3.   PROTEIN RELEASE FROM ENZYMATICALLY DEGRADING DEXTRAN HYDROGELS

Although dex-MA gels contain methacrylate esters in their crosslinks, the hydrolysis of these groups is very slow under physiological conditions. To obtain a degrading system, we incorporated dextranase in the gel[41]. The enzyme used (isolated from Penicillium Funiculosum) is an endodextranase which hydrolyzes α–1,6 glycosidic linkages[42]. The hydrogel degradation rate was quantified by determination of the release of reducing oligosaccharides, which are products of the degradation process. Figure 6 gives a representative plot of the release of reducing oligosaccharides from degrading dextran hydrogels which contained a varying concentration of dextranase. From this figure it appears that the initial degradation rate (defined as the slope of lines in Fig.6) is proportional to the enzyme concentration[41].

We demonstrated by electrospray mass spectrometry that even at high degree of substitution, non-crosslinked dex-MA was fully degradable with dextranase[43]. On the other hand, the degradation of dex-MA hydrogels catalyzed by dextranase strongly decreased with an increasing crosslink density[41,44]. Gels derived from dex-MA with a high degree of MA substitution (DS>17) did not show signs of degradation, even at high concentrations of dextranase incorporated.

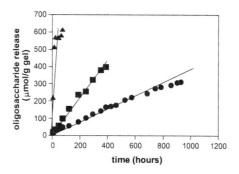

Figure 6. The cumulative release of reducing oligosaccharides from dex-MA hydrogels (DS 4, initial water content 70% w/w) containing dextranase (0.03 U/g gel (●), 0.1 U/g gel (■), 1 U/g gel (▲).

The release of a model protein (IgG) from degrading gels containing varying concentrations of dextranase was studied (Fig.7). In the absence of dextranase, only a marginal release of IgG was found (less than 10 % in 1000 hours). This indicates that the diameter of IgG (10 nm) is larger than the hydrogel mesh size. In the presence of a high concentration of dextranase, however, a fast release of IgG was observed. At lower dextranase concentrations, the observed release profiles were initially the same as for the gel in which no dextranase was incorporated. However, after a certain incubation time an abrupt increase in the release of IgG was found. It is likely that an abrupt increase in release occurs when the hydrogel is degraded to an extent at which the hydrogel mesh size equals the protein diameter. The time to reach this point is dependent on, as can be expected, the amount of dextranase present in the gel. Interestingly, the release of dextranase from the gels was slow as compared with the release of a protein with about the same size (ovalbumin). Most likely, the diffusivity of dextranase in dextran hydrogels is low due to binding (immobilization) of the enzyme to the hydrogel matrix [41].

4.   PROTEIN RELEASE FROM CHEMICALLY DEGRADING DEXTRAN HYDROGELS

In dex-MA (Fig.2) methacrylate esters are present which are sensitive to hydrolysis[45]. However, in hydrogels derived from dex-MA no significant hydrolysis of ester group occurs, even at extreme conditions (pH and temperature)[46]. As shown in the previous section, dex-MA can be rendered

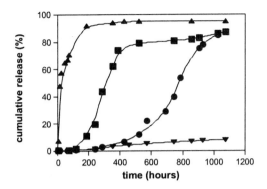

Figure 7. The cumulative release of IgG from dextran hydrogels (DS 4, initial water content 70% w/w) containing dextranase ((▼), 0.03 U/g gel (●), 0.1 U/g gel (■), 1 U/g gel (▲)).

degradable by the incorporation of dextranase (Fig.6), which is also an effective route to modulate the release of IgG from these matrices (Fig.7). An alternative approach to degradable, interpenetrating networks of dextran and oligomethacrylate, is the incorporation of hydrolytically labile spacers between the polymerized methacrylate groups and dextran. It has been reported that introduction of lactate esters induces degradability under physiological conditions in hydrogels derived from acrylated poly(ethylene glycol)-polylactate block-copolymers[47]. To develop hydrogels which degrade under physiological conditions, a new class of polymerizable dextrans with hydrolyzable groups was rationally designed by us[48].

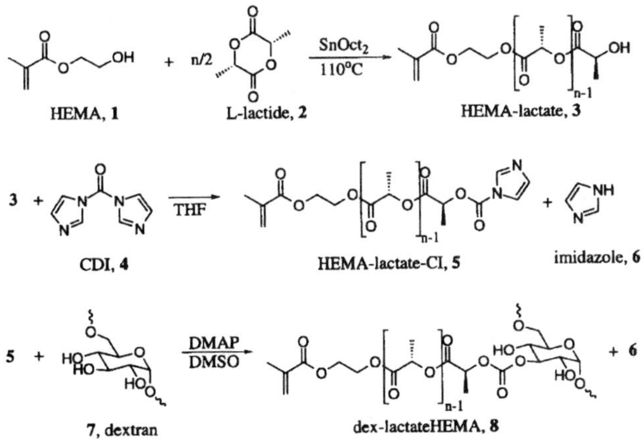

Figure 8. Reaction scheme for the synthesis of dex-lactateHEMA.

First, L-lactide (**2**) is grafted onto HEMA (**1**), yielding HEMA-lactate (**3**). After activation with *N,N'*-carbonyldiimidazole (CDI, **4**), the resulting

HEMA-lactate-CI (**5**) is coupled to dextran (**7**) to yield dex-lactateHEMA (**8**) (Fig.8). A comparable dextran derivative without lactate spacer between the methacrylate ester and dextran was also synthesized. For this compound, HEMA was activated with CDI, and the resulting HEMA-CI (**9**) is then coupled to dextran, yielding dexHEMA (**10**) (Fig.9).

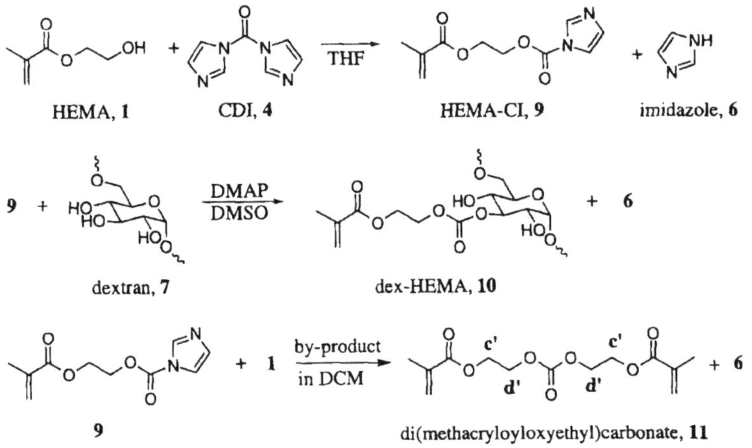

Figure 9. Reaction scheme for the synthesis of dex-HEMA.

The DS could be fully controlled by varying the molar ratio of HEMA-CI or HEMA-lactate-CI to dextran, with an incorporation efficiency of 60-80%[48]. Figure 10 shows the swelling behavior of a dex-MA hydrogel, a dex-HEMA hydrogel and a dex-lactateHEMA hydrogel, when exposed to an aqueous buffer (pH 7.2) at 37 °C. The dex-MA hydrogel reached an equilibrium swelling within 3 days; thereafter the weight of the gel remained constant, again demonstrating that no significant hydrolysis of ester groups in the crosslinks occurred. On the other hand, both the dex-HEMA and dex-lactateHEMA gels showed a progressive swelling in time, followed by a dissolution phase. Obviously, due to hydrolysis of esters, the molecular weight between the crosslinks (M_c) increases, resulting in an increased swelling.

Figure 11 summarizes the degradation times of all dex-(lactate)HEMA gels investigated. This figure shows that for both gel systems the degradation time increased with an increasing DS. It is obvious that an increasing DS at a fixed initial water content results in a network with a higher cross-link density. To dissolve the network, more cross-links have to be hydrolyzed, which requires more time. Figure 11 also shows that at a fixed DS, the degradation time increased with a decreasing initial water content.

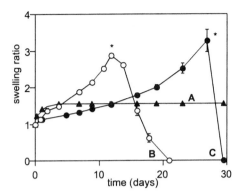

Figure 10. Influence of spacer on swelling behaviour of dextran hydrogels with initial water content 90%; dex-MA DS 4 (A), dex-lactateHEMA DS 10 (B) and dex-HEMA DS 9 (C). The asterisk indicates the time of maximum swelling, which is the transition from the swelling phase to the dissolution phase of the gel

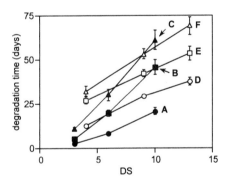

Figure 11. Degradation time of dex-HEMA (open symbols) and dex-lactateHEMA hydrogels (closed symbols) as a function of DS at various initial water contents; dex-lactateHEMA 90 % (A), 80 % (B), and 70 % (C); dex-HEMA 90 % (D), 80 % (E), and 70 % (F).

The average length of the oligolactate graft in HEMAlactate (Fig.8) can be fully controlled by the M/I ratio[48]. Besides, transesterification reactions occur during the grafting reaction[49] resulting in the presence of detectable amounts of HEMA-(lactate)$_{n=1-10}$ at a HEMA/lactide feed ratio of 1/1 (mole/mole)[50]. This means that the networks obtained after coupling of HEMA-lactate to dextran (Fig.8) and subsequent polymerization of the methacrylate groups, contain a varying number of lactate acid ester groups in their crosslinks. In order to control the properties of the hydroxy acid macromers as well as the polymers and networks derived hereof, it might be advantageous to use macromers with a fixed amount of hydroxy acid groups.

Using preparative HPLC, 2-(methacryloyloxy)ethyl-lactate and 2-(methacryloyloxy)ethyl-di-lactate were obtained[50]. Both compounds were coupled to dextran using the strategy shown in Figure 8 resulting in dex-(lactate)$_1$-HEMA and dex-(lactate)$_2$-HEMA, respectively. The degradation at pH 7.2 and 37 °C of hydrogels derived from dex-(lactate)$_{1,2}$-HEMA as well as a dex-HEMA hydrogel was studied (Fig.12). All hydrogels showed a progressive swelling in time, which is caused by hydrolysis of the crosslinks, followed by a dissolution phase. Interestingly, the degradation time of the hydrogel derived of dex-(lactate)$_2$-HEMA is shorter than for dex-(lactate)$_1$-HEMA as well as dex-HEMA hydrogels (6, 12 and 30 days respectively). This can be ascribed to the fact that the more hydrolytically sensitive groups are present in one crosslink, the greater the probability that one crosslink hydrolyzes and the shorter the degradation time of the hydrogel.

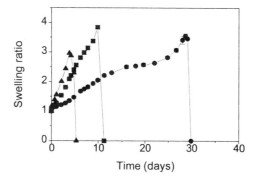

Figure 12. Swelling behavior of dex-HEMA (●), dex-lactate-HEMA (■) and dex-lactate$_2$-HEMA (▲) hydrogels in aqueous solution (pH 7.2, 37 °C). The initial water content of the hydrogels was 80%, the degree of methacryloyl substitution was approximately 6.

The results shown in Figures 11 and 12 demonstrate that the degradation time of dextran based hydrogels can be tailored by both the crosslink density (DS and initial water content; Fig.11) and the number of hydrolyzable groups in the crosslinks (Fig.12).

The release of a model protein (IgG) from degrading, cylinder-shaped (radius 0.23 cm, length 1 cm) dex-lactateHEMA hydrogels with varying DS and initial water content was investigated[46]. Representative release profiles are shown in Figure 13.

From this figure it appears that for gels with a high initial water content a first order release of the protein is observed (diffusion controlled release), whereas for gels with a lower initial water content an almost zero order release is observed for 35 days (degradation controlled release).

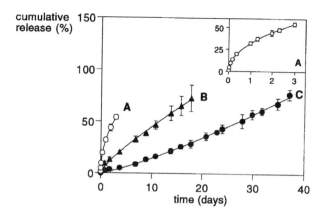

Figure 13. Cumulative release of IgG in time from dex-lactateHEMA hydrogels (DS 10) with initial water content 90 % (A), 80 % (B), and 70 % (C).

5. PROTEIN RELEASE FROM DEGRADING DEXTRAN MICROSPHERES

Figures 4, 5 and 13 show the release of proteins form macroscopic dextran-based hydrogels. For therapeutic applications of these protein-loaded gels, injectable dosage forms like microspheres are preferred. Extensive research has been done so far on the preparation of microspheres based on biodegradable polymers (e.g. PLGA). However, for the preparation of these microspheres, organic solvents are used which might cause denaturation of the encapsulated protein (see Introduction). Furthermore, the use of organic solvents has environmental and clinical drawbacks.

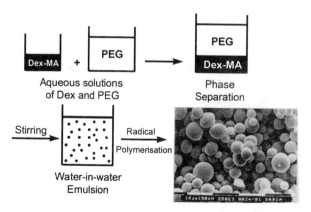

Figure 14. Schematic representation of the microsphere preparation process.

To overcome these problems, we have developed a technique to prepare microspheres in an all-aqueous system, avoiding the use of organic solvents[52,53]. This preparation method is based on the phenomenon that in aqueous two-polymer systems phase-separation can occur. The microsphere preparation process is schematically shown in Figure 14.

We have demonstrated that both the size and the initial water content of the microspheres can be fully controlled by the preparation conditions[53,54].

The release of a model protein, IgG was studied from both enzymatically degradable dex-MA microspheres[55] and chemically degrading dex-HEMA microspheres[56]. If an IgG solution was added to the dex-MA/PEG aqueous system prior to the polymerization reaction, the protein was encapsulated in the dextran microspheres with a high yield (around 90%). The high encapsulation efficiency can be ascribed to the favorable partitioning of proteins in the dextran phase of the dextran /PEG aqueous two-phase systems[57]. The release of IgG was studied as a function of the water content, the DS and the degradation rate of the microspheres. Dex-MA microspheres were rendered degradable by co-encapsulation of dextranase. Non-degrading microspheres mainly showed a burst release, which decreased with increasing crosslink density. By either a low water content (50% (w/w) or lower) or a high DS (DS 13), it was possible to reduce the burst release to about 10%, meaning that almost complete entrapment of the protein could be achieved. The release of IgG from degrading microspheres was predominantly dependent on the DS and the amount of encapsulated dextranase. No differences in release of IgG from microspheres with and without dextranase were observed at high DS (DS 13). This was ascribed to the inability of the enzyme to degrade these microspheres. On the other hand, the entrapped protein was completely released from enzymatically

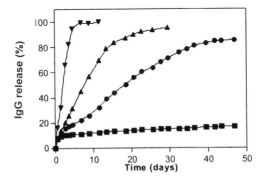

Figure 15. Cumulative release of IgG versus time from degrading dex-MA microspheres (DS 4, water content 50% (w/w)) as a function of the amount of incorporated dextranase: 2 U/g solid (▼), 0.7 U/g solid (▲), 0.2 U/g solid (●), 0 U/g solid (■).

degrading microspheres with a DS 4. Moreover, the release rate of IgG was proportional to the degradation rate of these microspheres (depending on the amount of co-encapsulated dextranase). Interestingly, an almost zero-order release of IgG from these microspheres was observed for periods up to 30 days (Fig.15).

Figure 16 gives the release of IgG from chemically degrading dex-HEMA microspheres as a function of the DS, at pH 7.0 and 37 °C. Interestingly, dex-HEMA microspheres showed a delayed release of the entrapped protein. The delay time increased with increasing DS, which demonstrates that the release of the protein was fully controlled by the degradation rate of the microspheres[56].

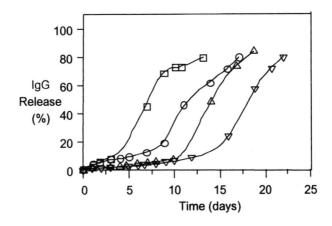

Figure 16. Cumulative release of IgG from degrading dex-HEMA microspheres (water content 50% (w/w)) in time DS 3(□), DS 6 (○), DS 8 (△) and DS 11(▽).

6.   IN VIVO BIOCOMPATIBILITY OF DEXTRAN BASED HYDROGELS

Both non-degradable dex-MA and degradable dex-lactateHEMA disc-shaped hydrogels varying in initial water content and degree of substitution were implanted subcutaneously in rats. The tissue reaction was evaluated over a period of 6 weeks[58].

The initial foreign body reaction to the dex-MA hydrogels was characterized by infiltration of granulocytes and macrophages and formation of fibrin, exudate as well as new blood vessels. This reaction depended on the initial water content as well as the DS of the hydrogel and decreased within 10 days. The mildest tissue response was observed for the gel with the highest water content and intermediate DS. At day 21, all dex-MA hydrogels were surrounded by a fibrous capsule and no toxic effects on the surrounding tissue were found. No signs of degradation were observed.

The initial foreign body reaction to the degradable dex-lactateHEMA hydrogels was less severe compared with the dex-MA gels. In general, the size of the dex-lactateHEMA hydrogels increased progressively in time and the gels finally completely dissolved. Degradation of the dex-lactate-HEMA hydrogels was associated with infiltration of macrophages and formation of giant cells, which both phagocytosed pieces of the hydrogel. A good correlation between the *in vitro* and *in vivo* degradation time was found. This suggests that the extra-cellular degradation is not caused by enzymes, but only depends on hydrolysis of the ester and/or carbonate bonds present in the crosslinks of the hydrogels. After 21 days, the degradable hydrogels as such could not be retrieved, but accumulation of macrophages and giant cells was observed, which contained particles of the gels intracellularly. As for the dex-MA hydrogels, no toxic effects on the surrounding tissue were found. These results demonstrate that dextran-based hydrogels can be considered biocompatible materials.

7.   CONCLUSION

A biodegradable and biocompatible dextran hydrogel system has been developed of which the degradation time can be tailored between 2 days and 3 months. This chapter shows that hydrogels based on crosslinked dextrans have unique properties as protein releasing matrices. Both the release pattern (first-order, zero-order or biphasic) as well as the duration of the release can be controlled by an appropriate selection of the characteristics and the geometry of the hydrogel.

ACKNOWLEDGMENTS

This research is supported by the Dutch Technology Foundation (STW, grant numbers UPR 66.4049 and UFA55.3931).

REFERENCES

1. D.J.A. Crommelin and R.D. Sindelar (Eds.), 1997, *Pharmaceutical Biotechnology*, Harwood Academic, Amsterdam.
2. T. Arakawa, S.J. Prestrelski, W.C. Kenney and J.F. Carpenter, 1993, *Adv. Drug Delivery Rev.*, 10: 1-28.
3. M.C. Manning, K. Patel and R.T. Borchardt, 1989, *Pharm. Res.* 6: 903-915.

4. T. Chen, Drug Dev. Indus. Pharmacy 1992, 18: 1311.
5. V.H.L. Lee and A. Yamamoto, 1990, *Adv. Drug Delivery Rev.* 4: 171-207.
6. K. Park, W.S.W. Shalaby and H. Park, 1993, *Biodegradable Hydrogels for Drug Delivery, Technomic*, Basel.
7. N.A. Peppas, 1986, *Hydrogels in Medicine and Pharmacy*, CRC, Boca Raton vol. 1-3.
8. J. Cheng, S. Jo and K. Park, 1995, *Carbohydrate Polymers*, 28: 69-76.
9. D.J.A. Crommelin and H. Schreier, 1994, *In: Colloidal Drug Delivery Systems*, J. Kreuter, (Ed.), Marcel Dekker, New York pp73-190.
10. D.J.A. Crommelin, G. Scherphof and G. Storm, Adv. Drug Delivery Rev. 1995, 17: 49-60.
11. L. Brannon-Peppas, 1995, *Int. J. Pharm.* 116: 1-9.
12. P. Couvreur, M.J. Blanco-Prieto, F. Puisieux, B.P. Roques and E. Fattal, 1997, *Adv. Drug Delivery Rev.* 28: 85-96.
13. L.M. Sanders and R.W. Hendren (Eds.), 1997, *Protein Delivery, Pharmaceutical Biotechnology*, Plenum, New York vol. 10: 1-43.
14. D.L. Wise, D.J. Trantolo, R.T. Marino and J.P. Kitchell, 1987, *Adv. Drug Delivery Rev.* 1: 19-39.
15. J.L. Cleland, 1997, *Protein Delivery, Pharmaceutical Biotechnology*, L.M. Sanders and R.W. Hendren (Eds.), Plenum, New York vol. 10, 1-43.
16. R. Arshady, 1991, *J. Controlled Release* 17: 1-22.
17. R. Jalil and J.R. Nixon, 1990, *J. Microencapsulation* 7: 297-325.
18. B. Conti, F. Pavanetto and I. Genta, 1992, *J. Microencapsulation,* 9: 153-166.
19. J. Cleland, 1998, *Biotechnology Progress* 14: 102-107.
20. T.G. Park, 1995, *Biomaterials* 16: 1123-1130.
21. A. Shenderova, T.G. Burke and S.P. Schwendeman, 1998, *Proceed. Int. Symp. Control. Rel. Bioact. Mater.* 25: 265-266.
22. A. Brunner, K. Mäder and A. Göpferich, 1998, *Proceed. Int. Symp. Control. Rel. Bioact. Mater.* 25: 154-155.
23. K. Mäder, S. Nitschke, R. Stösser, H.H. Borchert and A. Domb, 1997, *Polymer* 38: 4785-4794.
24. W. Lu and T.G. Park, 1995, *J. Pharm. Sci. Tech.* 49: 13-19.
25. V.N. Uversky, N.V. Narizhneva, S.O. Kirschstein, S. Winter and G. Löber, 1997, *Folding and design,* 2: 163-172.
26. K. Park, W.S.W. Shalaby and H. Park, 1993, *Biodegradable Hydrogels for Drug Delivery*, Technomic, Basel
27. N.A. Peppas, 1986, *Hydrogels in Medicine and Pharmacy*, CRC, Boca Raton vol. 1-3.
28. J. Cheng, S. Jo and K. Park, 1995, *Carbohydrate Polymers*, 28: 69-76.
29. P. Edman, B. Ekman and I. Sjöholm, 1980, *J. Pharm. Sci.* 69: 839-842.
30. N. Wang and X. Shen Wu, 1998, *Int. J. Pharm.*, 166: 1-14.
31. J.D. Andrade, 1973, *Med. Instrum.* 7: 110-121.
32. J.A. Hubbell, 1996, *J. Controlled Release.* 39: 305-313.
33. N.A. Peppas and J.E. Scott, 1992, *J. Controlled Release.* 18: 95-100.
34. M.T. am Ende and N.A. Peppas, 1997, *J. Controlled Release.* 48: 47-56.
35. R.L. Sidebotham, 1974, *Adv. Carbohydr. Chem.* 30: 371-444.
36. H. Brønsted, L. Hovgaard and L. Simonsen, 1995, *STP Pharm. Sci.* 5: 60-69.
37. K.R. Kamath and K. Park, 1995, *Polymer Gels and Networks.* 3: 243-254.
38. W.N.E. van Dijk-Wolthuis, O. Franssen, H. Talsma, M.J. van Steenbergen, J.J. Kettenes-van den Bosch and W.E. Hennink, 1995, *Macromolecules.* 28: 6317-6322.
39. W.N.E. van Dijk-Wolthuis, J.J. Kettenes-van den Bosch, A. van der Kerk-van Hoof and W.E. Hennink, 1997, *Macromolecules.* 30: 3411-3413.

40.W.E. Hennink, H. Talsma, J.C.H. Borchert, S.C. de Smedt and J. Demeester, 1996, *J. Controlled Release*. 39: 47-55.

41.O. Franssen, O.P. Vos and W.E. Hennink, 1997, *J. Controlled release*. 44: 237-245.

42.M. Sugiura, A. Ito, T. Ogiso, K. Kato and H. Asano, 1973, *Biochym. Biophys. Acta*. 309: 357-362.

43.O. Franssen, R.D. van Ooijen, D. de Boer, R.A.A. Maes, J.N. Herron and W.E. Hennink, 1997, *Macromolecules*. 30: 7408-7413.

44.O. Franssen, R.D. van Rooijen, D. de Boer, R.A.A. Maes and W.E. Hennink, 1999, *Macromolecules*. 32: 2896-2902.

45.W.N.E. van Dijk-Wolthuis, M.J. van Steenbergen, W.J.M. Underberg and W.E. Hennink, 1997, *J. Pharm. Sci.* 86: 413-417.

46.W.N.E. van Dijk-Wolthuis, M.J. van Steenbergen, C. Hoogeboom, S.K.Y. Tsang and W.E. Hennink, 1997, *Macromolecules*. 30: 4639-4645.

47.A.S. Sawhney, C.P. Pathak and J.A. Hubbell, 1993, *Macromolecules*. 26: 581-587.

48.W.N.E. van Dijk-Wolthuis, S.K.Y. Tsang, J.J. Kettenes-van den Bosch and W.E. Hennink, 1997, *Polymer*. 38: 6235-6242.

49.S.J. de Jong, W.N.E. van Dijk-Wolthuis, J.J. Kettenes-van den Bosch, P.J.W. Schuyl and W.E. Hennink, 1998, *Macromolecules*. 31: 6397-6402.

50.J.A. Cadée, M. de Kerf, C.J. de Groot, W. den Otter and W.E. Hennink, 1999, *Polymer*, in press.

51.M.L. Maniar, D.S. Kalonia DS and A.P. Simonelli AP., 1991, *J. Pharm. Sci.* 80: 778-782.

52.O. Franssen and W.E. Hennink, 1998, *Int. J. Pharm.* 168: 1-7.

53.R.J.H. Stenekes, O. Franssen, E.M.G. van Bommel, D.J.A. Crommelin and W.E. Hennink, 1998, *Pharm. Res.* 15: 555-559.

54.R.J.H. Stenekes and W.E. Hennink, 1999, *Int. J. Pharm.* 168: in press.

55.O. Franssen, R.J.H. Stenekes and W.E. Hennink, 1999, *J. Controlled Release*. 59: 219-228.

56.O. Franssen, L. Vandervennet, P. Roders and W.E. Hennink, 1999, *J. Controlled Release* in press.

57.P.-Å. Albertsson, 1986,*Partition of Cell Particles and Macromolecules*, Wiley, New York 3rd ed.

58.J.A. Cadée,, M.J.A. van Luyn, L.A. Brouwer, J.A. Plantinga, P.B. van Wachem, C.J. de Groot, W. den Otter and W.E. Hennink, submitted for publication.

BIODEGRADABLE NANOSPHERES: THERAPEUTIC APPLICATIONS

Jasmine Davda, Sinjan De, Wenzhong Zhou, and Vinod Labhasetwar
Department of Pharmaceutical Sciences , College of Pharmacy, University of Nebraska Medical Center, Omaha, NE 68198-6025

1.   INTRODUCTION

Drug delivery has become an integral part of drug development because it can significantly enhance the therapeutic efficacy of drugs [1]. Furthermore, newer drugs prepared by recombinant technology, though comparatively more potent and specific in their pharmacologic action, require efficient drug delivery systems. These drugs are either unstable in the biological environments [2] or are unable to cross the biological barriers effectively [3].

Our research is focused on biodegradable nanospheres as a drug carrier system. Nanospheres are submicron size colloidal particles with a therapeutic agent either entrapped in the polymer matrix or bound on to the surface. Nanocapsules consist of a drug core with a coating of a polymer film (Fig.1). The drug in the core of the nanocapsules is released by diffusion through the polymer coat [4]. In our studies, we have mainly used polylactic polyglycolic acid co-polymer (PLGA); a FDA approved biodegradable and biocompatible polymer to formulate nanospheres. The nanospheres were typically 100 to 150 nanometer in diameter with the drug entrapped into the nanosphere polymer matrix [5, 6].

Nanospheres as a drug delivery system could provide many advantages. Since nanospheres are submicron in size, they could be taken up more efficiently by the cells than the larger size particles. Furthermore, nanospheres could cross the cell membrane barriers by transcytosis. Drug molecules such as proteins and peptides and also DNA, which have a larger

Biomedical Polymers and Polymer Therapeutics
Edited by Chiellini *et al.*, Kluwer Academic/Plenum Publishers, New York, 2001

hydrodynamic diameter because of the charge and the bound water, could be entrapped into the nanosphere polymer matrix. Thus, the cellular uptake and transport of such macromolecules across biological membranes could be significantly improved by condensing them into nanospheres [1].

Nanospheres could protect the entrapped agent(s) from enzymatic and hydrolytic degradation. This is important because many drugs, such as oligonucleotides or DNA, and also proteins and peptides are susceptible to degradation due to nucleases and other enzymes present in the body [7]. Furthermore, the sustained release characteristics of nanospheres could be useful for many therapeutic agents that require repeated administration for their pharmacologic effects or to cure the disease completely. Such drugs could be delivered using nanospheres as a single-dose therapy. Most of the anticancer agents are substrates for the p-glycoprotein dependent efflux pump of the cell membrane. These drugs are either not taken up by the cells or are thrown out rapidly because of the efflux pump. Recent studies have demonstrated that the intracellular delivery of such drugs could be improved by encapsulating them into nanospheres.

Thus nanospheres could offer a solution to many problems related to the delivery of therapeutic agents. This book chapter will review the various pathophysiologic conditions where nanospheres could be used as an effective drug delivery system.

2. FORMULATION OF NANOSPHERES

Various polymers, both synthetic and natural have been investigated for the formulation of nanospheres. Most commonly used polymers are polyesters, polyacrylates, chitosans, gelatin, etc[5, 8]. The formulation of nanospheres depends on the polymer and the physicochemical properties of the drug molecules. Nanospheres using PLGA polymers are commonly formulated by an emulsion-solvent evaporation technique. The method involves formation of an emulsion, either a multiple (water-in-oil-in-water) emulsion for water-soluble drugs or a simple (oil-in-water) emulsion for oil-soluble drugs. The slow evaporation of the organic solvent from the emulsion results in the formation of nanospheres containing the drug (Fig.1).

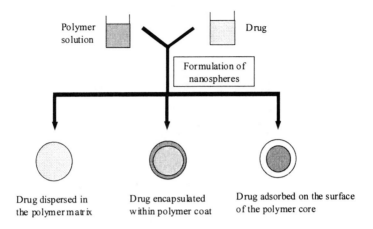

Figure 1. Types of nanospheres

3. ORAL DELIVERY

The research in oral delivery of nanospheres is focused on vaccines to induce mucosal and systemic immunity [9]. Nanospheres could gain entry into the lymphoidal tissue of the gut through the Peyer's patches, a group of specialised tissue in the gut consisting of the antigen presenting cells and the B cells and the T cells (Fig.2). The fate of the particles following their uptake by the lymphoidal tissue is mainly governed by their size [9]. Particles from 5-10 μm in size remain in the Peyer's patches, while smaller particles go to the mesenteric lymph node and then enter the systemic circulation where they are phagocytosed by the Kupffer cells of the liver (Fig.3). The particles, which enter the systemic circulation, are responsible for inducing systemic immune response. In our studies in an acute rat intestinal loop model, we observed a greater uptake of nanospheres by the Peyer's patch tissue as compared to larger size microspheres [10].

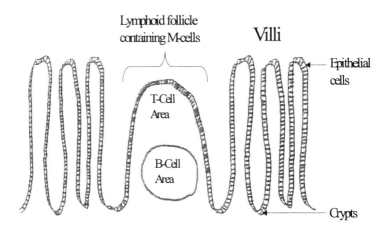

Figure 2. Schematic diagram of a Peyer's patch in the intestine

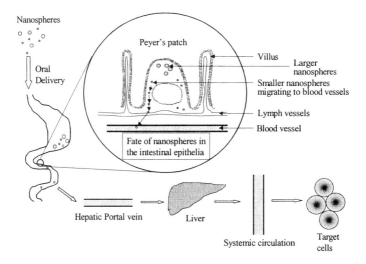

Figure 3. Schematic diagram of the fate of nanospheres following their oral delivery

Nanospheres containing antigen have been demonstrated to induce immune response following their oral administration. However, the important issue is whether the immune response generated by an oral administration of vaccines is strong enough to develop a protection against pathogens. An effective oral vaccine system will be more convenient and useful in preventing diseases against pathogens that gain entry into the body through mucosal contacts. Many infectious diseases are either water borne (e.g. cholera, typhoid) or air borne (e.g. tuberculosis) or communicable.

These diseases could be the target for oral vaccine system because it could generate mucosal immunity.

Nanospheres are also investigated to improve the oral bioavailability of the drugs that are not very well absorbed by the gastrointestinal tract [11], either due to their instability to the enzymes (cytochromes P) present in the intestinal tract or because they are the substrates of the efflux pump, P-glycoprotein that prevents their absorption [12]. In general, the uptake of drugs through the gastrointestinal tract could be via two different routes (Fig.4)
1. paracellular route (intercellular)
2. transcellular route (intracellular)

The paracellular route is probably less significant in the transport of nanospheres as compared to the transcellular route [13]. The tight junctions in between the cells may not allow nanospheres to cross the intestinal epithelial barrier via paracellular route. The transcellular route could allow a restricted passage of the nanospheres across the intestinal mucosal layer because of their moderate size range. In a tissue culture model using Caco-2 (enterocytes) cell monolayers, we have demonstrated transport of PLGA nanospheres across the cells. [14]

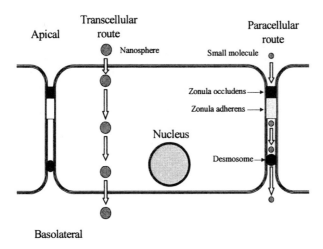

Figure 4. Routes of cellular transport

There are many issues with oral delivery of nanospheres, such as the efficiency of uptake, reproducibility, and the pathophysiologic condition of the patient (e.g. diarrhoea) that could affect the retention of orally administered nanospheres. To increase the efficacy of uptake of nanospheres by the gastrointestinal tract, investigators have been studying nanospheres which are surface modified with lectins and transferrin [15, 16]. The receptors for lectin are present on the intestinal epithelial cells. Therefore, lectin modified nanospheres could have greater uptake by the intestinal tissue. We have demonstrated greater transport across Caco-2 cell monolayers of PLGA nanospheres which were surface modified with transferrin [14]. Thus, the strategy of surface modification of nanospheres with specific ligands could be effective to improve the uptake of nanospheres by the intestinal tissue.

4.   TARGETED DELIVERY

Nanospheres could be used for targeting drugs to a particular tissue or cell population. Targeted drug delivery could increase the specificity of the pharmacologic action of the drug and also could reduce the dose and the toxic effects of the drug. Cell or tissue specific ligands could be coupled to the nanosphere surface to achieve site-specific delivery of the nanospheres. Ligands could be attached to the nanospheres either by physical adsorption through ionic interactions or by chemical means using spacers and modifiers such as dynacol, poly-L-lysine and, morpholineethanesulfonic acid buffer with 1-ethyl-3-(3-dimethylaminopropyl)carbodiimide [17, 18]. The most commonly used ligands are transferrin, folic acid, lectins, poly-L-lysine and antibodies for specific cells (Fig.5). Transferrin receptors are over expressed in intestinal cells, blood/brain endothelia and most cancerous cells. Other organs and cells could be targeted using conjugated nanospheres. For example, galactose can be used for targeting hepatocytes' because the receptors for galactose are present in these cells. Similarly folate receptors are over expressed in cancerous cells[17]. Lee et al. have demonstrated increased cellular uptake of liposomes which were conjugated to folic acid via a polyethyleneglycol spacer[19].

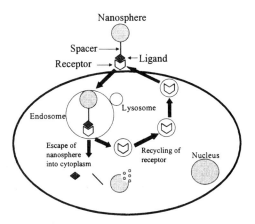

Figure 5. Proposed mechanism of receptor-mediated endocytosis of nanospheres in the cell

Nanospheres conjugated to ligands could act as a targeted reservoir drug delivery system. After reaching the target tissue, the nanospheres could provide a sustained release of the drug. However, many issues such as the stability of the conjugate and the increase in the particle size with conjugation of a ligand limiting the permeability of the nanospheres, are to be considered while developing such a system.

5. GENE DELIVERY

Gene therapy could be used to cure genetic diseases by adding a missing gene or correcting the defective gene[20]. However, one of the major limiting factors in gene therapy is the efficiency of gene transfer[21]. Many investigators are working on viral and nonviral methods of gene transfer. Nonviral methods are relatively less efficacious compared to viral methods of gene transfer but are considered safe. We have been investigating PLGA nanospheres as a gene delivery vector[22-24]. These nanospheres demonstrated sustained release of the encapsulated plasmid DNA under *in vitro* conditions. The sustained release of DNA is expected to produce a stable and prolonged *in vivo* gene expression. Using marker genes (luciferase or heat sensitive human placental alkaline phosphatase) in nanospheres, we have demonstrated gene expression in tissue culture[23]. In a bone osteotomy model in rat, in which the bone-gap was filled with DNA containing nanospheres, gene expression was observed in the tissue that was retrieved from the gap five weeks after the surgery[23]. This gene delivery strategy could be used to facilitate bone healing in bone fractures using therapeutic genes such as the gene encoding bone morphogenic protein. Nanospheres could act by facilitating the cellular uptake of DNA and protecting it from degradation due to nucleases and lysosomal enzymes. Leong's group has used cationic chitosan to formulate nanoparticles for gene transfer. A recent study from this group has demonstrated protection of animals against peanut allergy following oral administration of chitosan nanoparticles containing the gene for peanut allergen. Furthermore, they have higher secretory IgA antibody levels with chitosan-DNA nanoparticles as compared to a direct injection of naked DNA[25].

Many polymer-based systems have been recently investigated for gene delivery[26,27]. Shea et al. used polymer matrix containing gene to demonstrate greater gene expression in tissue culture[28]. In our recently published studies, we used a PLGA emulsion containing a marker gene (alkaline phosphatase) for coating a chromic gut suture. The gene-coated suture was used to close an incision in the rat skeletal muscles. Two weeks after the surgery, the tissue from the incision site demonstrated gene expression[29]. The gene-coated sutures encoding growth factors such as vascular endothelial growth factor could be useful to facilitate wound healing.

6.   VACCINE DELIVERY

A potent vaccine adjuvant could boost and sustain the immune response to the antigen. Nanospheres could be used as vaccine adjuvants because they could provide a sustained release of the antigen from the site of injection. In addition, due to the particulate nature of the nanospheres, they could attract macrophages and the antigen presenting cells to the site of administration of nanospheres. In our studies, we have demonstrated adjuvant properties of the nanospheres using staphylococcal enterotoxin B toxoid as a model antigen in rabbits. In this study, the systemic immune response with nanospheres (serum IgG levels) was comparable to alum[30]. Thus, nanospheres could act as an alternative to alum as a vaccine adjuvant to avoid the problem of toxicity associated with alum. The goal of vaccine delivery system is to combine the primary and the booster doses in a single-dose injection[31]. The advantage of a single-dose vaccine system is that it could prevent the fall-out in the number of people receiving the complete immunization with the increase in the number of injections[32].

7.   CANCER CHEMOTHERAPY

The major limitation in cancer chemotherapy is the poor uptake of anticancer agents by the cancer cells. The drug resistant cancer cells efflux out the drug due to the p-glycoprotein dependent cell membrane pump. By using nanospheres as a carrier system, researchers have demonstrated a higher cellular uptake of drugs and greater therapeutic effects[33,34]. The proposed mechanism is the drug in nanospheres (entrapped or bound) could bypass the efflux action of the p-glycoprotein pump[35].

The other therapeutic approach to treat cancers is to block the expression of certain oncogenes such as C-*raf* and C-*ras* and kinases[36]. The antisense oligonucleotide strategy is focused on blocking the activity of these genes. However, oligonucleotides are unstable in the body and are not taken up very efficiently by the cells. Nanospheres could facilitate the uptake of oligonucleotides and also stabilise them against nuclease attack[37]. In a recently published study, Tondelli et al. have demonstrated a 50-fold greater intracellular delivery of an antisense oligonucleotide in HL60 leukemia cells using nanospheres compared to free oligonucleotide[38]. In this study, the oligonucleotide was bound to the outer surface of the nanospheres via ionic interactions. The oligonucleotide bound to the nanospheres also demonstrated inhibition of cell growth at a dose 50-fold lower than the free oligonucleotide. In a study by Chavany et al., an oligonucleotide bound to nanoparticles demonstrated greater stability to nucleases than the free oligonucleotide[2]. Thus nanospheres could significantly improve the therapeutic efficacy of oligonucleotides by stabilising them against nucleases and facilitating their cellular uptake.

8. INTRAMUSCULAR DELIVERY

The intramuscular delivery of nanospheres could be used to provide sustained localized drug delivery or for sustained systemic drug effects. The nanospheres could form a depot at the site of injection from which the drug could be released slowly in the local tissue and then absorbed for the systemic effect. This strategy is useful for drugs which require repeated administration such as growth factors or hormones[39, 40].

9. LOCALISED DRUG DELIVERY

Nanospheres could be used as a colloidal injectable suspension for infusion into a certain diseased body compartment using a catheter. The infused nanospheres could provide a localised sustained drug effect with minimal toxicity. In our studies, we have successfully used nanospheres to localise therapeutic agents (e.g. dexamethasone, U-86983, a 2-aminochromone) into the arterial wall to inhibit restenosis in animal model studies[41-43].

Restenosis is the reobstruction of the artery following angioplasty procedure. Proliferation of the smooth muscle cells of the arterial wall in response to the injury due to balloon angioplasty is considered as the main reason for the pathophysiology of restenosis[44].

Intravenous or oral delivery of drugs is usually ineffective because it does not provide a therapeutic dose of the drug to the diseased artery for a sufficient period of time. Nanospheres could be infused into the arterial wall using infusion catheter following angioplasty (Fig.6). Nanospheres mobilised into the arterial wall could provide sustained drug levels in the diseased artery [41]. The localised delivery of drug using nanospheres could significantly reduce the total dose of the drug and also could prevent the toxic effects of the drug.

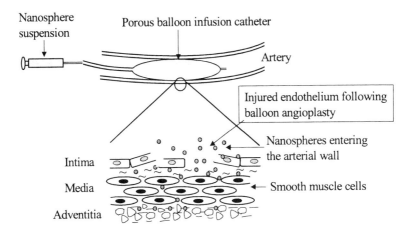

Figure 6. Schematic of arterial wall deposition of nanospheres after angioplasty

10.   CONCLUSION

Nanospheres could be used as an effective drug delivery tool to improve the therapeutic efficacy of drugs in various pathophysiologic conditions. It offers flexibility in terms of its composition and release kinetics. Furthermore, nanospheres could be surface modified to achieve drug targeting. It is important to understand the pathobiology of the disease so that one can determine the desired nanosphere formulation in terms of the release kinetics and the duration of drug release. At the same time, knowledge of the cell surface and receptors could help in developing a targeted drug delivery system using nanospheres. Gene delivery using nanospheres needs to be optimised to increase the transfection efficiency. In summary, nanospheres have potential applications as a drug carrier system in drug and gene therapy.

ACKNOWLEDGMENTS

The work from our laboratory was supported by grants from the National Institutes of Health (HL 57234) and the Nebraska Research Initiative-Gene Therapy Program.

REFERENCES

1. Davis, S. S., 1997, Biomedical applications of nanotechnology--implications for drug targeting and gene therapy. *Trends Biotechnol.* 15: 217-224.
2. Chavany, C., Saison Behmoaras, T., Le Doan, T., Puisieux, F., Couvreur, P., and Helene, C., 1994, Adsorption of oligonucleotides onto polyisohexylcyanoacrylate nanoparticles protects them against nucleases and increases their cellular uptake. *Pharm. Res.* 11: 1370-1378.
3. Bender, A. R., von Briesen, H., Kreuter, J., Duncan, I. B., and Rubsamen Waigmann, H., 1996, Efficiency of nanoparticles as a carrier system for antiviral agents in human immunodeficiency virus-infected human monocytes/macrophages in vitro. *Antimicrob. Agents Chemother.* 40: 1467-1471.
4. Morin, C., Barratt, G., Fessi, H., Devissaguet, J. P., and Puisieux, F., 1993, Biodegradable nanocapsules containing a lipophilic immunomodulator: drug retention and tolerance towards macrophages in vitro. *J. Drug Target.* 1: 157-164.
5. Labhasetwar, V., Song, C., and Levy, R. J., 1997, Nanoparticle drug delivery for restenosis. *Adv. Drug Del. Rev.* 24: 63-85.
6. Labhasetwar, V., 1997, Nanoparticles for drug delivery. *Pharmaceutical News* 4: 28-31.
7. Hedley, M. L., Curley, J., and Urban, R., 1998, Microspheres containing plasmid-encoded antigens elicit cytotoxic T-cell responses. *Nat. Med.* 4: 365-368.
8. Labhasetwar, V., and Dorle, A. K., 1990, Nanoparticles-a colloidal drug delivery system for primaquine and metronidazole. *J. Control. Rel.* 12: 113-119.
9. O'Hagan, D. T., Palin, K. J., and Davis, S. S., 1989, Poly(butyl-2-cyanoacrylate) particles as adjuvants for oral immunization. *Vaccine* 7: 213-216.
10. Desai, M. P., Labhasetwar, V., Amidon, G. L., and Levy, R. J., 1996, Gastrointestinal uptake of biodegradable microparticles: effect of particle size. *Pharm. Res.* 13: 1838-1845.
11. Sakuma, S., Suzuki, N., Kikuchi, H., and Hiwatari, K., 1997, Oral peptide delivery using nanoparticles composed of novel copolymers having hydrophobic backbone and hydrophilic brances. *Int. J. Pharmeutic.* 149: 93-106.
12. Damge, C., Vranckx, H., Balschmidt, P., and Couvreur, P., 1997, Poly(alkyl cyanoacrylate) nanospheres for oral administration of insulin. *J. Pharm. Sci.* 86: 1403-1409.
13. Thomas, N. W., Jenkins, P. G., Howard, K. A., Smith, M. W., Lavelle, E. C., Holland, J., and Davis, S. S., 1996, Particle uptake and translocation across epithelial membranes. *J. Anat.* 189: 487-490.
14. De, S., Huai-yun, H., Miller, D. W., and Labhasetwar, V., 1998, Transferrin surface modification enhances transcytosis of nanospheres across Caco-2 cell monolayers. *PharmSci. (Supplement)* 1: S-58.
15. Ritschel, W. A., 1991, Targeting in the gastrointestinal tract: new approaches. *Methods Find. Exp. Clin. Pharmacol.* 13: 313-336.
16. Naisbett, B., and Woodley, J., 1990, Binding of tomato lectin to the intestinal mucosa and its potential for oral drug delivery. *Biochem. Soc. Trans.* 18: 879-880.
17. Kranz, D. M., Patrick, T. A., Brigle, K. E., Spinella, M. J., and Roy, E. J., 1995, Conjugates of folate and anti-T-cell-receptor antibodies specifically target folate-receptor-positive tumor cells for lysis. *Proc. Natl. Acad. Sci. U S A* 92: 9057-9061.
18. Cristiano, R. J., and Roth, J. A., 1995, Molecular conjugates: a targeted gene delivery vector for molecular medicine. *J. Mol. Med.* 73: 479-486.
19. Lee, R. J., and Low, P. S., 1994, Delivery of liposomes into cultured KB cells via folate receptor-mediated endocytosis. *J. Biol. Chem.* 269: 3198-3204.
20. Dickler, H. B., and Collier, E., 1994, Gene therapy in the treatment of disease. *J. Allergy Clin. Immunol.* 94: 942-951.
21. Afione, S. A., Conrad, C. K., and Flotte, T. R., 1995, Gene therapy vectors as drug delivery systems. *Clin. Pharmacokinet.* 28: 181-189.

22. Labhasetwar, V., 1998, Sustained-release gene delivery systems:principles and perspectives. In *Self-Assebling Complexes for Gene Delivery* (A. Kabanov, P. Felger, and L.W. Seymour eds), *John Wiley & Sons, England, UK* , pp. 387-399.

23. Labhasetwar, V., Bonadio, J., Goldstein, S. A., and Levy, R. J., 1999, Gene transfection using biodegradable nanospheres: results in tissue culture and a rat osteotomy model. *Colloids and Surfaces B : Biointerfaces* 16: 281-290.

24. Labhasetwar, V., Chen, B., Muller, D. W. M., Bonadio, J., Ciftci, K., March, K., and Levy, R. J., 1997, Gene-based therapies for restenosis. *Adv. Drug Del. Rev.* 24: 109-120.

25. Roy, K., Mao, H. Q., Huang, S. K., and Leong, K. W., 1999, Oral gene delivery with chitosan--DNA nanoparticles generates immunologic protection in a murine model of peanut allergy. *Nat. Med.* 5: 387-391.

26. Bonadio, J., Smiley, E., Patil, P., and Goldstein, S., 1999, Localized, direct plasmid gene delivery in vivo: prolonged therapy results in reproducible tissue regeneration. *Nat. Med.* 5: 753-759.

27. Mathiowitz, E., Jacob, J. S., Jong, Y. S., Carino, G. P., Chickering, D. E., Chaturvedi, P., Santos, C. A., Vijayaraghavan, K., Montgomery, S., Bassett, M., and Morrell, C., 1997, Biologically erodable microspheres as potential oral drug delivery systems. *Nature* 386: 410-414.

28. Shea, L. D., Smiley, E., Bonadio, J., and Mooney, D. J., 1999, DNA delivery from polymer matrices for tissue engineering. *Nat. Biotechnol.* 17: 551-554.

29. Labhasetwar, V., Bonadio, J., Goldstein, S., Chen, W., and Levy, R. J., 1998, A DNA controlled release coating for gene trasnfer:transfection in skeletal and cardiac muscles. *J. Pharm. Sci.* 87: 1347-1350.

30. Desai, M., Hilfinger, J., Amidon, G., Levy, R. J., and Labhasetwar, V., 2000, Immune response with biodegradable nanospheres and alum:Studies in rabbits using staphylococcal entertoxin B-toxoid. *J. Microencapsulation* 17:215-225.

31. Cleland, J. L., Barron, L., Berman, P. W., Daugherty, A., Gregory, T., Lim, A., Vennari, J., Wrin, T., and Powell, M. F., 1996, Development of a single-shot subunit vaccine for HIV-1. 2. Defining optimal autoboost characteristics to maximize the humoral immune response. *J. Pharm. Sci.* 85: 1346-1349.

32. Aguado, M. T., 1993, Future approaches to vaccine development: single-dose vaccines using controlled-release delivery systems. *Vaccine* 11: 596-597.

33. de Verdiere, A. C., Dubernet, C., Nemati, F., Soma, E., Appel, M., Ferte, J., Bernard, S., Puisieux, F., and Couvreur, P., 1997, Reversion of multidrug resistance with polyalkylcyanoacrylate nanoparticles: towards a mechanism of action. *Br. J. Cancer* 76: 198-205.

34. Hu, Y. P., Jarillon, S., Dubernet, C., Couvreur, P., and Robert, J., 1996, On the mechanism of action of doxorubicin encapsulation in nanospheres for the reversal of multidrug resistance. *Cancer Chemother. Pharmacol.* 37: 556-560.

35. Couvreur, C., Roblot-Treupel, L., Poupon, M. F., Brasseur, F., and Puisieux, F., 1990, Nanoparticles as microcarriers for anticancer drugs. *Adv. Drug Deliv. Rev.* 5: 209-230.

36. Monia, B. P., Johnston, J. F., Geiger, T., Muller, M., and Fabbro, D., 1996, Antitumor activity of a phosphorothioate antisense oligodeoxynucleotide targeted against C-raf kinase [see comments]. *Nat. Med.* 2: 668-675.

37. Schwab, G., Chavany, C., Duroux, I., Goubin, G., Lebeau, J., Helene, C., and Saison Behmoaras, T., 1994, Antisense oligonucleotides adsorbed to polyalkylcyanoacrylate nanoparticles specifically inhibit mutated Ha-ras-mediated cell proliferation and tumorigenicity in nude mice. *Proc. Natl. Acad. Sci. U S A* 91: 10460-10464.

38. Tondelli, L., Ricca, A., Laus, M., Lelli, M., and Citro, G., 1998, Highly efficient cellular uptake of c-myb antisense oligonucleotides through specifically designed polymeric nanospheres. *Nucleic Acids Res.* 26: 5425-5431.

39. Hashimoto, T., Wada, T., Fukuda, N., and Nagaoka, A., 1993, Effect of thyrotropin-releasing hormone on pentobarbitone-induced sleep in rats: continuous treatment with a sustained release injectable formulation. *J. Pharm. Pharmacol.* 45: 94-97.

40. Ball, B. A., Wilker, C., Daels, P. F., and Burns, P. J., 1992, Use of progesterone in microspheres for maintenance of pregnancy in mares. *Am. J. Vet .Res.* 53: 1294-1297.

41. Guzman, L. A., Labhasetwar, V., Song, C., Jang, Y., Lincoff, A. M., Levy, R., and Topol, E. J., 1996, Local intraluminal infusion of biodegradable polymeric nanoparticles. A novel approach for prolonged drug delivery after balloon angioplasty. *Circulation* 94: 1441-1448.

42. Labhasetwar, V., Song, C., Humphrey, W., Shebuski, R., and Levy, R. J., 1998, Arterial uptake of biodegradable nanoparticles: effect of surface modifications. *J. Pharm. Sci.* 87: 1229-1234.

43. Humphrey, W. R., Erickson, L. A., Simmons, C. A., Northrup, J. L., Wishka, D. G., Morris, J., Labhasetwar, V., Song, C., Levy, R. J., and Shebuski, R. J., 1997, The effect of intramural delivery of polymeric nanoparticles loaded with the antiproliferative 2-aminochromone U-86983 on neointimal hyperplasia development in balloon-injured coronary arteries. *Adv. Drug Del. Rev.* 24: 87-108.

44. Fuster, V., Falk, E., Fallon, J. T., Badimon, L., Chesebro, J. H., and Badimon, J. J., 1995, The three processes leading to post PTCA restenosis: Dependence on the lesion substrate. *Thrombosis and Haemostasis* 74: 552-559.

APPLICATION TO CANCER CHEMOTHERAPY OF SUPRAMOLECULAR SYSTEM

[1]Katsuro Ichinose, [1]Masayuki Yamamoto, [2]Ikuo Taniguchi, [2]Kazunari Akiyoshi, [2]Junzo Sunamoto, and [1]Takashi Kanematsu
[1]*Department of Surgery II, Nagasaki University School of Medicine, 1-7-1 Sakamoto, Nagasaki 852-8501, Japan;* [2]*Department of Synthetic Chemistry & Biological Chemistry, Graduate School of Engineering, Kyoto University, Yoshida Hommachi, Sakyo-ku, Kyoto 606-8501, Japan*

1.   INTRODUCTION

Numerous studies have reported liposomes and microspheres to be a useful drug carrier in drug delivery systems. There are several problems including their instabilities and poor selectibilities for targeting. Since 1982, Sunamoto and his co-worker reported that hydrophobized polysaccharides, such as cholesterol conjugated pullulan (CHP), coated liposome showed an increased resistance to enzymatic destruction and CHP also formed hydrogel-nanoparticles after self-aggregation with various drugs in water[1-3]. In addition, recent in vitro studies have demonstrated that polysaccharides recognize lectin on the cell surface[4]. We synthesized CHP bearing galactose moiety (Gal-CHP) as a cell recognition element. In the present study, we evaluated the validity of the Gal-CHP coated liposome (Gal-CHP Lip) enclosing the anticancer drug, and the Gal-CHP self-aggregation complex with the anticancer drug.

Biomedical Polymers and Polymer Therapeutics
Edited by Chiellini *et al.*, Kluwer Academic/Plenum Publishers, New York, 2001   33

2.   METHODS AND RESULTS

2.1   Liposomes coated with Gal-CHP

In the first study, we tried to examine the validity of Gal-CHP liposome enclosing Adriamycin (Gal/CHP Lip-ADM). The radio-labeled CHP Lip or Gal/CHP Lip were added to cultured AH66 *rat liver cancer cells* in vitro. One hour later, the uptake of radio-labeled Gal/CHP Lip in AH 66 was the highest among all the tested liposomes. Athymic mice, which AH66 were subcutaneously transplanted, were intravenously administered free Adriamycin (ADM), conventional liposome enclosing

ADM, CHP coated liposome enclosing ADM (CHP Lip-ADM) or Gal/CHP coated liposome enclosing ADM (Gal/CHP Lip-ADM). The concentrations of ADM in the tumor were compared among these different groups of mice. The highest concentration of ADM in the tumor was observed in the Gal/CHP Lip-ADM group (Fig.1). The in vivo anti-tumor effect was investigated in the same manner as the former experiment. Tumor weight was estimated with the following fomula: Estimated tumor weight $=$ length x width^2 x 1/2. The results are expressed as relative tumor weights (W1/W0). Where W1= Estimated tumor weight at a given time; W0:=Tumor weight before treatment. The Gal/CHP Lip-ADM group was thus found to show a stronger effect (Fig.2). In Gal/CHP Lip-ADM group, the actual tumor weight was the least as compared to those in other groups (Table 1). Thus, Gal/CHP Lip-ADM was able to selectively and effectively deliver anti-cancer drug to the target.

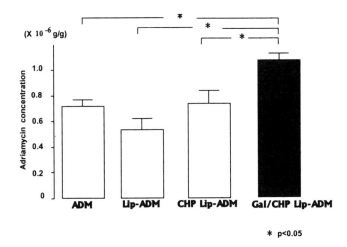

Figure 1. Concentration of adriamycin in AH66 tumor.

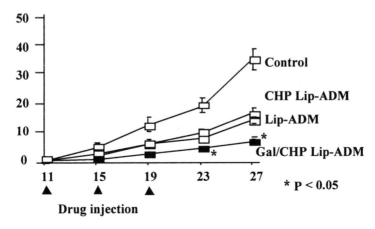

Figure 2. Therapeutic effect in various liposomal adriamycin on the growth of transplanted tumor.

Table 1. Actual weight of the excised tumors

Drugs	Weight (mg)	Tumor/Control (%)
Control	1590	100.0
ADM	890	55.9
CHP Lip/ADM	910	57.2
Gal-CHP Lip/ADM	410*	25.8*

* $P < 0.01$

2.2 Gal-CHP self-aggregation complex

In the second study, we tried to examine the validity of nanoparticles of Gal-CHP self-aggregation and neocarzinostatin chromophore (NCS-chr) complex (Gal-CHP/NCS-chr) as a drug carrier. The size of CHP self-aggregation complex was 25 nm. The galactose moiety, which is a cell recognition element, easily conjugates with the CHP self-aggregation complex. The radio-labeled CHP self-aggregation complex or Gal-CHP self-aggregation complex were then injected into the mice transplanted with 3'-mRLh-2 *rat liver cancer cells.* Three prime-mRLh-2 cell has a galactose-lectine in the cell surface. The activity rate for Gal-CHP self-aggregation complex was higher than for CHP self-aggregation complex. The stability of the anti-tumor effect of Gal-CHP/NCS-chr complex in the medium containing 10% Fetal bovine serum was investigated. NCS-chr is very instable, and loses anti-cancer effect immediately, but the strong anti-cancer effect of Gal-CHP/NCS-chr lasted for 120 minutes. Then, the therapeutic effect of these aggregation complex containing NCS-chr was examined. Among athymic mice transplanted 3'-mRLh-2 in the liver, mice injected Gal-CHP/NCS-chr demonstrated the longest survival.

3.   CONCLUSION

Cholesterol conjugated pullulan (CHP) allows drug carriers to easily conjugate with galactose moiety, thus resulting in the effective and selective targeting of anti-tumor drugs against cancer lesion

ACKNOWLEDGMENTS

This study was supported in part by a grant for general research (No. 02670582) from the Ministry of Education, Science and Culture, Japan, and Terumo Life Science Foundation, Japan.

REFERENCES

1. Sunamoto J. and Iwamoto K., 1986, *CRC Crit. Rev. Therapeutic Drug Carrier System* 2: 117-136.
2. Akiyoshi K. and Sunamoto J., 1996, *Supramolecular Science* 3: 157-163.
3. Taniguchi I., Fujiwara M., Akiyoshi K. and Sunamoto J., 1998, *Bull. Chem. Soc. Jpn.*, 71: 2681-2685.
4. Leonid M., Reuben L. and Avraham R., 1986,*Cancer Re*s. 46: 5270-5275.

DDS IN CANCER CHEMOTHERAPY
Clinical Study of Adriamycin Encapsulated PEG-immunoliposome and Development of Micelles Incorporated KRN5500

Yasuhiro Matsumura
Department of Medicine, National Cancer Center Hospital, 5-1-1 Tsukiji Chuo-ku, Tokyo, 104-0045, Japan

1.   INTRODUCTION

Although numerous anticancer agents have been developed, solid tumors in general respond poorly to treatment. Many people do not accept the current situation concerning cancer chemotherapy, especially in the field of gastrointestinal cancer. Is chemotherapy for gastrointestinal cancer meaningless?

My answer to that is no, even in the current situation.

Several GI oncology groups have conducted randomised control studies in which they have compared some combination chemotherapy with the best supportive care (BSC) alone in cases of advanced gastric cancer or advanced colorectal cancer[1-3]. As one such examples, in comparison with BSC, a regimen consisting of 5-FU, adriamycin and methotrexate appears to prolong survival in patients with advanced gastric cancer[1]. Similar results have been reported in a randomised study of advanced colorectal cancer by other groups[3].

Therefore current conventional chemotherapy for gastrointestinal cancer is not always useless. However, the situation is still far from satisfactory. In fact the median overall survival time of the treated group in the study for advanced gastric cancer was only 9 months[1].

Recently we have also obtained a similar median survival time, that is about 8 months, in advanced gastric cancer after the treatment with 5-FU and cisplatin[4].

Biomedical Polymers and Polymer Therapeutics
Edited by Chiellini *et al.*, Kluwer Academic/Plenum Publishers, New York, 2001

1.1   Why can we not eradicate solid tumors with anticancer agents?

Generally drugs can give patients proper benefits by adjusting a functional disorder. On the other hand, the target of an anticancer agent is cancer cells themselves, which originate from the patient's own cell. Unfortunately cancer cells are almost identical genetically to their corresponding normal cells. Although there are numerous reports concerning genetic and phenotype changes in tumors, there is as far as I know no pivotal change in tumor cells which distinguishes them from normal cells. I suppose that a subpopulation of dramatically changed tumors could be eradicated by the patients own immune system and then that subpopulation of tumors would be out of the question clinically. In general drugs, toxicity occurs at concentrations well above those needed to achieve maximum desired effect. So these drugs have a wide therapeutic window. On the other hand, in the case of anticancer agents, the therapeutic window is quite narrow because of a high degree of overlap between efficacy and toxicity.

1.2   We need DDS in cancer chemotherapy

Recently the concept of the molecular target has come into fashion in the development of new anticancer agents. I do not oppose the movement. However, people must be cautious, because they could possibly meet the same problems as with the current conventional cancer chemotherapy, even after exhaustive efforts for development. In the development of anticancer agent used conventionally, the drug was usually synthesized or extracted from a plant or bacteria and then screened using several cell lines, checked in in vivo systems and finally introduced into clinical trials. During the procedure, the macroscopic characteristics were not usually considered. On the other hand, the agents categorized in DDS have been developed with the macroscopic features of solid tumors in mind even in the manufacturing procedure.

1.3   EPR effect in solid tumor tissue

Figures 1 and 2 explain some examples of macroscopic derangement of solid tumor, which are now generally accepted as an effect named EPR (Enhanced Permeability and Retention) effect[5,6]. In the experiment shown in Figure 1, a blue dye, Evans blue was injected into the tail vein of tumor-bearing mice at a dosage of 10mg/kg. At this dose level there was no free dye in the plasma; it was mostly bound to albumin, as confirmed by molecular sieve chromatography. These four pictures illustrate tumor tissue

0, 6hr, 24hr and 72hr after iv injection of Evans blue; the tumor is blue compared with the normal tissue in the background. After extracting the dye, the concentration of Evans blue was determined spectrophotometrically.

Figure 2 shows quantification of Evans blue at different times in plasma, tumor and normal skin. A gradual increase in the intratumor concentration can be seen. What to be emphasized is that the intratumor concentration became much higher than that of the plasma 12hr after injection, while the plasma concentration progressively decreased. Another important thing is that the high intratumor concentration was retained over a prolonged period. This long retention of macromolecules in tumor tissue has been confirmed by many groups[7]. Then why do macromolecule including Evans blue-albumin as explained, accumulate effectively in tumor?

There were several data showing that vascular permeability occurred at the site of inflammation such as bacterial infection and was caused by bacterial proteases, which activate an endogenous kinin-generating cascade. Kinin is the most potent vascular permeability factor in the body[8].

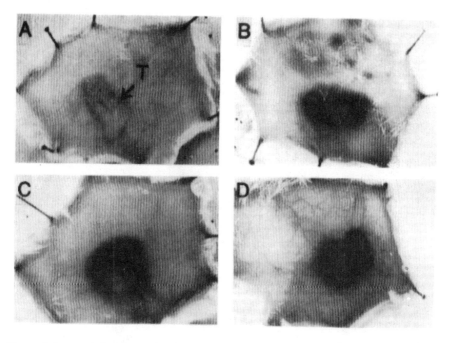

Figure 1. Accumulation of Evans blue-albumin complex in the tumor tissue. Tumor S-180 was injected into mice skin. A to D provide a macroscopic picture of the tumor in the skin taken at 0, 6, 24 and 72h, respectively, after i.v. injection of Evans blue (10mg/kg).

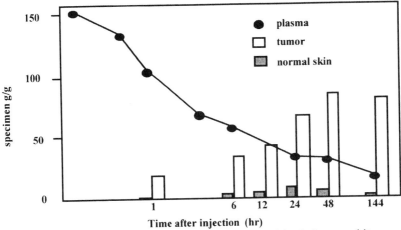

Figure 2. Clearance of Evans blue-albumin complex from blood plasma and its accumulation in tumor tissue and normal skin in tumor-bearing mice. Quantification of concentration of the Evans blue-albumin complex at different times for plasma (●), normal skin (▨), and tumor (□).

Then we hypothesized that both lesions, tumorous and inflammatory, shared a common cascade of kinin generation. In fact, we succeeded in purifying two types of kinin from the ascitic fluid of a patient with gastric cancer by gel filtration and reversed phase high-performance liquid chromatography (Fig.3).

Amino acid composition and amino acid sequence studies revealed that fraction I (panel 2) and fraction II (panel 3) were bradykinin and [hydroxyprolyl[3]] bradykinin, respectively.

At the same time we found that kinin content was relatively high in all tumorous exudates of patients with pancreatic, ovarian, pulmonary, hepatic and gastric cancer[9] (Table 1).

Table 1: Kinin Content in Various Tumor Effusions

Ascites	Kinin content (ng/ml)[a]	
	Bioassay	Enzyme immunoassay
1. S-180 (mice)[b]	1-4	0.625-2.5
2. AH-130 (rats)[b]	1-8	0.625-2.0
3. Pancreas cancer (human)[c]	1	0.625
4. Stomach cancer (human)[c]	30	8.0
5. Stomach cancer (human)[c]	40	10.0
6. Hepatoma with liver cirrhosis (human)[c]	2.5	—[d]
7. Ovarian cancer (human)[c]	2.5	—[d]
8. Lung cancer (human)[c]	40	20

a) Synthetic bradykinin was used as a standard; b) Values are from several rodent ascites. Different ascites and pleural effusion samples were used for bioassay and for enzyme immunoassay; c) Individual cases; d) Not done

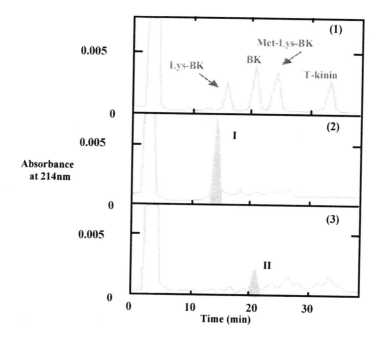

Figure 3. Comparison of elution profile of various authentic kinins (panel 1) and fractions with kinin oxytotic activity (panel 2 and 3) purified from ascitic fluid of a patient with gastric cancer.

Thus we suspected that ascitic fluid formation was brought about by enhanced vascular permeability caused by kinin.

By the way, kinin is generated by a kallikrein-dependent cascade (Fig.4).

We therefore examined whether ascitic formation could be blocked by the soy bean trypsin inhibitor SBTI, which is a specific inhibitor of kallikrein. SBTI was administered ip for 14 days beginning from the day of ip tumor inoculation. The results indicate significantly less accumulation of the fluid with SBTI (Fig.5). We concluded that enhanced vascular permeability occurred in tumor tissue partly because of the potent permeability factor, Kinin[9].

Figure 6 shows a diagram of normal and tumor tissue. It is well known that small molecules easily leak from normal vessels in the body and are rapidly filtered in the kidney glomeruli. Therefore they have a short plasma half life. On the other hand, macromolecules have a long plasma half life because they are too large to pass through the normal vessel walls, unless they are trapped by the reticuloendothelial system in various organs.

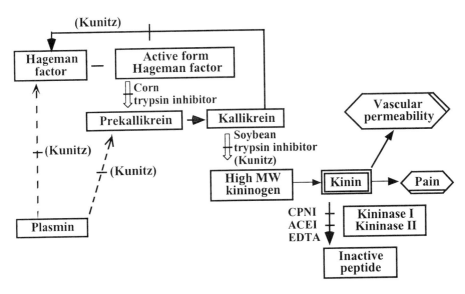

Figure 4. Kinin generation and degradation cascade and points of inhibitions by various inhibitors. ACEI, angiotensin converting enzyme inhibitor. Captopril and enalapril inhibit Kininase II (ACE). CPNI, carboxypeptidase N inhibitor. Dotted lines show possible activation.

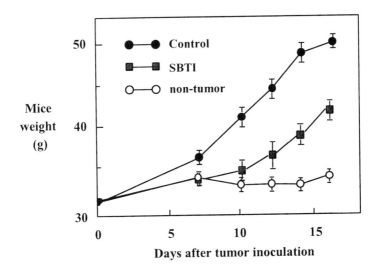

Figure 5. Effect of SBTI on ascitic fluid formation. SBTI was administered ip for 14 days beginning from the day of tumor inoculation. Ascitic volume can be estimated from the difference between the weight of controls with and without tumor. Tumor-bearing and nontumor-bearing mice (controls) are shown by ● and ■, respectively. SBTI (■) was given at 3mg/mouse per day.

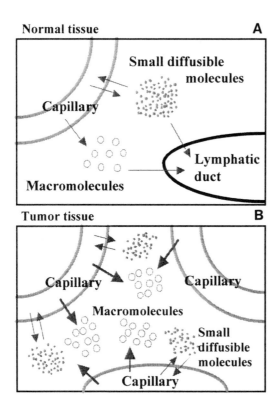

Figure 6. Diagrammatic presentations of normal tissue (A) and tumor tissue (B). Note that presence (A) and absence (B) of the lymphatic capillary, and much enhanced leakage of macromolecules are seen in tumor tissue. In tumor, macromolecules and lipids are not recovered via the lymphatic system but are readily accessible to the target tumor cells. Low molecular substances travese freely between the interstitial space and the blood capillary as well as the lymphatic duct.

In the solid tumor tissues shown in the lower panel, it was found from the pathological, pharmacological and biochemical studies of our or other group that solid tumors generally possess the pathophysiological characteristics: hypervasculature, incomplete vascular architecture, secretion of vascular permeability factors stimulating extravasation within the cancer, which was partly mentioned earlier, and immature lymphatic capillaries. These characteristics of solid tumors are the basis of the EPR effect, which was so named by Maeda[6]. Summarizing these findings, conventional low-molecular-weight anticancer agents disappear before reaching the tumor tissues and exerting their cell-killing effect. On the other hand, macromolecules and small particles should have time to reach and exit from tumor capillaries, by means of the EPR effect. To make use of the EPR effect, several techniques have been developed to modify the structure of drugs and make carriers.

2.   CLINICAL STUDY OF ADRIAMYCIN ENCAPSULATED PEG-IMMUNOLIPOSOME

2.1   Characteristics of the immunoliposome

Figure 7 shows the structure of MCC-465, which is the immunoliposome encapsulated adriamycin. The liposome is tagged with PEG and a monoclonal antibody.

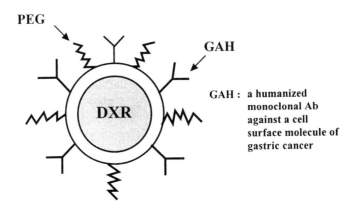

Figure 7. Schema of MCC-465, which is the immunoliposome encapsulated adriamycin. The liposome is tagged with PEG and the humanized F(ab')2 monoclonal antibody which mainly recognizes a cell surface protein of gastric cancer.

The antibody bound to liposome is humanized F(ab') 2.

At the present time, the epitope to be recognized by the antibody has not been well characterized, probably because the antibody recognizes the conformation of an epitope. Up to now, it has been found that the epitope is on the cell surface and 90% of gastric cancer tissue is positively stained while normal cells are always negative (data not shown).

We hope MCC-465 will be the "state of the Art" for the treatment of patients with advanced gastric cancer, because it may be able to kill the gastric cancer cells effectively through passive and active targeting.

2.2 The method of phase I study of the immunoliposome

The main purpose of this trial is to define the maximumtolerated dose (MTD), the dose-limiting toxicity and the recommended dose for phase II study of MCC-465, when used every 3 weeks for patients with metastatic or recurrent gastric cancer. We will also evaluate any indication of antitumor activity within a phase I setting.

Patients with metastatic or recurrent gastric cancers that are refractory to conventional chemotherapy or for which no effective therapy is available are eligible for entry, provided that the criteria described here are met.

Performance status must be 0 to 2, age is from 20 to less than 75 years, life expectancy at least 12 weeks. Patients must possess major organ function such as heart, lung, kidney, liver etc. Written informed consent must be obtained. A minimum of 4 weeks is required between prior chemotherapy or radiotherapy and entry in the study. Patients are ineligible if they have symptomatic brain metastasis or pre-existing cardiac disease including congestive heart failure, arrhythmia requiring treatment, or a myocardial infarction within the preceding 6months. Also ineligible are patients whose life time cumulative dose of doxorubicin exceeded $100mg/m^2$ or cumulative dose of epirubicin and pirarubicin exceeded $200mg/m^2$. Dose-limiting toxicity is defined according to the Japan Clinical Oncology Group Toxicity Criteria for the first cycle of drug administration; Grade 4 neutropenia not resolving within 5 days, febrile grade 4 neutropenia, grade 4 thrombocytopenia, or grade 4 nonhematologic toxicity (except nausea and vomiting). The maximum tolerated dose is defined as one dose level lower than the dose at which more than 50% of patients have experienced DLT.

Table 2. - Eligibility

1. Histologically confirmed adenocarcinoma of stomach.
2. Metastatic or recurrent disease.
3. After failure of standard chemotherapy.
4. Performance Status 0~2
5. Age: 20 to 75
6. Adequate organ function
7. Life oxpectancy $\geq$ 12 weeks
8. No radiotherapy within 4 weeks, No chemotherapy within 4 weeks
   (No MMC or Nitrosourea within 6 weeks)
   Prior ADM < 100 mg/m^2; Prior Epirubicin < 200 mg/m^2
9. No severe complication
10. Written informed consent

The immunoliposome is administered once every 3 weeks and the treatment is continued for up to 6 cycles.

The starting dose is 6.5 mg/m^2 which is equivalent to one tenth of the LD10 in rats.

Since many patients in phase I clinical trials are treated at doses of chemotherapeutic agents that are below the biologically active, they have a reduced the chance to get therapeutic benefit. Therefore we decided to adopt an accelerated dose escalation design followed by a modified Fibonacci method to reduce the number of such patients[10].

In this two stage design, the first stage allows for a single patient to be enrolled at each dose level. The dose of MCC-465 is doubled in each successive patient until grade2 toxicity is observed. If grade2 toxicity occurs in one patient, that dose level is given to another two patients. This is the time of starting the second stage design that is a modified Fibonacci method. Then, if one of the three patients develops DLT, that dose level is expanded to a total of six patients. If more than three of six patients experience DLT, the dose escalation is terminated (Fig.8).

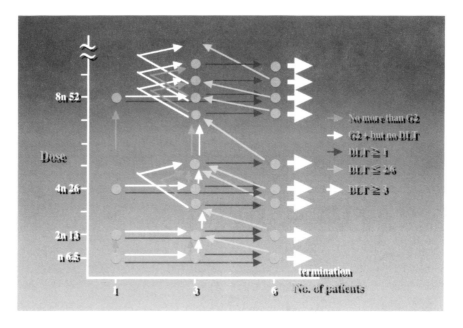

Figure 8. Dose escalation schema of MCC-465

3.   DEVELOPMENT OF MICELLES INCORPORATED KRN5500

This issue is not about a clinical trial but the story of how we made the polymeric micelles incorporating an anticancer agent based on the several inconvenient or adverse events resulting from a phase I clinical trial of the anticancer drug. KRN5500 is a newly semi-synthesized water-insoluble anticancer agent of which the main mechanism of anticancer activity is an inhibitory effect on protein synthesis[11].

The phase I clinical trial is now under way in our hospital and at NCI in the USA.

One of the problems in this clinical trial is that the injection of the drug is limited to a central vein. We can never inject the drug into a peripheral vein.

3.1   Structure and Characteristics of KRN5500

Figure 9 shows the chemical structure of KRN5500 which was semi-synthesized in an attempt to increase the therapeutic effects of spicamycin analogues. The main point is that the compound has a fatty acid chain. KRN5500 itself has little effect on the inhibition of protein synthesis in rabbit reticulocyte lysates[11].

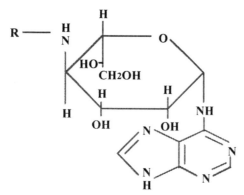

KRN5500   R; $CH_3(CH_2)_8CH=CHCH=CHCONHCH_2CO$
SAN-Gly   R; NH_2CH_2CO-
SAN   R; H

Figure 9. Chemical structure of KRN5500.

However, 4-N-glycylspicamycin aminonucleoside (SAN-Gly), which has no fatty acid chain and is thought be generated from KRN5500 by a cytosomal enzyme, has exhibited a marked inhibitory effect on protein synthesis in the cell-free system. Nevertheless, SAN-Gly has shown 1000-fold weaker cytotoxicity than KRN5500 in vitro because of its poor intracellular incorporation[11]. Therefore, to obtain an antitumor effect in vivo KRN5500 should be administered intravenously. Since it is insoluble in water, a mixture of organic solvents and chemicals must be used to dissolve the drug for iv injection.

The fatty acid chain of KRN5500 is pivotal for the drug internalization into cancer cells. 4-N-Glycyl spicamycin amino nucleoside (SAN-Gly), which has no fatty acid and is obtained after metabolization of KRN5500 by a cytosomal enzyme, exhibited a marked inhibitory effect.

More important thing is that several adverse events were seen after iv injection of even solvents alone in the preclinical experiment using rats (data not shown). This means that the solvents themselves has some toxic effects on rats. To overcome that problem, we have succeeded in incorporating the drug into polymeric micelles.

3.2   Incorporation of KRN5500 into Micelles

A-B block copolymer was dissolved in dimethyl sulfoxide and mixed with KRN5500 in DMSO. The mixture was then dialyzed against distilled water for at least 5hour using a cellulose membrane. Then sonication was carried out to obtain uniformly sized micelle particles. Finally we obtained these particles. The average diameter of the particles was 71nm[12] (Fig.10).

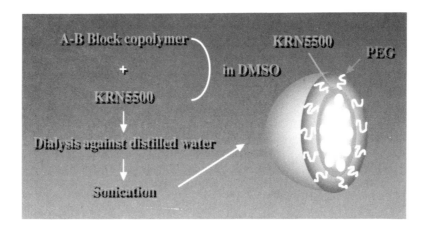

Figure 10. The method of incorporating KRN5500 into polymeric micelles. Block copolymer, PEG-P(BLA, C-16) was dissolved in dimethyl sulfoxide (DMSO) and mixed with KRN5500 in DMSO. The mixture was stirred at room temperature for 10 min, and then dialyzed against distilled water for at least 5h using a cellulose membrane. Sonication was then carried out to obtain uniformly sized micelle particles (approximate size, 70nm).

3.3 In vitro Antitumor Activity of KRN5500 and Micelles Incorporating KRN5500 (KRN/m)

IC50 values for KRN5500 and KRN/m in several human colonic, stomach and breast cancer cell lines are shown in Table 3. The level of antitumor activity was time-dependent for each drug. There was no remarkable difference between the IC50 values of KRN5500 and KRN/m at any exposure time[13].

Table 3. – ICS0 Value of KRN5500 and KRN/m in Various Cell Lines

	Exposure time					
	24 hr		48 hr		72 hr	
	KRN5500	KRN/m	KRN5500	KRN/m	KRN5500	KRN/m
Colonic cancer						
COLO 201	> 3.0	> 3.0	0.060	0.13	0.036	0.021
COLO 320	> 3.0	> 3.0	0.058	0.105	0.044	0.053
DLD-1	> 3.0	> 3.0	> 3.0	> 3.0	> 3.0	> 3.0
HT-29	> 3.0	> 3.0	0.046	0.050	0.038	0.051
LoVo	> 3.0	> 3.0	> 3.0	2.400	0.820	0.823
Stomach cancer						
MKN-28	> 3.0	> 3.0	0.122	0.250	0.045	0.041
MKN-45	> 3.0	> 3.0	0.037	0.042	0.019	0.021
MKN-72	> 3.0	> 3.0	0.085	0.156	0.065	0.062
TMK-1	> 3.0	> 3.0	0.035	0.059	< 0.01	0.010
KATO III	> 3.0	> 3.0	< 0.01	0.017	< 0.01	0.010
Breast cancer						
MCF-7	> 3.0	> 3.0	> 3.0	3.0	0.730	0.33
MDA-MB-435	> 3.0	> 3.0	> 3.0	2.45	> 3.0	0.40
T-47-D	> 3.0	> 3.0	> 3.0	> 3.0	0.160	0.17
SST	> 3.0	> 3.0	0.066	0.05	0.010	0.01

Each cell line was treated in triplicate for 24 h, 48 h and 72 h - MTT assay was used for obtaining IC_{50} value

3.4 In vivo Antitumor Activity of KRN5500 and KRN/m

The in vivo anti-tumor activity was evaluated with a human colonic cancer line, HT-29, which was inoculated into the abdominal skin of nude mice. 5.6mg/kg KRN5500 could be administered iv only once, because this dose of KRN5500 always induced irreversible inflammatory change in the tail of mice. We therefore evaluated the difference in anti-tumor activity between KRN5500 and KRN/m after a single injection of the two drugs.

One bolus injection of KRN5500 did not show any significant antitumor activity on HT-29 xenografts in comparison with the control, whereas the equivalent dose of KRN/m was significantly superior to the control. However, there was no statistically significant difference between the effects of these two drugs[13] (Fig.11).

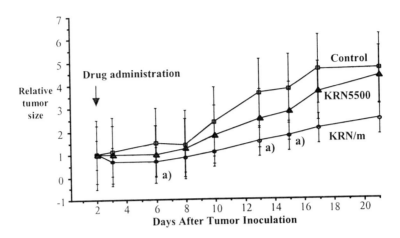

Figure 11. Changes in relative tumor (HT-29) size in abdominal skin of nude mice. After treatment with KRN5500 or KRN/m. KRN5500 at 5.6 mg/kg (▲), KRN/m in an equivalent amount to KRN5500 (●), or saline (■) was given iv on day 2 (arrow).
a) Significant difference between KRN5500 and control (P<0.05).

3.5   Toxicity of KRN5500 and KRN/m

Figure 12 shows body weight change of the nude mice. The data are from the same mice used in the treatment experiment. The weight of the mice injected with free KRN5500 was significantly lower than that of the controls or KRN/m-injected mice on day 7 after administration[13].

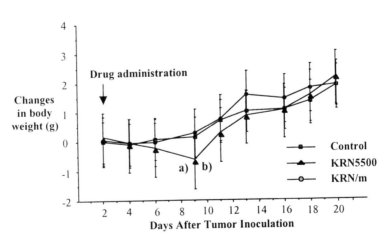

Figure 12. Body weight change of the nude mice. The data were from the same mice used in the treatment experiment (See Fig. 11). a) Significant difference between KRN5500 and control (P<0.05); b) Significant difference between KRN5500 and KRN/m (P<0.05).

Figure 13 shows pathological changes in the tail after iv injection of the two drugs. The tail of each mouse was cut in cross-sectional slices at 1cm proximal to the injection site one day after administration of the drugs. In the case of KRN5500, vascular necrosis with fibrin clot and skin degeneration was observed. On the other hand, in the case of KRN/m, there was no pathological change in the blood vessel or the skin of the tail[13].

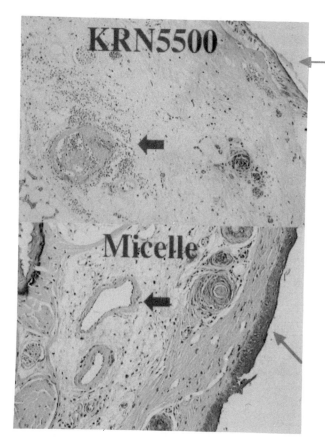

Figure 13. Vascular damage caused by iv injection of KRN5500, but not KRN/m. The tail of each mouse was cut in cross-sectional slices at 1cm proximal to the injection site after the iv administration of 5.6 mg/kg of KRN5500 or an equivalent amount of KRN/m.
A) KRN5500 : vascular necrosis with fibrin clot was observed (large arrow). Also, the skin of the tail had degenerated (small arrow).
B) KRN/m : there was no pathological change in the blood vessel (large arrow) or the skin of the tail (small arrow).

4.   CONCLUSION

From this study, we can conclude that incorporation of KRN5500 into micelles did not reduce its antitumor activity. The most remarkable result is that incorporation of KRN5500 into polymeric micelles showed a substantial reduction of toxicity caused by organic solvents or chemicals for dissolving KRN5500.

Therefore we expect that KRN/m will be superior to KRN5500 for clinical use[13].

More importantly the methodology of polymeric micelle drug carrier systems can be applied to other water-insoluble drugs used currently or in the future.

REFERENCES

1. Murad, AM. et al. 1993. Modified therapy with 5-fluorouracil, doxorubicin and Methotrexate in advanced gastric cancer. *Cancer*. 72: 37-41.
2. Glimelius, B. et al. 1994. Initial or delayed chemotherapy with best supportive care in advanced gastric cancer. *Ann. Oncol*. 5: 189-190.
3. Scheithauer, W. et al. 1993. Randomised comparison of combination chemotherapy plus supportive care with supportive care alone in patients with metastatic colorectal cancer. *Brit. Med. J*. 306: 752-755.
4. Shimada, Y. et al. 1999. Phase III study UFT+MMC versus 5-FU+CDDP versus 5-FU alone in patients with advanced gastric cancer. JCOG study 9205. *Proc Am Soc Clin Oncol*. 18: 272a (abst 1043).
5. Matsumura, Y and Maeda, H. 1986. A new concept for macromolecular therapeutics in cancer chemotherapy : mechanism of tumoritropic accumulation of proteins and the antitumor agent smancs. *Cancer Res*. 46: 6387-6392.
6. Maeda, H and Matsumura, Y. 1989. Tumoritropic and lymphotropic principles of macromolecular drugs. *Crit Rev Ther Drug Carrier sys*. 6: 193-210.
7. Duncan, R., Connors., and Maeda, H. 1996. Drug targeting in cancer therapy : the magic bullet, what next? *J Drug Targeting* 3: 317-319.
8. Matsumoto, K. et al. 1984. Pathogenesis of serratial infection : activation of the Hageman factor-prekallikrein cascade by serratial protease. *J Biochem*. 96: 739-746.
9. Matsumura, Y. et al. 1988. Involvement of the kinin generating cascade in enhanced vascular permeability in tumor tissue. *Jpn J Cancer Res*. 79: 1327-1334.
10. Simon, R. et al. 1997. Accelerated titration designs for phase I clinical trials in oncology. *J Natl Cancer Inst*. 89 : 1138-1147.
11. Kamishohara, M. et al. 1994. Antitumor activity of a spicamycin derivative, KRN5500 and its active metabolite in tumor cells. *Oncol Res*. 6: 383-390.
12. Yokoyama, M. et al. 1998. Incorporation of water-insoluble anticancer drug into polymeric micelles and control of their particle size. *J Controlled Release*. 55: 219-229.
13. Matsumura, Y. et al. 1999. Reduction of the side effect of an antitumor agent, KRN5500, by incorporation of the drug into polymeric micelles. *Jpn J Cancer Res*. 90: 122-128.

CELLULOSE CAPSULES – AN ALTERNATIVE TO GELATIN
Structural, Functional and Community Aspects

Shunji Nagata

Research Laboratories, Shionogi Qualicaps Co., Ltd, 321-5, Ikezawa-cho, Yamato-Koriyama, Nara 639-1032, Japan

1. INTRODUCTION

Hard gelatin capsules have been used in the pharmaceutical fields as an edible container for several decades. The development of mass-production facility and rapid capsule filling machine have made capsules one of the most popular oral dosage forms. However, gelatin capsules have some drawbacks derived from proteins. Gelatin capsule shells have 13-15 % water content and due to this gelatin capsule may not be suitable for readily hydrolyzed drugs. Furthermore when they were stored under severe conditions, some drugs reacted with amino groups of protein and crosslinked gelatin prolonged dissolution of drug[1]. Since gelatin for capsule is mainly derived from bovine, there is an implication of a potential risk posed by bovine spogiform encephalopathy (BSE)[2,3]. In addition, gelatin product from bovine and swine sources are sometimes shunned as a result of religious or vegetarian dietary restrictions.

For these reasons, trials to develop capsules free of proteins as an alternative to gelatin capsules were done in our company and the goals of new capsules are as follows:

1) To dissolve fast in water at 37°C,
2) To be manufactured by equipment used for gelatin capsules,
3) To overcome the drawbacks of gelatin capsules,
4) Materials should be approved as a pharmaceutical additive.

Biomedical Polymers and Polymer Therapeutics
Edited by Chiellini *et al.*, Kluwer Academic/Plenum Publishers, New York, 2001

Ultimately we came up with "cellulose capsules" which do not contain components of animal origin.

The composition of the capsules is shown in Table 1. The main material is hydroxypropylmethylcellulose (HPMC). HPMC alone does not gel at low temperature although it gels above 60°C. Hence small amounts (less than 1 % of HMPC) of carrageenan and potassium chloride are added as a gelling agent and a gelling promoter, respectively. Coloring agents can be added if necessary. The chemical structures of HPMC and kappa carrageenan are shown in Figure 1. Carrageenan is widely used in food industry as a natural gelling agent and derived from seaweed.

Table 1. Composition of HPMC capsules

Substance	%	Function
HPMC (USP)	q. s. 100	Base
Carrageenan (NF)	0.35-1.00	Gelling agent
Potassium chloride (USP)	0.35-0.50	Gelling promoter
Titanium oxide, Blue No.1, 2		Coloring agent
Yellow Ferric oxide, etc.		

HydroxyPropylMethylCellulose

Kappa Carrageenan

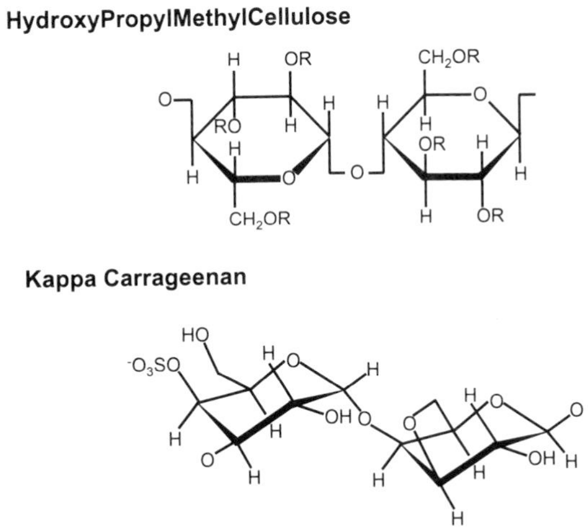

Figure 1. Chemical structures of hydroxypropylmethylcellulose and carrageenan

The cellulose capsules can be manufactured by the dipping and forming method, employed for the manufacture of hard gelatin capsules[4]. Shaped pins are dipped into HPMC solution of which temperature is maintained over room temperature. The pins are picked up from HPMC solution, cooled at room temperature, and dried by a warm air blow. HPMC base adhered to the pins gelled immediately to form the shape of capsule because carrageenan, a gelling agent, was added.

The typical pattern of the viscosity of HPMC solution against temperature is shown in Figure 2. HPMC is soluble in water below about 30°C but is not dissolved at higher temperature. The viscosity of HPMC has the shape of convexity when temperature falls from about 60°C. When the temperature of HPMC solution increased from about 30 to 40°C, the viscosity scarcely changed. Since the desirable viscosity of the base to manufacture capsules by the dipping method is about 4000 to 8000 mPas, cellulose capsules are manufactured using HPMC solution at 40 to 50°C.

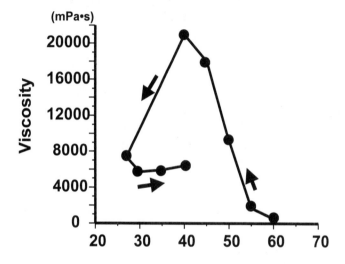

Figure 2. Viscosity of HPMC solution. 20% of HPMC solution was slowly refrigerated from 60°C to 27°C and slowly heated to 41°C. The viscosity was measured by Brookfield type viscometer.

2.   CHARACTERISTICS OF HPMC FILM

HPMC film may have looser structure compared to gelatin film according to the SEM observation of their section. We investigated the relation between the structure of film and the permeability of oxygen and water vapor through the films. The data of oxygen permeability are shown in Table 2. A diaphragmatic electrode was set on the one open side of a square stainless steel pipe of which volume was 104 cm^3. The opposite side was sealed by the films. Two stainless steel nozzles with valves were attached on the pipe and the inside air was replaced with nitrogen gas. Hence, at the beginning the inside concentration of oxygen is zero. Then the concentration of oxygen after 3 days was 0.1% in the case of gelatin film and 0.3% in the case of HPMC film[5].

Figure 3 shows the data of water vapor permeability through the films. Calcium chloride was placed in a cup, and the cup was sealed with the films and kept under the condition of 92%RH and 25°C. We checked the weight of calcium chloride increased linearly against time. The permeability rate of water vapor through gelatin and HPMC film was 446 and 263 $g/m^2/24hr$, respectively, indicating water vapor permeated more rapidly through gelatin film than HPMC film[5].

Oxygen permeability related to the looseness of film section but water vapor permeated more rapidly through gelatin film, which is tighter than HPMC film. Water vapor is considered not to permeate directly through the films and the permeability may relate to the water content of the films.

Table 2. Oxygen permeability

	Films		
	Gelatin	PEG/Gelatin *	HPMC
Conc. Of Oxygen in 3 days (%)**	0.1	0.0	0.3

*PEG/Gelatin ; gelatin including 5% of PEG4000
**Oxygen was determined by using a diaphragmatic electrode (n=3).

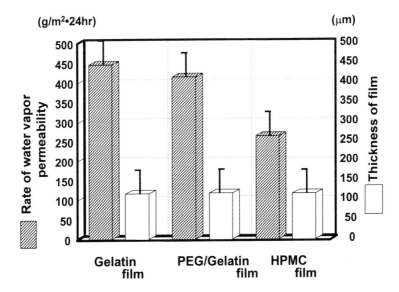

Figure 3. Rate of water vapor permeability through HPMC and gelatin films. Films of HPMC or gelatin, 22.4 cm^2, were covered over the cups where calcium chloride was placed. The cups were stored under the conditions of 25°C and 92%RH and the weight of calcium chloride was measured. Since the weight of calcium chloride was increased in proportion to the time, the rates were calculated at 24 hours as shown the below equation. The left scale is shown the rate of water vapor permeability and the right scale is shown the thickness of films.

$Q = 240 \cdot m/t \cdot a$, Q ; rate of water vapor permeability, m ; increased weight of $CaCl_2$ (mg), t ; time (hr), a ; effective area of films (cm^2)

3. CHARACTERISTICS OF HPMC CAPSULES

The equilibrium water contents of the capsules were measured by a Microbalance System (VTI, Florida, U.S.A.) at 25 and 40°C. Capsules were cut and about 10 mg of it was placed on a dish in Microbalance. After the piece was dried till the water content was 0%, the change of weight was continuously measured by changing relative humidity. The water contents of gelatin capsule were higher than that of HPMC capsule over the whole range of humidity (Fig.4). These data show that the water content of HPMC capsule is lower than gelatin capsule at 25°C and 40°C. In fact, the water contents of gelatin and HPMC capsule measured by Karl-Fisher were 13-15% and 4-6%, respectively when stored at 20-25°C and 40-60% RH.

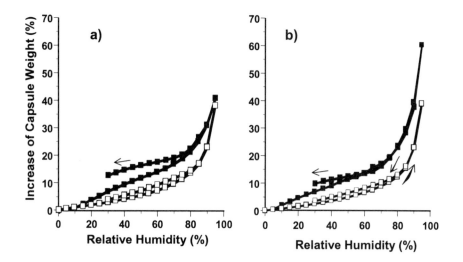

Figure 4. Equilibrium moisture content. a ; 25°C, b ; 40°C. Open symbols mean the data of HPMC capsules and closed symbols mean the ones of gelatin capsules.
(Courtesy of Formulation R & D Laboratories of Shionogi & Co., Ltd.)

It is well known that gelatin capsules become brittle when the capsules are dried and then the brittleness of capsules were examined. The capsules were kept under the various humidity conditions and the moisture content of them was adjusted at various levels. The brittleness of capsules was evaluated by the following two methods. One method reproduced a situation where the filled capsule was taken out of PTP (press through package) with fingers. The other reproduced a situation where empty capsules were handled in manufacturing or transporting. Figure 5 shows the respective equipment and the obtained data. In the former method (left figure), the capsules were filled with corn starch and pressed with 5 kg force. In the latter method (right figure), a 50 g weight was dropped on an empty capsule from 10 cm height. Gelatin capsules became brittle below 10% water content in both experiments. However, HPMC capsules were not broken even at 1% water content. HPMC capsules remain pliable even when their water contents become very low.

Generally, the lower the moisture content of a film is, the higher the static electricity is. The water content of HPMC capsules is lower than gelatin capsules but on the contrary the static electricity of HPMC capsules is lower than gelatin capsules as shown in Table 3[5]. One hundred of empty capsules were put into an acrylic vat of a friability tester and rotated for one minute at an angle of 50° and 26 rpm and the static electricity of capsules was immediately measured with STATIRON M2 (SHISHIDO Electrostatic Ltd., Tokyo, Japan). The static electricity of all capsules was positive. The static electricity of HPMC capsules was one-tenth of gelatin capsules regardless of the size of capsules.

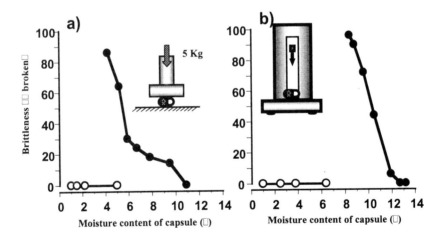

Figure 5. Brittleness test of capsules. a; pressure resistance test. b; hardness test. Capsules were filled with corn starch in case of a and empty capsules were tested in case of b. n=50. Open symbols mean the data of HPMC capsules and closed symbols mean the data of gelatin capsules.

Table 3. Static electricity of capsules. Each value in static column means the average and S.D. of three experiments.

Capsule	Size of capsules	Static (kV)
HPMC	1	0.17±0.29
	3	0.25±0.29
Gelatin	1	3.33±0.58
	3	3.67±0.58

4. LIQUID OR SEMI-SOLID FILLING

Capsules have traditionally been used to contain powder or granule formulations but in recent years have been adapted to contain oily liquid as soft gelatin capsules. However, hard gelatin capsules, stored under the high temperature and high humidity conditions, tend to be more brittle than soft capsules because the capsule shell is thinner.

We compared the appearance and the brittleness of gelatin and HPMC capsules, filling liquid and semi-solid excipients; propylene glycol, PEG400, labrasol and gelucire 44/14 (Gattefosse s.a.) , triacetine, triethyl citrate, medium chain fatty acid triglyceride (MCT), cotton seed oil, soybean oil, sesame oil, squalane, or emulsion. Gelatin capsules became brittle when stored in open state for one month at 45°C. However HPMC capsules did not become brittle except for the case of propylene glycol (Table 4). Then it is suggested that HPMC capsules should be suitable for many kinds of liquids or semi-liquids[6].

Table 4. Stability of liquid or semi-solid filling capsules after storage. The value in parenthesis means the broken % of the capsules after they were stored in glass bottles.

Excipients	Gelatin capsules		HPMC capsules	
	Appearance	% broken	Appearance	% broken
Propylene glycol	softening	100	softening	100
PEG400	good	100(100)	sweating	0
Labrasol	good	33(100)	deformation	0
Triacetine	good	66(0)	good	0
Triethyl citrate	good	0(0)	good	0
MCT	good	100(0)	good	0
Cottonseed Oil	good	100(0)	good	0
Soybean Oil	good	100(0)	good	0
Sesame Oil	good	100(0)	good	0
Squalane	good	33(0)	good	0
Emulsion	good	100(66)	sweating	0
Gelucire 44/14	good	66(0)	good	0

Emulsion; PEG400/water/MCT/BC-20TX = 74/13/10/3

5. DISINTEGRATION AND DISSOLUTION

Gelatin react with substances having aldehyde groups and gelatin capsules tend to be insoluble. After the capsules containing spiramycin, an antibiotic, were stored under 60°C and 75% RH for 10 days, the gelatin capsules did not disintegrate even at 30 min but the disintegration time of HPMC capsules treated by the same way did not change as shown in Table 5[7]. Spiramycin has an aldehyde group in its molecule and reacts with lysine residues in gelatin molecule, and then gelatin capsules became insoluble. As

HPMC has no reactive group like lysine in gelatin, HPMC capsules shell dose not react with spiramycin and its disintegration time is not changed.

Table 5. Disintegration time of capsules containing spiramycin after stored.

Preparation:	capsule (#2)	Apparatus: JP disintegration test with disk
	spiramycin 50 mg	Medium: JP 1st (pH 1.2)
	corn starch 200	Volume: 900 ml
Stored condition:	60°C, 75%RH	Temperature: 37°C

	Disintegration time (min)	
	Initial	after 10 days
HPMC capsule	4.6	4.5
Gelatin capsule	3.3	> 30

HPMC capsules filled with powder containing Vitamin C and B_2 and stored under the condition of 40°C and 75%RH for 2 months. Figure 6 shows the dissolution profiles of vitamin B_2. The dissolution rate of vitamin B_2 in all the mediums were fast and did not change after the storage under severe condition.

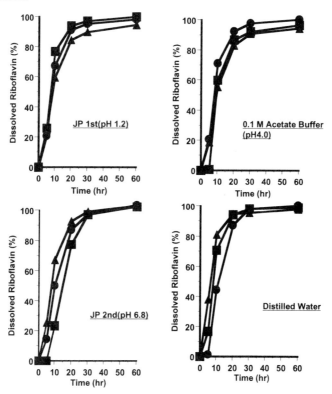

Figure 6. Dissolution profiles of HPMC capsules. Apparatus: Paddle, 50 rpm. Stored condition: 40°C and 75%RH. ■: initial, ●: one month, ▲: two months.
(Courtesy of Formulation R & D Laboratories of Shionogi & Co., Ltd.)

6. IN VIVO EVALUATION

HPMC capsules and gelatin capsules containing cephalexin were administered to six human volunteers. Plasma levels of cephalexin were determined by HPLC and the obtained pharmacokinetic parameters are shown in Table 6. All the parameters, AUC, Cmax, and Tmax are not significantly different between HPMC and gelatin capsules, suggesting that both capsules have equivalent biopharmaceutical performance[8].

Table 6. Pharmacokinetics parameters after administered cephalexin capsules with 100 ml of water according to crossover design test.

	AUC_{0-8} (mcg.hr/ml)	Cmax (mcg/ml)	Tmax (hr)
HPMC capsules	17.37±2.85	8.15±3.07	2.00±0.90
Gelatin capsules	16.86±3.56	9.56±3.70	1.50±0.80

7. CONCLUSION

Physical properties of capsule shells made of HPMC and gelatin are summarized in Table 7. Several merits of HPMC capsules stand out; low moisture content, low water vapor permeability, low static electricity, high tolerance to temperature, chemical inactivity, and solubility at room temperature. The features of HPMC capsules will break through the limited usability of traditional hard capsules.

Table 7. Physical characteristics of HPMC and gelatin capsules

Substance of Capsule	HPMC	Gelatin
Moisture content	3-7%	13-15%
Band sealing	Yes	Yes
Water vapor permeability	Low less than gelatin	Low
Oxygen permeabilty	High	Low
Light protection	Yes	Yes
Coloring	Yes	Yes
Maillard reaction with fillings	No	Yes
Degradation by light	No	Possible
Deformation by heat	Above ca. 80°C	Above Ca. 60°C
Static	Weak	Strong
Substrate for protease	No	Yes
Dissolution in water at room temperature	Soluble	Insoluble

ACKNOWLEDGMENTS

The development of HPMC capsules is achieved by three specialized companies. The first is Shin-Etsu Chemical Co., Ltd. which specialized in a field of polymer science. The second is Shionogi & Co., Ltd., an innovative pharmaceutical company and the last is Shionogi Qualicaps Co., Ltd., a hard capsule manufacturer. In particular, I very much appreciate of Dr. N. Muranushi's (Shionogi & Co., Ltd.) kindly supports.

REFERENCES

1. Digenis G. A., Gold T. B., and Shah V. P., 1994, Cross-Linking of Gelatin Capsules and Its Relevance to Their in Vitro-in Vivo Performance. *J. Pharm. Sci.*, 83: 915-921.
2. U.S. Department of Health and Human Services, Food and Drug Administration, Sep. 1997, Guidance for Industry "The Sourcing and Processing of Gelatin to Reduce the Potential Risk Posed by Bovine Spongiform Encephalopathy (BSE) in FDA-Regulated Products for Human Use."
3) The European Agency for the Evaluation of Medical Products, Human Medicines Evaluation Unit, Oct. 1998, "Note for Guidance on Minimising the Risk of Transmitting Animal Spongiform Encephalopathy Agents via Medicinal Products."
4) Jones B. E., Capsules, Hard. 1990, In Vol.2, "Encyclopedia of Pharmaceutical Technology" (J. Swarbrick and J.C. Boylan eds.), Marcel Dekker, Inc., New York and Basel.
5) Nishi K., Nagata S., 1999, Characteristics of HPMC capsules. *J. Pharm. Sci. Tech. Japan*, 59: Suppl. p.78, (Japanese).
6) Nagata S., Nishi N., and Matsuura S., 1999, Characteristics of HPMC capsules. Int. Symp. Strategies for Optimizing Oral Drug Delivery: Scientific to Regulatory Approaches (Kobe), Abst., p.161.
7) Matuura S. and Yamamoto T., 1993, New hard capsules prepared from water-soluble cellulose derivative. *Yakuzaigaku*, 53: 135-140, (Japanese).
8) Ogura T., Furuya Y., and Matsuura S., 1998, HPMC Capsules: An Alternative to Gelatin. *Pharm. Tech. Europe*, Nov.

POLYMERIC HYDROGELS IN DRUG RELEASE

[1]Federica Chiellini, [1]Federica Petrucci, [2]Elisabetta Ranucci, and [1]Roberto Solaro

[1] Dipartimento di Chimica e Chimica Industriale, Università degli Studi di Pisa, Via Risorgimento 35, 56126 – Pisa, Italy; [2] Department of Polymer Technology, Royal Institute of Technology, S-100 44 Stockholm, Sweden

1. INTRODUCTION

Among the several classes of biomedical materials, increasing attention is being devoted to polymeric hydrogels, which have the ability to swell in water or in aqueous solutions by forming a swollen gel phase that, in the case of crosslinked systems, will not dissolve regardless of the solvent[1]. An important feature of hydrogels is their biocompatibility, that can be attributed to their ability to simulate living tissue characteristics such as large water content[2], low interfacial tension with body fluids, permeability to metabolites, nutrients and oxygen[3].

At present, the most investigated hydrogels are those based on 2-hydroxyethyl methacrylate (HEMA), thanks to their ascertained non toxicity and widespread use in the production of soft contact lenses[4].

In the present paper we describe the synthesis and characterization of polymeric hydrogels based on HEMA to be used in the formulation of drug delivery systems. In particular HEMA hydrogels were developed to be used as components of dental implants amenable to the controlled release of antibiotics and in the preparation of polymeric scaffolds for tissue engineering application. Studies were carried out to assess a method for loading methronidasole, an antibiotic drug widely used in dentistry, into the hydrogel matrix. Attention was focused on the characterization of samples with different degree of crosslinking and of swelling in various aqueous solutions. Diffusion coefficients of water, inorganic salts and methronidasole were also investigated.

2.   PREPARATION OF HYDROGELS

Hydrogels were prepared by mixing HEMA with 1-8% by weight of crosslinking agents such as ethyleneglycol dimethacrylate (EGDMA) or tetraethyleneglycol diacrylate (TEGDA) and 0.75 % by weight of a water solution of ammonium persulphate/sodium pyrosulfite 1:1 mixture as a redox initiator.

The resulting solution was injected between two silanized glasses separated by a silicon rubber spacer (0.1–1 mm) and heated at 37 °C for 4 hours, to give clear transparent films (Fig.1).

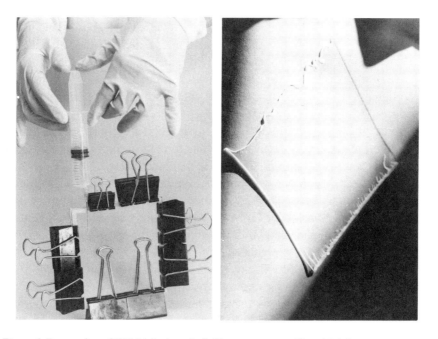

Figure 1. Preparation of HEMA hydrogels (left) as transparent films (right).

3. EVALUATION OF THE INTERACTIONS OF HEMA HYDROGELS WITH WATER

The degree of swelling of HEMA hydrogels was evaluated at 37 °C in deionized water, DMSO, phosphate buffer saline (pH 7.2), and cell culture medium (Dulbecco's modified eagles medium, 10 % foetal bovine serum, 4 mM glutamine).

The degree of swelling at equilibrium (*Se*) was evaluated as:

$$S_e = 100 \ \frac{W_s - W_d}{W_d}$$

where W_d and W_s are the weights of dry and swollen sample, respectively.

The degree of swelling of the investigated hydrogels in water and in aqueous solutions resulted included between 30 and 70 % and was not significantly influenced by the type and amount of crosslinking agent, while the degree of swelling in DMSO was almost ten times higher (100-400 %) and closely related to hydrogel structural parameters (Fig.2).

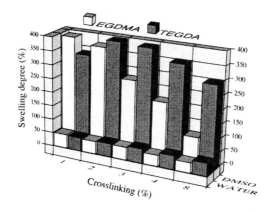

Figure 2. Swelling in water and DMSO of HEMA hydrogels containing different type and amount of crosslinking agents.

Water in swollen hydrogels can be classified as bound, interfacial and free water depending on the extent of interaction with the polar portion of polymeric material. Bound water is strongly associated to polymer chains by hydrogen bonding or dipolar interaction, interfacial water has weak interactions with polymer macromolecules and free water exhibits the same characteristic as pure water and does not interact with polymer chains (Fig.3).

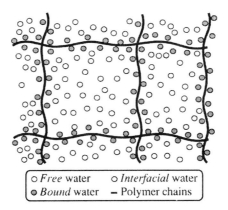

Figure 3. Schematic representation of the water structure in hydrogels.

The structure of water in HEMA hydrogels and its interactions with the polymer chains was investigated by differential scanning calorimetry (DSC)[5,6] and nuclear magnetic resonance (NMR)[7].

DSC analysis of water structure in HEMA hydrogels showed that both the total content of water and the percent of freezing water decrease with increasing the degree of crosslinking (Table 1).

The presence of several overlapping endothermic peaks connected to the melting of water indicates the existence of different types of non-freezing water in the hydrogels (Fig.4).

Table 1. Evaluation of non-freezing water in HEMA hydrogels.

Crosslinking agent		Type of water		
Type	(%-weight)	Total (%-weight)	Freezing (%-weight)	Non-freezing[a] (moles per m.u.)
EGDMA	1	52	24	5.5
	8	37	12	3.8
TEGDA	1	57	31	6.6
	8	50	30	5.1

[a] m.u. = monomeric unit

Evaluation of water mobility was performed also by measuring NMR T_1 and T_2 relaxation times to further investigate the water structure in HEMA hydrogels. The biexponential decay of NMR signals resulted in accordance with the presence of two different types of water molecules, one external to the polymer network representing about 60 % of the total water, with T_1 and T_2 values of about 2 and 1 s, and one internal (~ 40 %) with T_1 and T_2 values of about 100 and 20 ms, respectively (Tables 2 and 3).

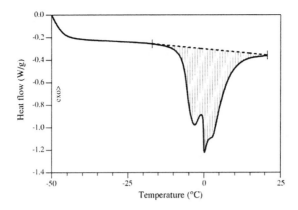

Figure 4. Heating curve of a HEMA hydrogel crosslinked with 8% of TEGDA and containing 49,9% by weight of water.

Table 2. Water relaxation times (T_1) of HEMA hydrogels[a]

EGDMA (%-weight)	Component I (%)	(s)	Component II (%)	(s)
0	60,9	2,39	39,1	0,121
1	51,7	1,65	48,3	0,110
2	54,6	1,39	45,4	0,108
4	57,6	2,02	42,4	0,098
8	72,5	2,16	27,5	0,105

[a] Evaluated by biexponential fit of the magnetization recovery at 37 °C.

Table 3. Water relaxation times (T_2) of HEMA hydrogels[a]

EGDMA (%-weight)	Component I (%)	(s)	Component II (%)	(s)
0	62,8	1,18	37,2	0,023
1	49,1	0,75	50,9	0,066
2	73,8	0,35	26,2	0,018
4	70,2	0,76	29,8	0,015
8	75,3	1,31	24,7	0,168

[a] Evaluated by biexponential fit of the magnetization recovery at 37 °C.

These data can be associated with the presence of two types of water having a different degree of interaction with the polymer network. No simple relationship between hydrogel structure and relative contribution of these two components can be established. Moreover, even the most mobile water component exhibited relaxation times smaller than those of pure water (T_1 = 4,5 s and T_2 = 2,8 s) indicating that some degree of interaction of water molecules with the polymer occurs in all cases.

The different results obtained by DSC and NMR analysis must be attributed to the different phenomena that are measured by these techniques[7].

DSC analysis showed that also the glass-transition temperature (Tg) of HEMA hydrogels is related to both polymer structure and hydrogel hydration extent. In particular the value of Tg decreased with the degree of hydration due to the plasticizing effect of water (Table 4).

Table 4. Dependence of the glass transition temperature (Tg) of HEMA hydrogels on their water content

Crosslinking agent		Hydration	Tg	ΔCp
Type	(%-weight)	(%-weight)	(°C)	(J/g· K)
EGDMA	1	0	109,4	0,24
EGDMA	1	0,6	99,7	0,24
EGDMA	1	5,2	72,4	0,12
EGDMA	1	15,1	39,9	0,05
EGDMA	8	0	108,7	0,25
TEGDA	1	0	102,4	0,24
TEGDA	8	0	97,8	0,25

The diffusion of water in HEMA hydrogels was investigated at 37 °C by measuring the kinetics of water absorption (Fig.5).

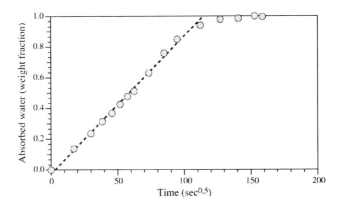

Figure 5. Variation of the water content *vs.* the square root of time in HEMA hydrogels crosslinked with 8 % EGDMA

The linear trend of water absorption *vs.* the square root of time is in accordance with a Fickian diffusion mechanism[8]. The diffusion coefficient D was evaluated by the following equation[9], where M_t is the mass of the water absorbed at time t, M_0 is the mass of the water at equilibrium, t is time in seconds, L is the sample thickness in centimeters and D is the diffusion coefficient in $cm^2 \cdot sec^{-1}$.

$$\frac{M_t}{M_0} = \frac{4}{\pi^{1/2}} t^{1/2} \, D^{1/2} \, L^{-1}$$

The diffusion coefficients (D ~ 1,4-1,6·10⁻⁷ cm²·sec⁻¹) resulted almost independent of the hydrogel structure (Table 5).

Table 5. Water diffusion coefficients (D) of HEMA hydrogels at37 °C

| Crosslinking agent | | $D \cdot 10^7$ |
Type	(%-weight)	$(cm^2 \cdot sec^{-1})$
EGDMA	1	1,65
EGDMA	3	1,55
EGDMA	8	1,43
TEGDA	1	1,59
TEGDA	3	1,65
TEGDA	8	1,75

4. METHRONIDASOLE LOADED HEMA HYDROGELS

Hydrogels loaded with methronidasole, an antibiotic specific for periodontal infections, were prepared according to two strategies, that is either by soaking preformed HEMA hydrogels in methronidasole solutions or by direct polymerization of monomers/methronidasole mixtures.

Methronidasole

In the first case preformed HEMA hydrogels were soaked in a methronidasole solution in water (10 g/l) or in DMSO (124 g/l). The amount of absorbed drug from the water solution was rather low (5,5 % by weight) while a 48 % drug absorption was obtained from the DMSO solution. However, when the soaked hydrogel was washed with diethyl ether to remove DMSO, most of the absorbed drug was removed as well.

Hydrogels with an appreciable content of methronidasole were obtained by direct polymerization of HEMA/EGDMA mixtures containing 15 % by weight of the drug, in the presence of ammonium persulphate/sodium pyrosulfite as redox initiator. Experiments were carried out at both at 37 and 80 °C. When the polymerization was performed at 37 °C, macroscopic phase separation of methronidasole crystals was observed. Rather homogeneous films were obtained at 80 °C, indeed the solubility of the drug in the monomer mixture is close to 15 % at this temperature.

The release of methronidasole from HEMA hydrogels water solution was monitored by measuring the UV absorption at 320 nm (Fig.7 and 8).

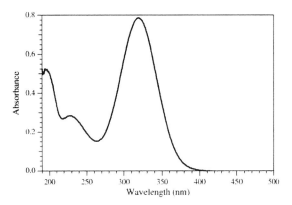

Figure 6. UV spectrum of methronidasole in water.

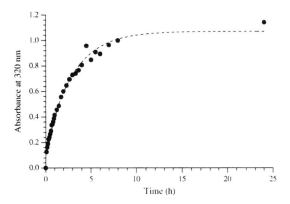

Figure 7. Release kinetics of methronidasole from HEMA hydrogels containing 3 % by weight of EGDMA.

The linear trend of methronidasole release *vs.* the square root of time (Fig.8) indicates that this process is diffusion controlled according to a Fickian II mechanism.

The methronidasole diffusion coefficient, $D_M \sim 1,0 \cdot 10^{-7}$ cm^2·sec^{-1}, resulted almost independent of polymer structure (Table 6) and somewhat lower than that of water, very likely due to the larger size of the methronidasole molecule. The much larger D_M value of the uncrosslinked sample can be attributed the synergistic effect of its linear structure and low molecular weight.

It is worth noting that in all cases the total amount of released methronidasole is significantly lower than that of the loaded drug. At present, no explanation of this unexpected behavior can be proposed.

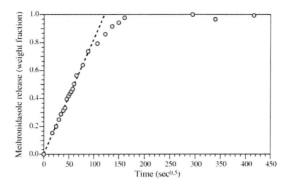

Figure 8. Diffusion plot of methronidasole from HEMA hydrogels containing 3 % by weight of EGDMA.

Table 6. Release of methronidasole from HEMA hydrogels in water

EGDMA (%-weight)	Temp.[a] (°C)	Released methronidasole[b] (%-weight)	$D_M \cdot 10^{7}$ [c] (cm$^2 \cdot$ sec^{-1})
0 [d]	80	59	34,7
1	37	82	0,7
1	80	85	7,5
3	37	65	1,2
8	37	75	1,1

[a] Polymerization temperature. [b] Referred to the hydrogel methronidasole content. [c] Methronidasole diffusion coefficient. [d] Low molecular weigh linear polymer.

5. MODIFIED HEMA HYDROGELS FOR TISSUE ENGINEERING APPLICATIONS

It is well known that hydrogels are characterized by a very good biocompatibility[10], mostly because of the low interfacial tension between the hydrogel surface and biological fluids that minimizes protein adsorption and cell adhesion[11].

In particular, HEMA hydrogels have found widespread biomedical use mainly for the production of soft contact lenses[12]. However, their surface properties must be modified to improve cell adhesion and proliferation in order to produce polymeric scaffolds suitable for tissue engineering applications. This objective can be achieved by introduction of charged groups onto the hydrogel surface. Accordingly, preformed HEMA hydrogels were reacted for 24 hours with succinic anhydride (SA) (Scheme 1) or 1,4-diaminobutane (DAB) in DMF solution (Scheme 2), at 50 and 100 °C, respectively.

Hydrogels containing 3-8 % carboxylated monomeric units or 3-15 % of aminated repeating units were obtained by using HEMA/SA and HEMA/DAB molar ratios of 5 and 0.5, respectively.

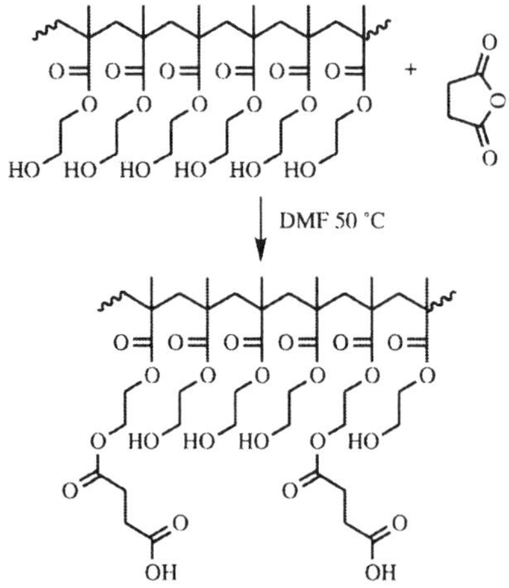

Scheme 1. Reaction of preformed HEMA hydrogels with succinic anhydride.

The alternative route based on the polymerization of mixtures of HEMA, 1-8 % of crosslinking agent, and 1-10 % of either methacrylic acid or 2-(N,N-dimethylamino)ethyl methacrylate was also investigated. Polymerization experiments were performed at 37 °C for 4 hours. By this procedure, HEMA hydrogels containing 1-10 % by weight of either carboxylic or amino groups were obtained as thin clear films.

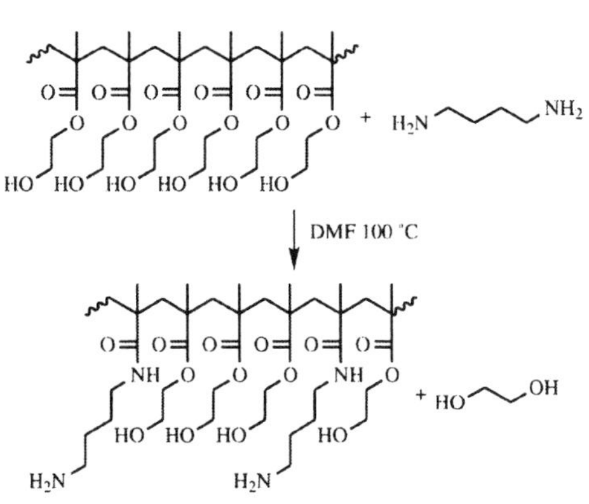

Scheme 2. Reaction of preformed HEMA hydrogels with 1,4-diaminobutane.

Surface charge measurements confirmed the presence of negative and positive charges on the surface of hydrogels containing either carboxylic or amine groups. Preliminary *in vitro* experiments seems to suggest that the presence of a small amount of amino groups on the hydrogel surface may improve cell adhesion. This indication needs however more substantial evidence.

6.   CONCLUSIONS

Biocompatible polymers with specific shape and tailored hydrogel properties can be obtained by polymerization of mixtures of HEMA with different amounts and type of crosslinking agent by using a redox initiator.

Investigation of the swelling behavior of the prepared hydrogels evidenced an appreciable dependence on both solvent type and polymer chemical structure. Additionally, the solvation process resulted to be controlled by solvent diffusion, according to a Fickian II mechanism.

The presence of several types of water characterized by different melting behavior and NMR relaxation times was observed in fully swollen hydrogels.

Controlled release systems can be obtained by loading HEMA hydrogels with an antibiotic drug, either by soaking preformed polymers in methronidasole solutions or by direct polymerization of monomer/drug mixtures. Also the release of methronidasole in water obeyed diffusive type kinetics, albeit at a slower rate than water and it was completed within 10 hours.

Introduction of charged positive and negative groups can be easily achieved by direct polymerization of appropriate monomer mixtures and by chemical transformation of preformed hydrogels.

The reported results indicate that HEMA hydrogels are endowed with suitable properties for their use in pharmaceutical and biomedical applications.

ACKNOWLEDGMENTS

The present work was performed with the partial financial support of CNR Progetto Finalizzato "Materiali Speciali per Tecnologie Avanzate II". The authors are grateful to Prof. M. Delfini (University "La Sapienza", Rome-Italy) for carrying out relaxation time measurements.

REFERENCES

1. Kim, S. W., Bae Y.H., Okano T., 1992, Hydrogels:Swelling, Drug Loading, and Release. *Pharmaceutical Research*, 9 (3) pp. 283-290.
2. Ratner B., Hoffman A., 1967,.in *Hydrogels for medical and related applications* (J. Andrade, Ed.), ACS, Washington DC.
3. Peppas N.A, 1996, in *Biomaterials Science, An Introduction to Materials in Medicine,* (Ratner B.D., Hoffman A.S., Schoen F.J, Lemons J.E. , Eds.), Academic Press, San Diego p. 60.
4. Davis P.A., Huang S.J., Nicolais L., Ambrosio L., 1991, in *High Performance Biomaterials,* (Michael Szycher, Ed.), Technomics, Basel, p.343.
5. Khare A.R., Peppas N.A., 1993, *Polymer*, 34 (22).
6. Jhon M.S., Amdrade J.D., 1973, *J. Biomed. Mater. Res.*, 7 (509).
7. Barbieri R., Quaglia M., Delfini M., Brosio E., 1998, *Polymer*, 39 (1059).
8. Kaneda Y., 1998 in *Biorelated Polymers and Gels, Controlled Release and Applications in Biomedical Engineering,* (T. Okano, Ed.), Academic Press, San Diego, p. 32
9. Migliaresi C., Nicodemo L., Nicolais L., Passerini P., 1984, *Polymer* 25 (686)
10. Saltzman W.M., 1997, in *Principles of Tissue Engineering*, (Lanza R., Langer R., Chick W.L., Eds.), Landes Bioscience, Academic Press, San Diego, p.225.
11. Folkman J., Moscona A., 1978, *Nature* 273 p. 345-349.
12. Wichterle O., Lim D., 1960, *Nature* 185 (117)

BIODEGRADABLE POLYROTAXANES AIMING AT BIOMEDICAL AND PHARMACEUTICAL APPLICATIONS

Tooru Ooya and Nobuhiko Yui
School of Materials Science, Japan Advanced Institute of Science and Technology, 1-1 Asahidai, Tatsunokuchi, Ishikawa 923-1292, Japan

1.   INTRODUCTION

Since 1978, to supramolecular chemistry has been given the most attention with regard to the chemistry of molecular assemblies including rotaxanes and polyrotaxanes[1]. A family of polyrotaxanes has been recognized as a molecular assembly in which many cyclic compounds are threaded onto a linear polymeric chain capped with bulky end-groups[2]. The name of rotaxane was given from the Latin words for wheel and axle, and thus it refers to a molecular assembly of cyclic and linear molecules. Polypseudorotaxanes are defined as inclusion complexes in which many cyclic molecules are threaded onto a polymeric chain (Fig.1a). Further, bulky blocking-groups are introduced at the ends of the pseudo-polyrotaxanes (Fig.1b) to prevent dethreading of the cyclic molecules. This is known as a polyrotaxane which is a family of newly categorized molecular assemblies. Our recent studies have focused on the design of biodegradable polyrotaxanes aiming at biomedical and/or pharmaceutical applications.

Biomedical Polymers and Polymer Therapeutics
Edited by Chiellini *et al.*, Kluwer Academic/Plenum Publishers, New York, 2001

(a)

(b)

Figure 1. Schematic illustration of polypseudorotaxane (a) and polyrotaxane (b).

1.1 Studies on Polyrotaxanes using Cyclodextrins

Cyclodextrins (CDs) are frequently used as building blocks for supramolecular structure because they have a hydrophobic cavity that can encapsulate a guest molecule[3]. Also, CDs are of great importance in the pharmaceutical sciences, since CDs and certain drugs form water-soluble and low-toxic complex molecules. Rotaxanes, polypseudorotaxanes and polyrotaxanes have been studied as a new molecular architecture in comparison with homopolymers and/or copolymers. In 1976, Ogata *et al.* observed a polycondensation reaction of a pseudorotaxane consisting of a β-CD and an α, ω-diaminohydrocarbon with a diacid chloride to form a polypseudorotaxane[4]. This paper was the first one describing the synthesis and characterization of a polypseudorotaxane using CDs. Recent progress in polypseudorotaxane preparation using CDs has been extensively accelerated by Harada *et al.*, who found polypseudorotaxanes consisting of an α- or γ-CD and poly(ethylene glycol) (PEG)[5-7], a β- or γ-CD and poly(propylene glycol) (PPG)[8,9], a α-CD and polyesters[10,11], and a γ-CD and polyisobutylene[12]. Further, they developed a molecular necklace in which many α-CDs are threaded onto an α, ω-diamino PEG chain capped with dinitrophenyl groups[6,13,14]. Their findings gave us a great opportunity to design polyrotaxane-based drug carriers.

1.2 Considerable Functions Based on The Polyrotaxane Structure

According to the previous studies on polyrotaxanes, significant characteristics of the polyrotaxanes involve (i) non-covalent bonds between CDs and linear polymeric main chains and (ii) the chemical modification of CDs in the polyrotaxanes. Such the chemical characteristic of non-covalent bonds gives the following two functions: one is supramolecular dissociation if terminal blocking-groups are eliminated (Fig.2a), and the other is the sliding function of CDs along linear polymeric main chains (Fig.2b). Rather, the hydroxyl groups of the threaded CDs can be chemically modified to alter the physicochemical properties[15,16].

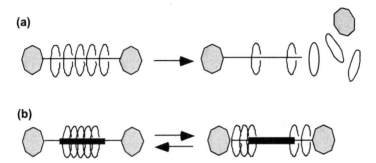

Figure 2. Proposed two functions of the polyrotaxanes: (a) supramolecular dissociation by terminal hydrolysis and (b) sliding of CDs along the liner polymeric chain.

2. BIODEGRADABLE POLYROTAXANES FOR DRUG DELIVERY SYSTEMS

We have proposed the design of biodegradable polyrotaxanes as a novel candidate for drug carriers. As mentioned above, polyrotaxanes exhibit their specific supramolecular structure with bulky blocking-groups. We proposed biodegradable polyrotaxanes that could be dissociated by hydrolysis of the terminal moieties. α-CD and PEG are used as main components of the polyrotaxane, because i) this combination is known to form a polypseudorotaxane that was reported by Harada *et al.*[5,6], ii) α-CD and PEG as degradation products have been used frequently in the pharmaceutical and biomedical fields[17,18], and iii) α-CDs in the polyrotaxanes can be chemically modified in order to incorporate drugs and to improve physicochemical properties such as solubility[15,16]. α-CD release accompanied by supramolecular dissociation via terminal hydrolysis may lead to a new mechanism of drug release from a polymeric drug carrier.

During the last two decades, water-soluble and/or biodegradable polymers have been widely demonstrated as polymeric carriers for biologically active agents[19]. The degradable bonds in biodegradable polymers can be arranged in many ways. The type of degradation is classified as illustrated in Figure 3 [20,21] In Type I, degradable bonds in a part of polymer backbone is cleaved. In Type II, the polymer is chemically crosslinked. Crosslinks of the polymers are soluble by backbone cleavages.

2.1   Types of Biodegradation and Drug Release

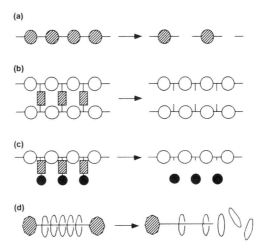

Figure 3. Classification of biodegradable polymers: (a) Type I, (b) Type II, (c) Type III, and (d) Type IV.

In Type III, active agents are directly immobilized to the polymer backbone. Such chemical immobilization of any drug onto water-soluble polymeric backbone via biodegradable spacer has been a general strategy to design polymeric drug carriers since Rigsdorf proposed his idea on a polymeric drug carrier in 1975[22]. In the case of biodegradable polyrotaxanes, the degradation type (Type IV in Fig.3) will be quite different: supramolecular dissociation by terminal hydrolysis. Generally, the drug in polymer-drug conjugates based on Rigsdorf's model was covalently bound to the polymeric backbone via biodegradable moieties. Those polymer-drug conjugates are to provide prolonged release action of the drugs by the hydrolysis or biological scission of the covalent bonds[23]. In our approach, drugs can be introduced to α-CDs via another biodegradable spacers in the polyrotaxane. Thus, regulating the degradability of the terminal degradable moiety and the spacers, it may be likely that the terminal moieties of the polyrotaxanes are firstly degraded by hydrolytic enzymes to release the drug-immobilized α-CDs. The covalent bonds between the drug and the α-CDs are then subjected to hydrolysis to release the active drugs.

2.2 Synthesis of Biodegradable Polyrotaxanes and Drug-Polyrotaxane Conjugates

We have synthesized a series of biodegradable polyrotaxanes in which many α-CDs are threaded onto a PEG chain capped with an L-phenylalanine (L-Phe) moiety via a peptide linkage[24-27]. The synthesis was carried out by i) preparing a polypseudorotaxane consisting of α-CD and amino-terminated PEG as previously reported by Harada *et al.*[5,6], and ii) capping both terminals of amino-groups in the polypseudorotaxane with benzyloxycarbonyl (Z-) L-Phe-succinimide in dry DMSO. The obtained polyrotaxanes were hydroxypropylated using propylene oxide in a 1N NaOH solution, and Z-groups were deprotected with palladium-carbon. The hydroxypropylation of the α-CDs improved the solubility of the polyrotaxanes in a phosphate-buffered saline. The purity of the polyrotaxanes was evaluated by ^{1}H-NMR and GPC[26]. The threading of α-CDs onto a PEG chain has been confirmed by detecting the hydrophobic cavity of α-CDs when the terminal peptide linkages were cleaved by papain[24,25].

The conjugation of the polyrotaxane with a model drug (theophylline) was carried out[28]. Hydroxyl groups of the hydroxypropylated (HP-)α-CDs in the polyrotaxane (ca. 20-22 α-CDs were threaded onto PEG, Mn=4000, with the degree of HP substitution being 8-9/α-CD) were activated by 4-nitrophenyl chloroformate in DMSO/pyridine at 0°C. The coupling of N-aminoethyl-theophylline-7-acetoamide with the activated polyrotaxane was carried out using DMSO/pyridine (1:1) as a solvent. Z-groups were deprotected to obtain a drug-polyrotaxane conjugate (Fig.4). The synthesis was confirmed by ^{1}H-NMR and GPC. The number of HP-α-CDs in the drug-polyrotaxane conjugate was determined to be ca. 13-14 by ^{1}H-NMR spectrum. From the GPC peaks, one theophylline molecule was found to be introduced into every two α-CD molecules in the drug-polyrotaxane conjugate.

R = H or

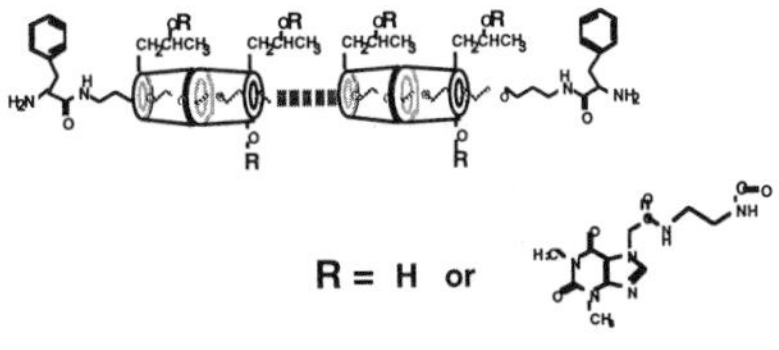

Figure 4. Schematic representation of the drug (theophylline) polyrotaxane conjugate.

2.3   Drug Release from Biodegradable Polyrotaxanes via Supramolecular Dissociation and its Relation to Solution Properties

2.3.1   Solution Properties of The Polyrotaxanes and Drug-Polyrotaxane Conjugates

Solution parameters of the polyrotaxane and the drug-polyrotaxane conjugate, which were determined by static light scattering measurements, were summarized in Table 1. Association number of the polyrotaxane and the drug-polyrotaxane conjugate was found to be much smaller than that of L-Phe-terminated PEG. This result indicates that hydroxyl groups in HP-α-CDs participate in preventing the association of such a supramolecular-structured polymer. The second virial coefficient (A_2) value is a measure of the extent of polymer-polymer or polymer-solvent interaction. Polymers are avoiding one another as a result of their van der Waals "excluded volume" when the A_2 value is large, and they are attaching one another in preference to the solvent when it is small or negative. Taking the characteristics of the A_2 value into account, it is suggested that the polyrotaxanes are considered to exhibit a loosely packed association, presumably due to their supramolecular structure. On the other hand, L-Phe-terminated PEG seems to exhibit a tightly packed association, because the terminal hydrophobic moieties can associate easily due to the flexibility of the PEG chains[29]. In addition, the drug-polyrotaxane shows less association state with strong molecular interaction of theophylline[28]. Furthermore, the ratio of hydrodynamic radius and radius of gyration (R_h/R_g) of the polyrotaxane revealed an anisotropic shape, indicating its rod-like structure[29]. The R_h/R_g value of the polyrotaxane was calculated to be 0.46. This value indicates that the polyrotaxanes do not have a spherical shape since the R_h/R_g value of spherical particles was ca. 1.29. The rod-like structure of the polyrotaxane is likely to affect the reduced association state.

Table 1. Solution parameters determined by light scattering measurements

Name	The Number of HP-α-CDs	Association Number	$A_2 \cdot 10^4$ (ml$\cdot$ mol$\cdot$ g^{-2})
Drug-polyrotaxane conjugate	13 - 14	6	-8.03
Polyrotaxane	20 - 30	2	3.40
L-Phe-terminated PEG	0	46	0.25

2.3.2   In vitro degradation of the drug-polyrotaxane conjugates and theophylline-immobilized HP-α-CD release in relation to their solution properties

As for the polyrotaxane, HP-α-CD was found to be released by the terminal hydrolysis in the presence of papain as a model enzyme (Fig.5). The terminal hydrolysis in the polyrotaxanes proceeded to over 85 %, in contrast to the limited hydrolysis in L-Phe-terminated PEG (~50%)[29]. Taking the A_2 values of the polyrotaxane into account (Table 1), the complete dissociation of the polyrotaxanes by hydrolysis may be due to the less association state related to the rod-like structure. On the other hand, the high association state of L-Phe-terminated PEG with strong molecular interaction is considered to decrease the accessibility of papain to the terminals. *In vitro* degradation of the drug-polyrotaxane conjugate by papain was examined in order to clarify the terminal hydrolysis of the drug-polyrotaxane conjugate and drug-immobilized HP-α-CDs (drug-HP-α-CDs) release. As shown in Figure 6, the release of drug-HP-α-CD from the drug-polyrotaxane conjugate was found to be completed around 200 h. HP-α-CD release from the polyrotaxane proceeded to reach 100% at 320 h under a similar experimental condition. Considering the solution properties of the drug-polyrotaxane conjugate as mentioned above, these results indicate that the association of the conjugate does not induce the steric hindrance but rather enhances the accessibility of enzymes to the terminal peptide linkages. Presumably, the terminal peptide linkages in the drug-polyrotaxane conjugate are likely to be exposed to the aqueous environment as a result of the specific association nature.

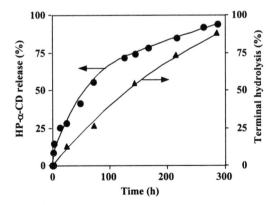

Figure 5. HP-α-CD release and terminal hydrolysis of the polyrotaxane in the presence of papain as a model enzyme.

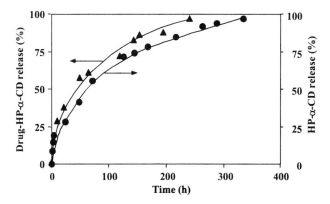

Figure 6. Drug-HP-α-CD release from the drug-polyrotaxane conjugate (▲) in comparison with HP-α-CD release from the polyrotaxane (●).

The most important aspect in relation to the solution properties is complete degradation of the drug-polyrotaxane conjugates. Previously, Ulbrich *et al.* reported that the rate of drug release via enzymatic hydrolysis of the spacers is reduced with an increase in the association of the conjugates[23]. This is due to the steric hindrance of enzymatic accessibility to the spacers. Therefore, special interest should be given to our results that the association of our designed conjugates did not hinder but rather enhanced the accessibility of the enzyme to the terminal peptide linkages.

3.   BIODEGRADABLE POLYROTAXANES FOR TISSUE ENGINEERING

Biodegradable polymers have been studied as implantable materials for cell growth and tissue regeneration[30]. Aliphatic polyesters such as poly(L-lactic acid) (PLA) is a representative of implantable materials for tissue engineering. However, high crystallinity of PLA reduces water intrusion into the crystalline regions resulting in incomplete hydrolysis; crystalline oligomers remain in the tissue for a long time. Such a complex situation causes chronic inflammatory reactions at the implanted sites[31]. Our next concern of biodegradable polyrotaxanes is in the design of implantable materials for tissue engineering. A novel design for biodegradable polymers has been proposed by constructing a supramolecular structure of a polyrotaxane; α-CDs threaded onto a PEG chain are capped with benzyloxycarbonyl (Z-) L-Phe via ester linkages[32]. The most favorable characteristics of the polyrotaxanes as implantable materials will involve their excellent biocompatibility, mechanical properties and instantaneous dissociation properties. The polyrotaxane may have high crystallinity which is due to intermolecular hydrogen bonds between the hydroxyl groups of the α-CDs. This high crystallinity is postulated to disappear due to terminal hydrolysis, although the crystallinity of aliphatic polyesters usually increases as a result of their hydrolysis. The hydrolysis of the ester linkage in a certain PEG terminal will trigger rapid dissociation of the supramolecular structure into α-CDs, PEG and Z-L-Phe. This specific structure of the polyrotaxane may enable us to maintain the appropriate mechanical properties until tissue regeneration and to be degraded completely. Further, hydroxyl groups of α-CDs may be available to introduce biologically active molecules such as RGD peptide for enhancing cell interactions and hydrophobic groups for controlling the supramolecular dissociation. Such hydrophobization of α-CDs in the polyrotaxane may make it possible to delay the time to complete the supramolecular dissociation, although the terminal hydrolysis may proceed completely.

3.1 Preparation of a Biodegradable Polyrotaxane with Ester Linkages and its Chemical Modification

In order to ester linkages at both terminals of PEG (*Mn*=3300), terminal hydroxyl groups were carboxylated by using succinic anhydride. The carboxy-terminated PEG was activated using *N*-hydroxynimide and dicyclohexylcarbodiimide. An amino-terminated PEG was prepared by using the activated PEG and ethylenediamine. Polypseudorotaxane (inclusion complex) was prepared from α-CDs and the amino-terminated PEG, according to the method reported by Harada *et al* [5]. The end-capping reaction of polypseuorotaxane was carried out using Z-L-Phe succinimide ester in DMSO as mentioned in the section **2.2**. Acetylation of the polyrotaxane was carried out using acetic anhydride in pyridine. From the ^{1}H-NMR spectrum of the acetylated polyrotaxane, the peaks of α-CDs, CH_2 of PEG, aromatic of Z-L-Phe and CH_3 of acetyl groups were confirmed. Further, the polyrotaxane structure was confirmed by differential scanning calorimetry (DSC): the endothermic peaks at 50, 140 and 244 °C relating to the melting points of the amino-terminated PEG, Z-L-Phe and the acetylated α-CD were not observed. The average number of α-CDs in the acethylated polyrotaxane was determined to be *ca.* 20 by the ^{1}H-NMR spectra. Table 2 summarized the reaction condition and the result of the acetylation. When the feed ratio was one and three, the degree of acetylation was calculated to be 15 and 32%, respectively. In order to obtain much higher degree of acetylation, it was considered to feed higher concentration of acetic anhydride or to prolong higher reaction time. In this case, the reaction time was prolonged to obtain much higher degree of acetylation (**ACRX-100** in Table 2) because the reaction mixture was heterogeneous. These results suggest that the degree of acetylation can be controlled by both the feed ratio and the reaction time.

Table 2. The reaction condition [a] and the result in the acetylation of the polyrotaxane

Code	Acetic anhydride in feed (mmol)	Reaction time (h)	Degree of acetylation (%) [b]	Number of α-CDs in polyrotaxane	
				Before acetylation	After acetylation
ACRX-15	7.2	0.25	15	20	20
ACRX-32	21.6	0.25	32	20	18
ACRX-100	21.6	48	100	20	19

a Hydroxyl groups in the polyrotaxane were 7.2 mmol.
b Determined by ^{1}H-NMR spectra.

3.2   In vitro hydrolysis of the acetylated polyrotaxanes

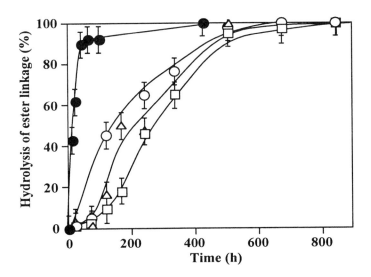

Figure 7. Cumulative hydrolysis of ester group in the polyrotaxane. non-acethylated polyrotaxane:($\bullet$), ACRX-15: ($\square$), ACRX-32: (Δ), ACRX-100: ($\triangle$), (n=3, Mean $\pm$ S. E. M.).

In vitro hydrolysis of the acethylated polyrotaxanes was performed, and the acethylated α-CD release via terminal hydrolysis was evaluated by HPLC using a reverse phase column.   Figure 7 shows the hydrolytic behavior of ester linkage of non-acethylated polyrotaxne and the acethylated polyrotaxanes. The time to complete the terminal ester hydrolysis was significantly delayed by the acetylation. This result suggests that the rate of the terminal hydrolysis can be controlled by chemical modification of α-CD in the polyrotaxane.   Interestingly, the time to complete the terminal ester hydrolysis was found to be shorter than that of α-CD release in any acetylated   polyrotaxanes.   If   the   terminal   ester   linkages   are   cleaved statistically, the percentage of acetylated α-CD release can be calculated to 75% at the terminal ester hydrolysis of 50%[26,29]. Based on this assumption, delay time of α-CD release (Δt) in relation to the degree of acetylation was calculated as follows:

$$\Delta t = T_{obs} - T_{theo}$$

where T_{theo} is theoretical time to reach 75% α-CD release (observed time to reach 50% hydrolysis in Fig.7), and T_{obs} is observed time to reach 75% α-CD release. The relationship between Δt and the degree of acetylation is shown in Figure 8. Δt reached 400 h when the degree of acetylation was over 30%. This result indicates that a degree of acetylation of at least 30% is required to prolong the supramolecular dissociation. Thus, it is suggested that supramolecular structure of the ordered α-CDs based on the polyrotaxane structure remains over 30% of the degree of acetylation, although the terminal ester hydrolysis proceeds completely.

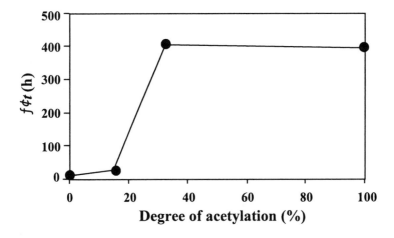

Figure 8. Delay-time of the supramolecular dissociation as a function of the degree of acetylation.

4.   POLYROTAXANES FOR STIMULI-RESPONSIVE MATERIALS

New synthetic methods have been investigated over the past two decades in order to design "intelligent" or "smart" materials which exhibit large property changes in response to small physical or chemical stimuli. Stimuli-responsive polymers have been studied as smart materials aiming at the applications in biomedical fields or chemomechanical devices. The representatives of the stimuli-responsive polymers are poly(*N*-

isopropylacrylamide) and its copolymers, in which the coil-globule transition based on hydration-dehydration behavior is induced in response to temperature. Such a stimuli-responsive mechanism involves long relaxation time for the transition of the polymeric chain into solvent. The chemomechanical devices have been designed as hydrogels crosslinked with the temperature-responsive polymers[1]. In these hydrogels, the characteristic time of swelling-deswelling is typically governed by diffusion-limited transport of the polymeric components of the network in water. Since the polymer chain in the hydrogels is fixed at the both terminals, rate-determining step for the swelling-deswelling involves the relaxation of the polymer chains. This physicochemical property is likely to limit the advances in the design of actuators for chemomechanical devices. One of the ideal chemomechanical systems will be a natural architecture of muscle contraction where myosin headpieces slide along actin filaments to initiate the contraction process. Such a sliding function is considered to be advantageous for energy balance compared with any other stimuli-responsive mechanisms.

Our alternative concern regarding polyrotaxanes is how such a molecular assembly can be utilized as a material with molecular dynamic functions: threading of many CDs onto a polyrotaxane might change the location along a linear polymeric chain in response to external stimuli which would be perceived as the action of mechanical pistons[34]. Because the driving force for such inclusion complexation of CDs with a polymeric chain is due to intermolecular hydrogen bonding between neighboring CDs as well as steric fittings and hydrophobic interaction between host and guest molecules, several stimuli such as temperature and dielectric change may be used to control the assembled state of the CDs onto a polyrotaxane (Fig.9).

dispersed   **assembled**

Figure 9. Stimuli-responsive change in CD location in a polyrotaxane.

4.1   Synthesis and Characterization of a Polyrotaxane Consisting of β-Cyclodextrins and a Poly(ethylene glycol)-Poly(propylene glycol) Triblock-Copolymer

A polyrotaxane, in which many β-CDs are threaded onto PEG-*block*-PPG-*block*-PEG triblock-copolymer (Mn = 4200, PPG segment Mn = 2250, PEG segment Mn = 975 x 2) named Pluronic® P-84 capped with fluorescein-4-isothiocyanate (FITC), was synthesized as a model of stimuli-responsive molecular assemblies[35]. Terminal hydroxyl groups of the triblock-copolymer were modified to obtain amino-terminated triblock-copolymer. The polypseudorotaxane between the amino-terminated triblock-copolymer and β-CDs was prepared in 0.1 M phosphate buffered saline at 40 ˚C. The terminal amino-groups in the polypseudorotaxane were allowed to react with FITC in DMF at 5 °C to obtain the polyrotaxane (Scheme 1). In this condition, a side reaction of β-CDs and FITC was prevented[35,36]. The threading of β-CDs in the polyrotaxane was confirmed by GPC, ¹H-NMR and 2D NOESY NMR spectroscopies using a 750 MHz FT-NMR spectrometer[35]. The number of β-CDs in the polyrotaxane was determined to be ca. 7 from the ¹H -NMR spectrum.

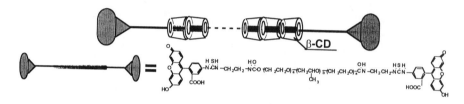

Scheme 1. A polyrotaxane in which many β-CDs are threaded onto the triblock-copolymer capped with fluorescein-4-isothiocyanate (FITC).

4.2   Change in the Location of β-CDs in Response to Temperature

The interaction of β-CDs with terminal FITC moiety was analyzed by means of induced circular dichroism (ICD) measurement in 0.01 N NaOH at 25°C[35]. A positive ellipticity [θ] of the polyrotaxane around 490 nm decreased when the temperature ncreased from 20 to 40 °C. This result ndicates that the interaction between β-CD and the terminal FITC moiety is considered to decrease with increasing temperature. The interaction of β-CD hydrophobic cavity with the PPG segment of the triblock-copolymer was analyzed by means of 750MHz ¹H-NMR spectroscopy. With increasing temperature, the peak of the methyl protons in the PPG segment shifted to the lower field and slightly broadened for the polyrotaxane[35]. The methyl proton peak shifted to the lower field is considered to be due to the interaction between the methyl protons of propylene glycol (PG) units and the cavity of the β-CDs. Assuming that the shifted methyl proton peak is the included methyl protons with β-CDs, the number of β-CDs located on the PG units can be estimated[35]. As a result, it is considered that ca. 5.2 β-CD molecules were localized on the PPG segment at 50 ˚C, although ca. 1.3 β-CD molecules existed on the PPG segment at 10 °C. Based on this number, the assembled ratio of β-CDs onto the PPG segment can be estimated, and

76% of β-CDs in the polyrotaxane was found to be assembled to the PPG segment at 50 °C. (Fig.10). Presumably, the number of β-CD molecules on the PPG segment will be determined by the balance of the following two forces: the enhanced hydrophobic interaction between the β-CD cavity and the PPG segment: the repulsive forces between the ionized hydroxyl groups in β-CD in the alkaline condition. Therefore, it is imagined that the majority of β-CDs move toward the PPG segment with increasing temperature although some β-CDs may reside on the PEG segments.

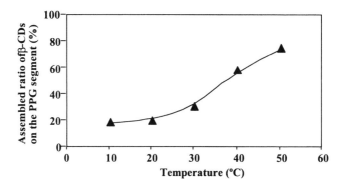

Figure 10. The assembled ratio of β-CDs onto the PPG segment as a function of temperature.

It was found that the association number of the polyrotaxane was decreased although that of the model triblock-copolymer was slightly increased from 35 °C to 50 °C[37]. These results indicate that the dissociation behavior of the polyrotaxane may be related to the localization of β-CDs along the triblock-copolymer. Stimuli-responsive polyrotaxanes can be one of candidates to create novel smart materials acting in a similar manner to natural chemomechanical proteins of myofibrils where myosin headpeices slide along actin filaments to initiate the contraction process.

5.   CONCLUSION

It is concluded that our designed biodegradable polyrotaxanes were demonstrated to be promising as a drug carrier as well as an implantable materials for tissue engineering[38,39]. Stimuli-responsive polyrotaxanes will provide new fields of smart materials that are constructed by mimicking natural supramolecular assemblies. The supramolecular structure and dissociation of the polyrotaxanes will be the most unique characteristic when considering biomedical and pharmaceutical applications. These polyrotaxanes are designed as to be degraded after meanwhile or by completing their medical functions in living body, and thus, to be excreted as soluble compounds. In near future, biodegradable and/or stimuli-responsive polyrotaxanes will enable us to achieve novel biomedical and pharmaceutical properties.

ACKNOWLEDGMENTS

The authors are grateful to Dr. Fujita, H., and Mr. Watanabe, J. (JAIST) for their help in the experiments on biodegradable polyrotaxanes. A part of this study was financially supported by a Grant-in-Aid for Scientific Research (B) (No.851/09480253) and a Grant-in-Aid for Scientific Research on Priority Areas "New Polymers and Their Nano-Organized System" (No. 277/09232225), from The Ministry of Education, Science, Sports and Culture, Japan.

REFERENCES

1. Lehn, J.-M., 1995, *Supramolecular Chemistry*, VCH, Weinheim.
2. Gibson, H. W., and Marand, H., 1993, *Adv. Mater.*, 5: 11-21.
3. Wenz., G., 1994, *Angew. Chem. Int. Ed. Engl.*, 33: 803-822.
4. Ogata, N., Sanui, K., and Wada, J., 1976, *J. Polym. Sci., Polym. Lett. Ed.*, 14: 459-462.
5. Harada, A., and Kamachi, M., 1990, *Macromolecules*, 23: 2821-2823
6. Harada, A., Li, J., and Kamachi, M., 1992, *Nature*, 356: 325-327.
7. Harada, A., Li, J., and Kamachi, M., 1994, *Nature*, 370: 126-128
8. Harada, A., and Kamachi, M., 1990, *J. Chem. Soc., Chem. Commun.*, 1322-1323
9. Harada, A., Okada, M., Li, J., and Kamachi, M., 1995, *Macromolecules*, 28: 8406-8411.
10. Harada, A., Kawaguchi, Y., Nishiyama, T., and Kamachi, M., 1997, *Macromol. Rapid Commun.*, 18: 535-539.
11. Harada, A., Nishiyama, T., Kawaguchi, Y., and Kamachi, M., 1997, *Macromolecules*, 30: 7115-7118.
12. Harada, A., Li, J., Suzuki, S., and Kamachi, M., 1993, *Macromolecules*, 26: 5267-5268
13. Harada, A., Li, J., Nakamitsu, T., and Kamachi, M., 1993, *J. Org. Chem.*, 58: 7524-7528
14. Harada, A., Li, J., and Kamachi, M., 1994, *J. Am. Chem. Soc.*, 116: 3192-3196
15. Hirayama, F., Minami, K., and Uekama, K., 1995, *J. Pharm. Pharmcol.*, 48: 27-31.
16. Pitha, J., Milecki, J., Fales, H., Pannell, L., and Uekama, K., 1986, *Int. J. Pharm.*, 29: 73-82.
17. Brewster, M. E., Simokins, J. W., Hora, M. S., Stern W. C., and Bodor, N., 1989, *J. Parent. Sci. & Tech.*, 43: 231-240.
18. Zalipsky, S., 1995, *Adv. Drug Deliv. Rev.*, 16: 157-182
19. Okano, T., Yui, N., Yokoyama, M., Yoshida, R., 1994, *Advances in Polymeric Systems for Drug Delivery*, Gordon & Breach Sci. Pub., Yverdon.
20. Heller, J., 1980, *Biometerials*, 1: 51-57.
21. Baker, R. W., 1987, *Controlled Release of Biologically Active Agents*, John Wiley, NY
22. Ringsdorf, H., 1975, *J. Polym. Sci. Symp*, 51: 135-153.
23. Ulbrich, K., Konak, C., Tuzar, Z., and Kopecek, J., 1987, *Makromol. Chem.*, 188: 1261-1272.

24. Ooya, T., Mori, H., Terano, M., and Yui, N., 1995, *Macromol. Rapid Commun.*, 16: 259-263.

25. Yui, N., and Ooya, T., 1996, *Advances Biomaterials in Biomedical Engineering and Drug Delivery* (Ogata, N., Kim, S. W. and Okano, T. eds.), Springier, Tokyo, pp.333-334.

26. Ooya, T., and Yui, N., 1997, *J. Biomater. Sci. Polym. Edn.*, 8: 437-456.

27. Yui, N., Ooya, T., and Kumeno, T., 1998, *Bioconjugate Chem.*, 9: 118-125.

28. Ooya, T., and Yui, N., 1999, *J. Controlled Release*, 58: 251-269.

29. Ooya, T., and Yui, N., 1998, *Macromol. Chem. Phys.*, 199: 2311-2320.

30. Langer, R., Vacanti, J. P., 1993, *Tissue Engineering, Science*, 260: 920-926.

31. Tsuruta, T., 1993, *Biomedical Applications of Polymeric Materials*, CRC Press, NY - USA

32. Watanabe, J., Ooya, T., Yui, N., 1998, *Chem. Lett.*, 1031-1032.

33. Ichijo, H., Hirasa, O., Kishi, R., Oosawa, M., Sahara, K., Kokufuta, E., and Kohno, S., 1995, *Radiat. Phys. Chem.*, 46: 185.

34. Fujita, H., Ooya, T., Kurisawa, M., Mori, H., Terano, M., and Yui, N., 1996, *Macromol. Rapid Commun.*, 171: 509-515.

35. Fujita, H., Ooya, T., and Yui, N., 1999, *Macromolecules*, 32: 2534-2541.

36. Fujita, H., Ooya, T., and Yui, N., 1999, *Macromol. Chem. Phys.*, 200: 706-719.

37. Fujita, H., Ooya, T., and Yui, N., 1999, *Polym. J.,* in press.

38. Ooya, T., and Yui, N., 1999, *Crit. Rev. Ther. Drug Carrier Syst.*, 16: 289-330.

39. Ooya, T., and Yui, N., 1999, *s.t.p. Pharma. Sci.*, 9: 129-138.

LIPOSOMES LINKED WITH TIME-RELEASE SURFACE PEG

Yuji Kasuya, Keiko Takeshita, and Jun'ichi Okada
Product Development Laboratories, Sankyo Co., Ltd. 1-2-58 Hiromachi, Shinagawa-ku, Tokyo 140-8710, Japan

1. INTRODUCTION

Retention of liposomes in the blood circulation and the resulting accumulation into pathological tissues like tumours can be increased by suppressing the interaction between the liposomes and biocomponents, such as plasma proteins and phagocytic cells. Modification of the liposome surface with poly (ethylene glycol) is one of the most efficient methods to obtain the long retention in the blood circulation [1-7].

Once the liposomes reach the target site, the pharmaceutical activity of the liposome-encapsulated drug can be expressed by one of the following ways: 1) drug is released from stably existing liposome, 2) drug is liberated by the erosion of the liposome, and 3) internalisation of drug-encapsulating liposome into the target cell. In the case of 2 or 3, strong interaction between liposomes and biocomponents is favourable. However, this feature is incompatible with the long retention in the blood circulation. To answer this problem, we focused on the preparation of liposomes whose surface can change from bioinert into bioreactive so that the accumulation at the target site could be followed by the interaction with biocomponents. We designed liposomes linked on the surface with PEG which releases the PEG with time due to unstable linkage between the liposome and PEG. In this study, basic physicochemical and biological properties of the PEG-releasing liposomes were investigated to confirm their applicability.

Biomedical Polymers and Polymer Therapeutics
Edited by Chiellini *et al.*, Kluwer Academic/Plenum Publishers, New York, 2001

2.   MATERIALS AND METHODS

2.1   Preparation of the Liposomes

Large multi-lamellar liposomes were prepared by the usual thin film hydration method[8] using phosphate buffered saline (PBS, 20 mM NaH_2PO_4, 135 mM NaCl, pH 7.4 adjusted with 1M NaOH) as the aqueous phase. The liposomes were composed of L-α-dipalmitoylphosphatidylcholine (DPPC), cholesterol (Chol), and another lipid, such as monomethoxy-poly (ethylene glycol)$_{2000}$-succinyl phosphatidylethanolamine (MPEG-DSPE, NOF Corporation, Japan), stearic hydrazide (SHz), and stearylamine (SA). The Chol content was fixed at 40 mol% of the total lipid content. The liposome dispersion was extruded through poly-carbonate membranes with a pore diameter of 50 nm, according to the method of MacDonald *et al*[9]. For the experiments in which the liposome amount was to be determined, liposomes labelled with the fluorescent marker, 1,1'-dioctadecyl-3,3,3',3'-tetramethyl-indocarbocyanine perchlo-rate[10] (DiI, Lambda Probes & Diagnostics, USA, 0.2 mol% of the total lipids) were prepared. Liposomes containing hydrazide groups on the surface (SHz liposomes) were prepared by adding 16 mol% of SHz in the lipid mixture. To 50 μL of the SHz liposome dispersion (lipid concentration: 100 mM) was added 450 μL of 100 mg/mL monomethoxypoly (ethylene glycol)$_{5000}$ -aldehyde (MPEG-CHO, NOF Corporation, Japan) in PBS. This mixture was incubated for at least 2 days to form a hydrazo bond between the hydrazide group on the liposome surface and the aldehyde group at the end of MPEG-CHO. The dispersion of SHz liposomes linked to MPEG-CHO (MPEG-SHz liposome) thus obtained was stored at 4°C and tested for its physicochemical or biological properties within 2 weeks. As a comparison to MPEG-SHz liposomes, other kinds of PEG-modified liposomes were prepared. Liposomes containing stearylamine (SA liposomes) were prepared in the same manner as SHz liposomes but by using stearylamine instead of stearic hydrazide. SA liposomes were also linked to MPEG-CHO, but the reaction was performed in the presence of 10 mg $NaCNBH_3$ to reduce the imino bond to a secondary amine so that the linkage could become irreversible (stabilised MPEG-SA liposomes).

Abbreviations: Chol, cholesterol; DiI, 1,1'-dioctadecyl-3,3,3',3'-tetramethylindocarbocyanine perchlorate; DPPC, L-α-dipalmitoylphosphatidylcholine; MPEG-CHO, mono-methoxypoly (ethylene glycol)$_{5000}$-aldehyde; MPEG-DSPE, monomethoxypoly (ethylene glycol)$_{2000}$-succinyl phosphatidyl-ethanolamine; PBS, phosphate buffered saline; PMN, polymorphonuclear leukocyte; SA, stearylamine; SHz, stearic hydrazide.

In addition, conventional PEG-linked liposomes (MPEG-DSPE liposomes) were prepared by adding MPEG-DSPE in the lipid mixture before hydration or after preparation of the bare liposome dispersion. In the latter method, the bare liposome dispersion was mixed with MPEG-DSPE and incubated at 50 °C for 3h to allow the anchor (i.e. distealoyl) moiety to be inserted into the liposome membrane. The final lipid concentration was maintained at 37.5 mM.

2.2   Physicochemical Evaluations

The partition coefficient of liposomes between immiscible aqueous biphases was determined as by Senior et al.[1] with some modification. Briefly, to 50 µL of DiI-labelled liposome dispersion (lipid concentration: 37.5 mM) was added 2 mL of PBS containing 5 w/v% PEG 6000 and 5 w/v% dextran T-500 (Pharmacia Biotech, Sweden). After vortexing, the mixture was allowed to stand for 30 min to achieve a phase separation (upper layer, PEG rich; lower, dextran rich). Then, the concentration of DiI in each layer was fluorophotometrically measured (exitation, 550 nm; emission, 565 nm) after solubilising the liposomes and DiI with the addition of 2 w/v% aqueous solution of Triton to 50 µL of each layer. The partition coefficient was calculated by dividing the DiI concentration in the upper layer by that in the lower layer.

The hydrodynamic size of the liposomes in PBS was measured by dynamic light scattering using Nicomp Particle Sizer Model 370 (Nicomp Particle Sizing Systems, USA).

In the case of MPEG-DSPE liposome, amount of MPEG-DSPE incorporated was determined as by Allen et al.[7]. Briefly, the liposome dispersion was ultracentrifuged (150000 G, 20 min) to obtain a supernatant containing MPEG-DSPE free of liposomes. Then, the MPEG-DSPE concentration in the supernatant was determined by a Bio Rad Protein assay method. From this value, the amount of liposome-incorporated MPEG-DSPE was calculated.

2.3   Evaluation of the Affinity of Liposomes for Cells

Human polymorphonuclear leukocytes (PMNs) were isolated and purified from human peripheral blood by the dextran-ficoll method[11], and finally suspended in PBS containing of 152 mM NaCl, 2.7 mM KCl, 8 mM Na_2PO_4, 1.5 mM KH_2PO_4, 5 mM glucose, 1 mM $CaCl_2$ and 1 mM $MgCl_2$ (PBS(+)) at a concentration of 10^7 cells / mL. The isolated PMN suspension was mixed with the liposome dispersion within 2 h after isolation as described below.

DiI-labelled MPEG-SHz liposome dispersion (25 µL, lipid concentration: 100 mM) was diluted up to 500 µL with PBS (+) and incubated at 37 °C for a predetermined period to allow PEG to be released from the liposome surface. This dispersion was mixed with 500µL of the above-obtained PMN suspension. After incubation at 37 °C for the predetermined period, the mixture was centrifuged to sediment PMN-associated liposomes, leaving the free liposomes in the supernatant. To the resultant pellet of PMNs (approx. 50 µL) was added 1 mL of PBS containing 2 w/v% Triton. The mixture was sonicated for 1 min with a probe sonicator (35 W) to dissolve DiI in the PMN-associated liposomes.

The amount of liposome associated with PMNs (A) was calculated by the following equation:

$$A = \{(I / I_0) \times C \times V\} / N \times 10^7 \text{ (nmol lipids / } 10^7 \text{ cells)}$$

where I and I_0 are, respectively, the fluorescence intensities measured from the solution of PMN-associated liposomes and the original DiI-labelled liposome dispersion with lipid concentration C (50 nmol /mL); V is the volume of the eluent (1.05 mL), and N is the number of cells (0.5×10^7 cells).

3.   RESULTS AND DISCUSSION

3.1   Preparation of Liposomes Linked with Time-release Surface PEG

The release of PEG from the liposome surface can be achieved by linking PEG to the liposome surface via an unstable bond. Among several kinds of unstable bonds, we selected the hydrazo bond, a kind of reversible bond, as the formation and cleavage of the bond could be controlled easily; MPEG-SHz liposome can be prepared by mixing a high concentration of SHz liposomes and MPEG-CHO as the equilibrium shifts towards the formation of the hydrazo bond. Also, a hydrazo bond would cleave with time to release PEG gradually from the liposome surface when the liposome dispersion is diluted with plasma after injection or with a medium in vitro.

Liposome formulation prepared from SHz, DPPC, cholesterol (16:44:40, molar ratio) resulted in a fine dispersion. In this study, we did not determine if the hydrazide groups were exposed on the liposome surface. However, we did confirm that SHz alone could not by a fine dispersion (i.e. micelles) as could be obtained from the liposomal formulation. This observation strongly suggested that SHz was stoichiometrically introduced in liposomes.

The partition coefficient of liposomes in the aqueous immiscible biphases is widely known to evaluate their surface properties[1]. This has also been utilised to evaluate the PEG surface coverage of liposomes. It is known that the higher the PEG coverage on the liposome surface, the larger the partition coefficient. We further performed a quantitative investigation of the relationship between the PEG content on the liposome surface and the partition coefficient quantitatively by using conventional PEG-modified liposomes (MPEG-DSPE liposomes) with different MPEG-DSPE content. When MPEG-DSPE was formulated in the lipid mixture or added to the preformed liposome dispersion, the MPEG-DSPE content in the liposome was controlled between 0 and 10 mol% or 0 and 2 mol%, respectively. Using these two samples, we were able to confirm the linear correlation between the PEG-content and the logarithm of the partition coefficient (data not shown).

On the basis of these results, the progressive hydrazo bond formation introducing MPEG-CHO to the liposome surface was estimated by the partition measurement. The effects of different reaction conditions such as temperature, time, nominal concentration and molar ratio of the liposomes over MPEG-CHO, on the partition coefficient of MPEG-SHz liposomes were investigated (data not shown). It was found that the conditions for hydrazo bond formation between SHz liposomes and MPEG-CHO could be optimised (as in *MATERIALS AND METHODS*).

3.2   Physicochemical Evaluations of PEG-releasing Liposomes

Changes in the physicochemical properties of MPEG-SHz liposomes incubated in PBS at 37°C were evaluated with time, and compared with those of liposomes with stably linked PEG, namely, stabilised MPEG-SA liposomes and MPEG-DSPE liposomes. The lipid composition and linkage between PEG and the liposome surface of the liposomes tested here are shown in Table 1.

Figure 1 shows the time course of the partition coefficient change of MPEG-SHz liposomes incubated in PBS at 37°C. The partition coefficient of MPEG-SHz liposomes decreased with time. This indicates that the PEG content on the liposome surface decreased with time. Figure 2 shows the time course of the hydrodynamic size changes of MPEG-SHz liposomes incubated in PBS at 37°C. The hydrodynamic size of MPEG-SHz liposome decreased over several hours. This suggests that the thickness of the hydrated layer on the liposome surface decreased as the PEG content on the liposome surface decreased. In contrast, both the partition coefficient and hydrodynamic size of the liposomes with stably linked PEG were maintained almost constant (Fig.3 and 4).

These results indicate that the decrease in the PEG content on the MPEG-SHz liposomes is attributable to the unstable bonding between PEG and the liposome surface, but not to the physical extraction of the PEG-lipid conjugate from the liposomal membrane.

Table 1. Liposomes tested for changes in the physicochemical properties with time in PBS at 37°C

Liposome	Lipid composition (molar ratio) and surface modification	linkage between PEG and liposome
MPEG-SHz	SHz / DPPC / cholesterol = 16 / 44 / 40	$-CH=N-NH_2-CO-CH_2-$ (hydrazone)
DPPC	DPPC / cholesterol = 60 / 40	-
Stabilised-MPEG-SA	SA / DPPC / cholesterol = 16 / 44 / 40 Irreversibly modified with MPEG-CHO in the presence of $NaCNBH_3$	$-CH_2-NH-CH_2-$ (secondary amine)
MPEG-DSPE	MPEG-DSPE / DPPC / cholesterol =5 / 55 / 40	$-CH_2-OCO-NH-CH_2-$ (carbamate)

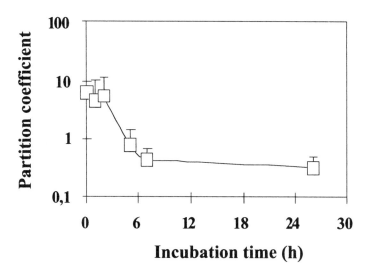

Figure 1. Time course of the partition coefficient changes of MPEG-SHz liposomes incubated in PBS at 37 °C.

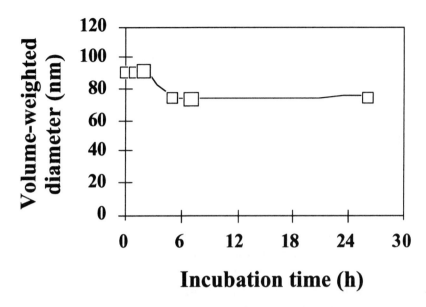

Figure 2. Time course of the hydrodynamic size changes of MPEG-SHz liposomes incubated in PBS at 37 °C.

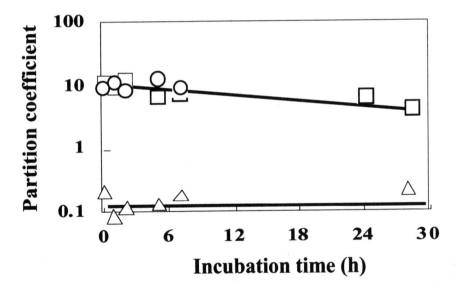

Figure 3. Time course of the partition coefficient of the liposomes with stably linked PEG. □, Stabilised MPEG-SA liposomes; O, MPEG-DSPE liposomes; Δ DPPC liposomes (as reference)

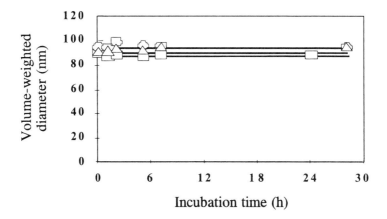

Figure 4. Time course of the hydrodynamic size of the liposomes with stably linked PEG. □, Stabilised MPEG-SA liposomes; O, MPEG-DSPE liposomes; Δ DPPC liposomes

3.3 Evaluation of the Affinity of PEG-releasing Liposomes for Cells

Changes in the affinity of MPEG-SHz liposomes for cells was investigated using PMNs as the model phargocytic cell. MPEG-SHz liposomes were incubated in PBS at 37 °C for a predetermined period to allow MPEG-CHO to be released from the liposome surface. Then this liposome suspension was mixed with a PMN suspension. Figure 5 is the time course of the adherence of MPEG-SHz liposomes to PMNs over the total period of the preincubation and incubation time with the cells.

The amount of liposomes that adhered to PMNs immediately after mixing, increased with preincubation time. The rate of further adherence after mixing appeared to follow the relationship. This result suggests that the affinity of MPEG-SHz liposomes for the cells increased as the PEG was released from the liposome surface with time. In other words, liposome surface substantially changed from being bioinert into bioreactive with time.

It was found that the time-dependency of the physicochemical properties and that of the affinity for the cells were different from each other. The affinity of the liposomes for PMN gradually increased for an incubation time of up to 20h, in contrast to the changes in the partition coefficient and the hydrodynamic size, in which case both drastically changed within 6 h and remained constant thereafter.

This discrepancy may be due to the aggregation tendency of the PEG-releasing liposomes, especially in the evaluation system of the affinity for PMNs. It is plausible that divalent cations and/or proteins secreted from PMNs could induce the aggregation of MPEG-SHz liposomes as the incubation time is prolonged, resulting in an increase in the particle size. It is known that the recognition of particles by leukocytes increases with increasing particle size[12,13].

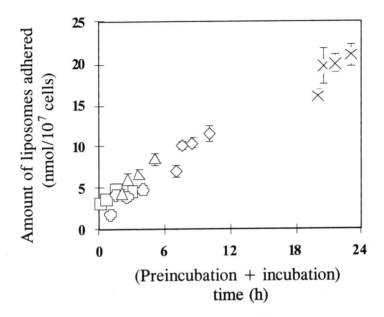

Figure 5. Time course of the liposome adhesion to PMNs. MPEG-SHz liposomes were preincubated in PBS (+) at 37 °C for 0 (□), 1 (○), 2 (Δ), 7 (◆), or 20 (×) h, then mixed with PMNs and incubated. The incubation with PMNs was stopped just after mixing (1st plot of each symbol) and continued for 0.5 (2nd), 1.5 (3rd) and 3.0 (4th) h.

Thus, we consider that the gradual liposome aggregation occurs in parallel with the changes in the surface properties, and they synergistically enhance the interaction with the cells.

3.4   Possible Applicability of PEG-releasing Liposomes

From the results of the physicochemical and *in vitro* biological evaluations, it was confirmed that surface-linked PEG could be released from liposomes by cleavage of the hydrazo bond between PEG and the liposome surface.

Several attempts have already been made on releasing surface-linked PEG from liposomes. Holland et al.[14] prepared PEG-releasing liposomes by controlling the length of the PEG-lipids. In this case, the PEG-lipids themselves were released by physical extraction by the host lipid or by a physical stimuli such as blood flow. Kirpotin[15] et al linked PEG on the liposome surface via a disulfide bond. In this case, PEG-SH could be released by a disulfide exchange with SH-containing compounds in the body. In these liposomes, the amount and body distribution of biocomponents (e.g. lipidic or SH compound) affect the release mechanism. In contrast, in our case, PEG is mainly released by a change in the chemical equilibrium and without any requirement for the presence of biocomponents. Therefore, a smaller variation in the PEG release rate can be expected among individuals and animal species.

In the case of utilising the biointeractive ligands attached to the liposome surface, we believe the concealed ligands is gradually exposed with time, as shown in Figure 6. At the beginning of incubation, surface-linked PEG conceals the ligand so that the passive targeting can occur. After accumulation at the target site, the PEG is released from the liposome surface and the ligand can interact with the biocomponent. This mechanism may be applicable to diverse kinds of ligands, such as adhesion molecules, receptor substrates, and antibodies.

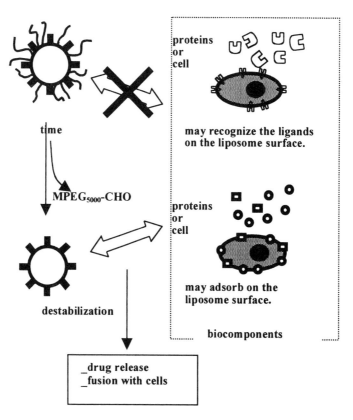

Figure 5. Schematic illustration of the time-dependent interaction between MPEG-SHz liposomes and biocomponents.

4.   CONCLUSION

Liposomes modified with PEG via a hydrazo bond (MPEG-SHz liposome) were designed as a PEG-releasing liposome. From physico-chemical evaluations, it was confirmed that PEG chains were released gradually from the liposome surface with time. From biological evaluations *in vitro*, pre-incubation of MPEG-SHz liposomes in PBS resulted in an increase in cell adherence. The MPEG-SHz liposomes are characterised by both effective passive targeting from the blood circulation system and high availability of encapsulated drugs at the target site.

ACKNOWLEDGMENTS

The authors are grateful to Drs. M. Ikeda and A. Kusai of Product Development Laboratories, Sankyo Co., Ltd. for helpful supervision.

REFERENCES

1. Senior, J., Delgado, C., Fisher, D., Tilcock, C., and Gregoriadis, G., 1991, Influence of surface hydrophilicity of liposomes on their interaction with plasma protein and clearance from the circulation: studies with poly (ethylene glycol)-coated vesicles. *Biochim. Biophys. Acta* 1062: 77-82.
2. Klibanov, A.L., Maruyama, K., Torchilin, V.P., and Huang. L., 1990, Amphipathic polyethyleneglycols effectively prolong the circulation time of liposomes. *FEBS Lett.* 268: 235-237.
3. Klibanov, A.L., Maruyama, K., Beckerleg, A.M., Torchilin, V.P., and Huang, L., 1991, Activity of amphipathic poly (ethylene glycol) 5000 to prolong the the circulation time of liposomes depends on the liposome size and is unfavorable for immunoliposome binding to target. *Biochim. Biophys. Acta*, 1062: 142-148.
4. Blume, G., and Cevc, G., 1990, Liposomes for the sustained drug release in vivo. *Biochim. Biophys. Acta*, 1029: 91-97.
5. Maruyama, K., Yuda, T., Okamoto, A., Kojima, S., Suginaka, A., and Iwatsuru, M., 1992, Prolonged circulation time in vivo of large unilamellar liposomes composed of distearoyl phosphatidylcholine and cholesterol containing amphipathic poly (ethylene glycol). *Biochim. Biophys. Acta*, 1128: 44-49 (1992).
6. Maruyama, K., Yuda, T., Okamoto, Ishikura, C, Kojima. S., and Iwatsuru, M., 1991, Effect of molecular weight in amphipathic polyethyleneglycol on prolonging the circulation time of large unilamellar liposomes. *Chem. Pharm. Bull.*, 39: 1620-1622.
7. Allen, T.M., Hansen, C., Martin, F., Redemann, C., and Yau Young, A., 1991, Liposomes containing synthetic lipid derivatives of poly (ethylene glycol) show prolonged circulation half lives in vivo. *Biochim. Biophys. Acta*, 1066: 29-36.
8. Bangham, A.D., and Horn, R.W., 1964, Negative staining of phospholipids and their structured modification by surface active agents as observed in the electron microscope. *J. Mol. Biol.*, 8: 660-668.

9. MacDonald, R.C., MacDonald, R.I., Menco, B.P.M., Takeshita, K., Subbarao, N.K., and Hu, L.-R., 1991, Small-volume extrusion apparatus for preparation of large, unilamellar vesicles. *Biochim. Biophys. Acta*, 1061: 297-303.

10. Claassen, E., 1992, Post-formation fluorescent labelling of liposomal mambranes. In vivo detection, localisation and kinetics. *J. Immunol. Methods*, 147: 231-240.

11. Boyum, A., 1968, A one-stage procedure for isolation of granulocytes and lymphocytes from blood. Scand. *J. Clin. Lab Invest., Suppl.*, 97: 51-76.

12. Tabata, Y., and Ikada, Y., 1988, Effect of the size and surface charge of polymer microspheres on their phagocytosis by macrophage. *Biomaterials*, 9: 356-362.

13. Kawaguchi, H., Koiwai, N., Ohtsuka, Y., Miyamoto, M., and Sasakawa, S., 1986, Phagocytosis of latex particles by leucocytes. ?Dependence of phagocytosis on the size and surface potential of particles. *Biomaterials*, 7: 61-66.

14. Holland, J. W., Madden, T., and Cullis, P., 1996, Bilayer stabilizing components and their use in forming programmable fusogenic liposomes. *PCT WO 96/10392*.

15. Kirpotin, D., Hong, K., Mullah, N., Papahadjopoulos, D., and Zalipsky, S., 1996, Liposomes with detachable polymer coating: destabilization and fusion of dioleoylphosphatidylethanolamine vesicles triggered by cleavage of surface-grafted poly(ethylene glycol). *FEBS Letters,* 388: 115-118.

CARRIER DESIGN: INFLUENCE OF CHARGE ON INTERACTION OF BRANCHED POLYMERIC POLYPEPTIDES WITH PHOSPHOLIPID MODEL MEMBRANES

[1]Ferenc Hudecz, [1]Ildikó B. Nagy, [1]György Kóczán, [2]Maria A. Alsina, and [3]Francesca Reig
[1]*Research Group of Peptide Chemistry, Hungarian Academy of Science, Eötvös L. University, P.O. Box 32, Budapest 112, H-151 Hungary,* [2]*Peptide Department, Research Council, CID. CSIC. Jordi Girona 16. 08034 Barcelona, Spain;* [3]*Physicochemical Unit, Faculty of Pharmacy, Barcelona, Spain*

1.   INTRODUCTION

Synthetic biodegradable polymers represent a new group of compounds with a promising potential for biomedical application. Such polymers as therapeutic agents has been used with success in experimental cancer chemotherapy[1] and in the treatment of human multiple sclerosis[2]. Alternatively, polymeric compounds can be utilised as macromolecular component of bioconjugates in the development of target specific delivery systems as well as preparation of long acting drug - macromolecule constructs. Improved pharmacological properties (extended blood circulation, elevated accumulation in tissue) diminished cytotoxicity and/or decreased immunogenic properties of biologically active compounds (e.g. drugs, radionuclides, and enzymes) have been achieved by their covalent attachment to polymers.[3-8]

The second field of application of biocompatible polymers is the construction of efficient synthetic antigens containing peptide epitope(s) of microbial (mainly viral) origin and non-immunogenic carrier moiety. This approach could lead to the development of the new generation of synthetic subunit vaccines.[9-11]

The common difficulty of the above described biomedical utilisation is the selection of polymeric macromolecule with appropriate biological properties. Besides empirical experimental work there is an increasing need to establish structure-function type correlation applicable for the rational design of polymeric compounds with required structural and functional parameters. In order to contribute to this line of research we have prepared new groups of branched polypeptides with the general formula poly[Lys(X_i-DL-Ala$_m$)] (XAK[†]) or poly[Lys(DL-Ala$_m$-X_i)] (AXK) where i<1, m ~ 3, and X represent an additional optically active amino acid residue.[12,13] These polymeric polypeptides were characterised by their size, chemical (primary structure, solution conformation) and biological (*in vitro* cytotoxicity, pyrogenicity, biodegradation, immunoreactivity and biodistribution) properties.[13-15]

Based on these data various conjugates with antitumor agents (daunomycin[16], methotrexate[13], GnRH antagonist[17,18]), boron compounds[19] or radionuclides[20,21] were prepared. Antitumor activity of Ac-[D-Trp[1-3], D-Cpa[2], D-Lys[6], D-Ala[10]]-GnRH - poly[Lys(Ac-Glu$_i$-DL-Ala$_m$)] (AcEAK) conjugate has been clearly documented.[17,18] The coupling of acid labile derivative of daunomycin to poly[Lys(Glu$_i$-DL-Ala$_m$)] (EAK) resulted in compensation for the immunosuppressive effect of the drug and this polymeric conjugate *in vivo* was very effective against L1210 leukaemia producing 66-100 % long-term survivors (> 60 days) in mice.[22]

Several groups of branched polypeptide based synthetic compounds were also prepared by introduction of B-cell or T-cell epitope peptides onto the side end of the branches.[23] We have demonstrated that the efficacy of antibody response specific for a *Herpes Simplex virus* glycoprotein D epitope is highly dependent on the chemical structure (sequence and conformation) of branched polypeptide carrier.[24,25] Recently a fully synthetic prototype conjugate of EAK with two independent T cell epitope peptides of

---

[†] Abbreviations for amino acids and their derivatives follow the revised recommendation of the IUPAC-IUB Committee on Biochemical Nomenclature, entitled "Nomenclature and Symbolism for Amino Acids and Peptides" (recommendations of 1983). Nomenclature of branched polypeptides is used in accordance with the recommended nomenclature of graft polymers (IUPAC-IUB recommendations, 1984). For the sake of brevity codes of branched polypeptides were constructed by us using the one-letter symbols of amino acids (Table I). The abbreviations used in this paper are the following. AK, poly[Lys-(DL-Ala$_m$)]; AXK, poly[Lys-(DL-Ala$_m$-X_i)]; XAK, poly[Lys(X_i-DL-Ala$_m$)]; X = Ser (SAK), Orn (OAK), Glu (EAK), or Ac-Glu (Ac-EAK). All amino acids are of L-configuration unless otherwise stated. DPH, 1,6-diphenyl-1,3,5-hexatriene; ANS, sodium anilino naphthalene sulfonate; DPPC, dipalmitoyl phosphatidyl choline; PG, phosphatidyl glycerol; Z, benzyloxycarbonyl; Pcp, pentachlorophenol; P, polarisation.

M.tuberculosis proteins (16 kDa and 38 kDa) were prepared. *In vivo* data clearly suggest that both mycobacterial epitopes preserved their capability to induce specific T-cell proliferation.[26]

These studies have also demonstrated that the biological properties (e.g. epitope-related immunogenicity or drug-related antitumor effect) and pharmacokinetics (e.g. blood survival, tissue distribution) of conjugates are greatly influenced by the chemical structure of the carrier.[13,23,27]

In view of the fact that polymeric polypeptides as well as their conjugates act on biological membranes we have initiated model experiments to study the correlation between structural characteristics and membrane activity of selected polymeric polypeptides using phospholipid mono- and bilayer models.[28]

In this communication we describe the results of a systematic study in which

1. the surface activity of polypeptides at the air/water interface was analysed and

2. the effect of branched polypeptides with polycationic, polyanionic or amphoteric nature was investigated on phospholipid membranes with different compositions.

New set of data will be provided on the characterisation of these polymers by RP-HPLC approach developed in our laboratory. Data presented on polymers demonstrate that the side composition of these compounds could markedly influence their membrane activity by altering the relative hydrophobicity and the charge properties of the polymer.

2.   SYNTHESIS AND CHEMICAL STRUCTURE

Branched chain polymeric polypeptides used in these studies are poly[L-Lys] derivatives substituted by short (3-6 amino acid residue) branches at the ε-amino groups. The side chains are composed of about three DL-Ala residues and one other amino acid residue (X) at the N-terminal end of the branches (XAK) (Fig.1). It should be noted that similar polymeric polypeptides with much longer branches (20-30 residues) and copolymeric N-terminal portions were described by Sela et al.[29] Experimental details of the synthetic procedures of the XAK polymers were described previously.[30,31] Briefly, poly[L-Lys] was prepared by the polymerisation of N^α-carboxy-N^ε-benzyloxycarbonyl-lysine anhydride under conditions that allowed a number average degree of polymerisation ($\overline{DP_n}$) approximately 60. Protecting groups were cleaved and DL-alanine oligomers were grafted to the ε-amino groups of polylysine by polymerisation of N-carboxy-DL-alanine anhydride to produce poly[Lys(DL-Ala$_m$)] (AK). The suitably protected and activated Glu[30], Ser[31] or Z-Orn(Z)-OPcp residue was coupled to the end of branches of AK by amide bond. After the removal of protecting groups samples were dialysed against distilled water, freeze-dried. From sedimentation equilibrium measurements of polylysine, the number average degree of polymerisation ($\overline{DP}_n$), the average molar mass, and the relative molar mass distribution ($\overline{M}_z / \overline{M}_w$) were calculated.[13] The average molar mass of the branched polypeptides was estimated from $\overline{DP}_n$ of polylysine and from the amino acid composition of the side chains determined by quantitative amino acid analysis. Characteristics of polymeric polypeptides used for the present experiments are summarised in Table 1.

Table I. Characteristics of branched polypeptides

Code [a]	Amino acid composition [b]			M_w [c]	R_t [e]
	Lys	Ala(m)	X[d](i)	(±5 %)	(min)
polylysine	1.00			12800	25.1
AK	1.00	4.00	-	26000	26.9
SAK	1.00	4.20	0.75	29100	27.0
OAK	1.00	4.22	0.85	32000	26.4
EAK	1.00	4.30	0.95	33900	26.1
Ac-EAK	1.00	4.30	0.95	37500	28.7

[a] Code of branched polypeptides, based on one letter symbol of amino acids. [b] Determined by amino acid analysis after hydrolysis in *6 M* HCl at 105 °C for 24 h. [c] Calculated from the number average degree of polymerisation of polylysine ($\overline{DP}_n = 61$) and from the amino acid composition of branches. [d] X=Ser/Orn/Glu [e] RP-HPLC retention time.

Schematic structure of branched chain polymeric polypeptides is shown in Figure 1. Polylysine with free ε-amino groups, AK or poly[Lys(Ser$_i$-DL-Ala$_m$)], (SAK) containing α-amino groups and poly[Lys(Orn$_i$-DL-Ala$_m$)], (OAK) possessing both α- and ε-amino groups can be considered as polycations. Side chains of poly[Lys(Glu$_i$-DL-Ala$_m$)], (EAK) contains glutamic acid at the end of the branches. Therefore this polymer has not only free α-amino, but also free γ-carboxyl group in the side chain, consequently this compound has amphoteric character. Acetylation of EAK resulted in a polyanionic derivative poly[Lys(Ac-Glu$_i$-DL-Ala$_m$)], (Ac-EAK).

The size and architecture of all these polymeric polypeptides are rather similar, but differences in amino acid X in their side chains resulted in altered relative hydrophobicity and electrical charge. These features provide a logical basis for comparative studies of their membrane interaction.

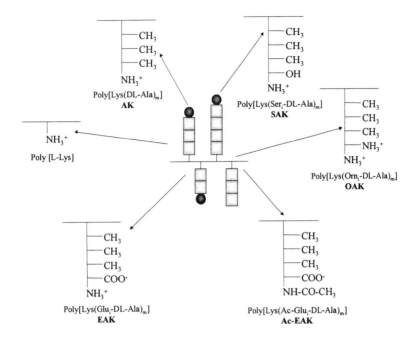

Figure 1. Simplified structure of the poly[Lys(X_i-DL-Ala$_m$)], XAK type branched polymeric polypeptides

3. REVERSED PHASE HPLC CHARACTERISATION

The analysis of the molecular mass distribution of branched chain polymeric polypeptides was performed earlier by gel-filtration using Sephadex packing materials.[32] Preliminary data have been reported on the heterogeneity of drug-conjugates studied by capillary electrophoresis.[33] Here we describe a RP-HPLC based analytical approach developed by us for the highly charged branched polymeric polypeptides. Polymer samples were run on a Waters (Milford, USA) HPLC system using a Delta Pak C_{18} column (300 x 3.9 mm I.D.) with 15 µm silica (300 Å pore size) (Mulligan, Milford, MA, USA) as a stationary phase, and 0.1 % TFA in water was used as eluent A and 0.1% TFA in acetonitrile-water (80:20, v/v %) as eluent B.

Linear gradient from 10% v/v of B to 65% v/v of B in 45 minutes was used at a flow rate of 2 ml/min at ambient temperature and UV detection at 208 and 214 nm, respectively. Polymer samples were dissolved in eluent A at a concentration of 3mg/ml. Depending on the UV absorbance of the polymer 20 to 100µl aliquots were injected.

Both the basic and amphoteric polymers form highly charged TFA salt under the acidic elution conditions. Wide elution profile was observed only with slight differences in retention times. It is interesting to note that the polyanionic polymer (Ac-EAK) produced a relatively sharp peak. It might be due to the lack of salt formation with TFA.

Due to the good recovery of the polymers from the column, RP-HPLC was used with success for the detection of small molecular mass impurities and for the small-scale purification of branched chain polypeptides. Chromatograms are illustrated in Figure 2.

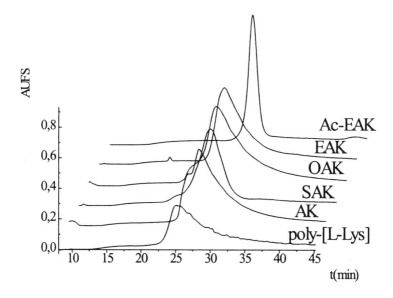

Figure 2. RP-HPLC chromatograms of branched polypeptides

4.   SURFACE ACTIVITY OF POLYMERS

The behaviour of polymeric polypeptides at the air/water interface was studied. Relative hydrophobicity was quantified using a Langmuir balance device designed and constructed in our laboratory (FR) according to Verger and de Haas.[34]

Polymers dissolved in *0.1M* sodium acetate buffer (pH 7.4) at different concentrations were injected in the aqueous subphase. Surface pressure values as a function of time were recorded. Data are summarised in Figure 3.

In case of branched polypeptides the increase of surface pressure was dependent on the polymer concentration and on time.

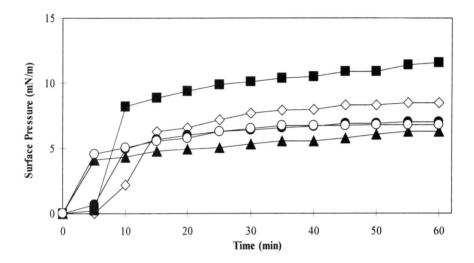

Figure 3. Surface activity of branched polypeptides at the air/water interface. Surface pressure values represented as a function of time. Polypeptides AK (—◇—), SAK (—■—), OAK (—▲—), EAK (—●—) and Ac-EAK (—○—) dissolved in 0.1M acetate buffer (pH 7.4) were injected into the subphase at $c=4.50\times10^{-8}$ M.

The maximum value was recorded at polymer concentrations higher than 1.8×10^{-8} M after 60 min (Fig.3 and 4). It is interesting to note that while no delay of incorporation was observed for OAK and Ac-EAK, a certain period of time was needed for the appearance of polymers AK, SAK and EAK at the interface (Fig.3.). This induction time was markedly longer at low concentrations (e.g. 15 min at $c = 0.9\times10^{-8}$ M and only 5 min at $c = 4.5\times10^{-8}$ M for SAK) (data not shown). Similar findings were described in the literature for proteins and interpreted as a need of time for an ordered structure to rearrange at the air/water interface.[35]

The surface pressure values for the branched polymers as a function of concentration are compared in Figure 4.

In case of polylysine the lack of change in surface pressure even at high concentration (10^{-6} M) was observed (data not shown). However data indicate that all branched polypeptides are able to form monolayers at the air/water interface. Maximal surface pressure (π_{max}) achieved has increased with polymer concentration and after saturation no further alteration occurred. SAK showed the highest surface activity (π_{max}=11.6 mN/m), while π_{max} was nearly the same (6.3-8.5 mN/m) for other polypeptides. These values were lower than those observed for linear polymers with similar molecular weight [poly[Glu(OMe)][36] and polyethylene glycol[37]].

Taken together these finding suggest that the surface activity of polymeric polypeptides at the air/water interface was dependent on the side chain terminal amino acid residue of polymers and can be described by the SAK>AK>EAK>Ac-EAK>OAK>>polylysine order.

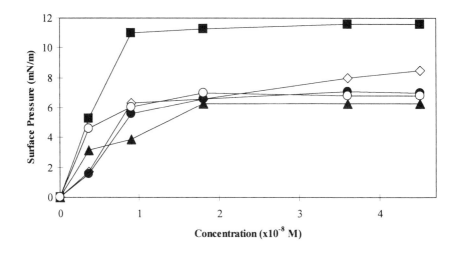

Figure 4. The effect of concentration of branched polypeptide on the surface pressure. Surface pressure values were determined after 60 min of injection of polymers in the subphase: AK (–◇–), SAK (–■–), OAK (–▲–), EAK (–●–) and Ac-EAK (–○–).

5. PENETRATION OF POLYMERS IN MONOLAYERS

Phospholipid monolayers are excellent model membranes into which the penetration of macromolecules can be sensitively monitored through changes in surface pressure. In order to establish correlation between chemical structure and penetration properties we prepared monolayers composed of DPPC, DPPC/PG (95/5 mol/mol) and of DPPC/PG (80/20 mol/mol). Insertion experiments were carried out in the same Teflon cuvette as described above. Lipid solutions were spread on the aqueous subphase to form a monolayer with a given initial surface pressure. Once monolayers were established, polymer solutions were injected in the subphase and pressure increases were recorded as a function of time.

The interaction of polylysine, AK, SAK, OAK as well as EAK, Ac-EAK with monolayers was studied at different initial surface pressures (5, 10, 20 and 32 mN/m) and constant area. The increases in surface pressure values are presented as histograms on Figure 4. In the case of DPPC monolayer, the presence of SAK resulted in the highest surface pressure increase (6.8 mN/m at 5 mN/m initial pressure) (Fig.5a). The surface pressure increase was somewhat lower when other polypeptides were in the subphase (6.3 mN/m for AK, 2.5 mN/m for OAK, 4.7 mN/m for EAK and 4.4 mN/m for Ac-EAK at 5 mN/m). Essentially the same tendency was observed at higher initial pressures (at 10, 20 or 32 mN/m), but penetration decreased with elevated phospholipid packing density.

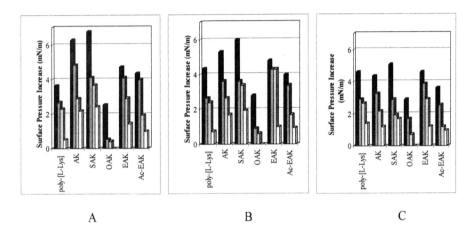

Figure 5. Penetration of branched polypeptides into monolayers of DPPC (A) DPPC/PG (95/5 mol/mol) (B) and DPPC/PG (80/20 mol/mol) (C). Effect of initial surface pressure (5 mN/m ■, 10 mN/m ▨, 20 mN/m ☐ and 32 mN/m ☐)

In the presence of monolayer composed of DPPC and 5 or 20 % PG (Fig.5b, c), the surface activity of polycationic SAK and AK (6.0 mN/m for AK, 5.3 mN/m for SAK) was less pronounced. However, no significant changes were observed with EAK and Ac-EAK (4.8 mN/m for EAK and 4.0 mN/m for Ac-EAK at 5 mN/m in DPPC/PG, 95/5 mol/mol) as compared to values of the DPPC monolayer. Similar observation was interpreted as barrier effect of the monolayer that prevent the incorporation of the relatively more hydrophobic molecules into the surface of the phospholipid monolayer.[38] Presence of negatively charged PG in the monolayer resulted in increased penetration ability of free ε-amino group containing OAK and polylysine (e. g. surface pressure increase for polylysine at 5 mN/m initial pressure was 3.6 mN/m in the presence of DPPC, while it was 4.6 mN/m in the presence of DPPC/PG (80/20 mol/mol).

Based on surface pressure increase values observed, penetration of polypeptides into DPPC, DPPC/PG (95/5 mol/mol) and DPPC/PG (80/20 mol/mol) monolayers can be described with SAK>AK>EAK≅Ac-EAK>polylysine>OAK order. Differences in π_{max} values indicate that polymeric polypeptides could be classified as follows.

1. For polymers containing ε-amino group (polylysine, OAK) the negative phospholipid content resulted in increased electrostatic interaction which facilitated their penetration at low surface activity.
2. Polypeptides possessing only free α-amino group at the end of the side chains were less capable to penetrate into monolayers with negatively charged PG. It is in good correlation with results of surface activity measurements where AK and SAK had the highest relative hydrophobicity. This indicates that in the interaction of AK or SAK with phospholipid monolayers electrostatic and hydrophobic forces might be involved.
3. The amphoteric EAK and its polyanionic derivative, Ac-EAK showed low penetration, which was almost independent from the composition of the monolayer suggesting the predominant role of hydrophobic attraction.

The maximum initial surface pressure we have used in these experiments was 32 mN/m and represents a value described for cellular membranes of biological systems.[39] Therefore differences in surface pressure increases observed with branched polypeptides might be indicative for their potential interaction with plasmamembranes.

The insertion of the polymers into monolayers was also studied by comparing the compression isotherms. Lipid monolayers were formed on *0.1M* sodium acetate buffer (pH 7.4) subphase in the presence or absence of polymers in a rectangular Teflon cuvette of 28.5 cm x 17.5 cm. After 10 min stabilisation period, a Teflon barrier compressed the monolayer at a speed of 4.2 cm/min and surface pressure vs. area (π vs. A) isotherms were recorded. Data are summarised in Table II. We observed no changes in the shape of area vs. pressure curves obtained in the presence of polymers in the subphase (data not shown), but polymers induced an expansion of the monolayer. These changes were detected at various surface pressures (10, 20, 30, 40 and 50 mN/m) and are expressed as area/molecule of phospholipid values (Table 2). These values indicate significant differences in the interaction of polycationic and amphoteric/polyanionic polypeptides. Marked expansion of DPPC monolayer occurred in the presence of SAK or AK (ΔA=0.23-0.51), while EAK and Ac-EAK initiated only moderate changes in this parameter (ΔA=0.01-0.13). The effect of polylysine and OAK was negligible (ΔA=0.01-0.03).

In the case of DPPC/PG monolayers similar tendency could be detected but only at the lowest surface pressures. These results underline the side/chain dependent involvement of electrostatic and/or hydrophobic forces in the polymer - monolayer interaction. In accord with the surface activity measurement at the air/water interface we have observed pronounced effect of SAK on monolayers at low compression pressures.

Table 2. Changes in the area (nm^2) per molecule of DPPC or DPPC/PG (95/5, 80/20 mol/mol) values induced by the polymeric polypeptides measured at different surface pressures

	π mN/m	ΔA [a]					
		Polymer					
		poly[L-Lys]	AK	SAK	OAK	EAK	Ac-EAK
DPPC	10	0.01	0.40	0.51	0.03	0.10	0.13
	20	0.01	0.31	0.31	0.02	0.01	0.05
	30	0.02	0.30	0.28	0.02	0.02	0.03
	40	0.02	0.23	0.25	0.02	0.02	0.03
	50	0.00	0.23	0.23	0.02	0.01	0.01
DPPC/PG	10	0.06	0.25	0.34	0.05	0.13	0.17
	20	0.05	0.17	0.39	0.03	0.24	0.05
(95/5)	30	0.06	0.14	0.12	0.03	0.25	0.01
	40	0.02	0.13	0.07	0.01	0.23	0.01
	50	0.02	0.14	0.07	0.01	0.19	0.01
DPPC/PG	10	0.08	0.23	0.26	0.05	0.18	0.10
	20	0.07	0.14	0.11	0.03	0.15	0.02
(80/20)	30	0.03	0.11	0.07	0.03	0.11	0.02
	40	0.00	0.08	0.06	0.01	0.08	0.02
	50	0.00	0.03	0.04	0.01	0.05	0.02

[a] Changes in area/molecule values calculated at 10, 20, 30, 40 and 50 mN/m surface pressure from compression isotherms of DPPC and DPPC/PG (95/5, 80/20 mol/mol) monolayers formed on polymer polypeptide solution and buffer subphase. ($\Delta A = A_{polymer} - A_{buffer}$).

6. INTERACTION OF POLYMERS WITH PHOSPHOLIPID BILAYERS

The effect of branched polypeptides on phospholipid membranes was further investigated using lipid bilayers with DPPC/PG (95/5, 80/20 mol/mol). Two fluorescent probes of different character were used to analyse the effect of polymers on the outer surface (negatively charged, sodium anilino naphthalene sulfonate, ANS) and on hydrophobic core (hydrophobic, 1,6-diphenyl-1,3,5-hexatriene DPH) of bilayers. For these studies small unilamellar vesicles were used.

6.1 Preparation of liposomes

Lipids dissolved in chloroform were mixed in a round bottom flask and solvent was eliminated by rotary evaporation from the clear solution at 55°C in vacuum. The liposome preparation was dried in high vacuum for 2 h. Lipid film was hydrated with 2 ml of 0.1 M sodium acetate buffer (pH 7.4), and multilamellar vesicles thus formed were submitted to sonication in an ultrasounds bath until the solution was only slightly turbid. Size of liposomes measured in Malvern Autosizer and as described earlier[40] was lower than 80 nm in diameter. In some cases an external sonication probe was used to reduce the size of liposomes. In this case samples were submitted to centrifugation.

6.2   Fluorescence studies

Fluorescence studies were carried out by measuring changes in fluorescence intensity and polarisation of ANS and DPH probes located in the bilayers, using a PE-LS50 spectrofluorometer provided with four cuvettes thermostated bath. Fluorescence intensity changes could be due to variations in the environmental conditions of the fluorophore, while polarisation is mainly related to its motion.

In the first set of experiments liposomes were incubated with fluorescent probes at various concentrations in the dark at 55°C for 60 min and the saturation curves were recorded. The optimal probe/phospholipid molar ratios determined from the saturation curves were 1/26 (mol/mol) for ANS/phospholipid and 1/240 (mol/mol) for DPH/phospholipid.

At these ratios, liposome preparations were mixed either with polymer solution or with buffer (0.1 M sodium acetate, pH=7.4) and fluorescence intensity was measured as a function of temperature. Excitation and emission wavelengths were λ=380-480 nm for ANS and λ=365-425 nm for DPH, respectively. The degree of fluorescence polarisation was calculated according to Shinitzky and Barenholz applying equation 1.[40]

$$P = (I_\| - I_\perp \cdot G) / (I_\| + I_\perp \cdot G) \qquad\qquad \text{eq. 1}$$

where $I_\|$ and $I_\perp$ are the intensities measured with its polarisation plane parallel ($\|$) and perpendicular ($\perp$) to that of the exciting beam. G is a factor used to correct polarisation of the instrument and is given by the ratio of vertically to horizontally polarised emission components when the excitation light is polarised in the horizontal direction. From these data transition temperature (T_c) - characteristic for the transformation from gel to liquid crystalline state of the bilayer - was calculated. In the absence of polymers these values were in the range of $T_c \approx$40-41.5°C.

First the interaction of polymeric polypeptides with the polar surface of liposomes was analysed. Changes in fluorescence polarisation of ANS in

DPPC/PG (95/5, 80/20 mol/mol) liposomes in the presence of polycationic polypeptides are documented in Figure 6.

For liposomes of DPPC/PG (95/5 mol/mol) in $T<T_c$ temperature range small increase in polarisation ($\Delta P \sim 0.05$) was observed in the presence of polylysine and OAK (Fig.6a), but in the $T>T_c$ domain only OAK built up from double positive charged monomers induced changes. These data indicate that polylysine interacts with phospholipid membrane mainly in gel phase. In case of other polycations with single α-amino group at the end of the branches (AK, SAK) and of amphoteric (EAK) or polyanionic (Ac-EAK) polypeptides no changes in polarisation was observed.

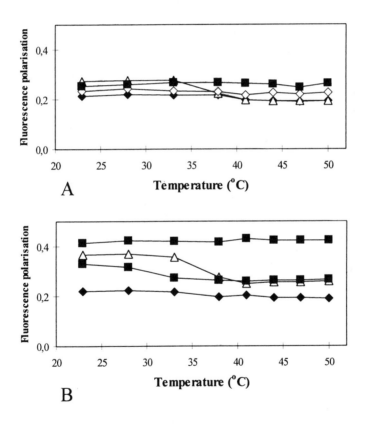

Figure 6. Interaction of polymeric polypeptides with liposomes composed of DPPC/PG (95/5 mol/mol) (A) or (80/20 mol/mol) (B). Fluorescence polarisation of ANS as a function of temperature in the presence of polylysine (—△—), SAK (—■—), AK (—▲—) or *0.1M* sodium acetate buffer (pH 7.4) (—◆—).

The influence of polypeptides on ANS saturated liposomes containing 20% PG was also investigated (Fig.6b). ΔP values obtained in $T<T_c$ range for OAK ($\Delta P \sim 0.2$) and for polylysine ($\Delta P \sim 0.15$) suggest strong interaction

with phospholipid bilayers. Presence of AK and SAK containing only α-amino group caused less pronounced increase in polarisation referring to weaker interaction. At T>T$_c$ temperature liposomes incubated with OAK indicated the highest increase in polarisation (ΔP~0.23). For polylysine, AK and SAK, ΔP values obtained were similar (ΔP~0.07). In the presence of EAK and Ac-EAK a relatively small increase (ΔP~0.07) was observed in T<T$_c$ as well as in T>T$_c$ temperature ranges (data not shown). ΔP values calculated indicate that these polypeptides possessing free carboxyl group interact with liposomes containing 20% PG. This observation could be explained by the pH-shifting effect of negatively charged lipids to the acidic range in the close vicinity of the liposomes. Consequently protonation of the carboxyl groups of the polymers is favoured and resulted in relative increase in hydrophobicity and accompanied by marked ability to penetrate into phospholipid membrane.[41,42]

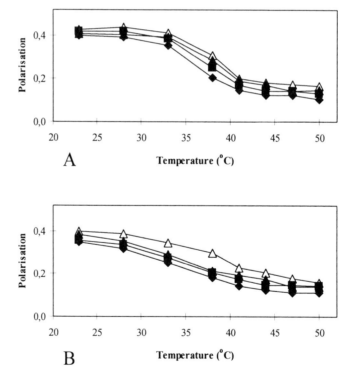

Figure 7. Interaction of polymeric polypeptides with liposomes composed of DPPC/PG (95/5 mol/mol) (A) or (80/20 mol/mol) (B). Fluorescence polarisation of DPH as a function of temperature in the presence of polylysine (—△—), SAK (—■—), AK (—▲—) or *0.1M* sodium acetate buffer (pH 7.4) (—◆—).

Interaction of polymeric polypeptides with liposomes was further studied using DPH localised in hydrophobic core region of the bilayer. Polarisation values for DPH in DPPC/PG (95/5, 80/20 mol/mol) bilayers were dependent on temperature (Fig.7a, b).

Taken together data obtained indicate that the effect of polymeric polypeptides on DPPC (95/5, 80/20 mol/mol) liposomes saturated with ANS was strongly influenced by the PG content of the bilayer. In both liposome compositions polycationic OAK with α- and ε-amino groups caused the highest increase of polarisation. Similar effect was observed at $T<T_c$ in the presence of polylysine with ε-amino group, but at $T>T_c$ temperatures its behaviour was similar to that observed for AK, SAK, EAK or Ac-EAK polypeptides.

$P\sim0.4$ values were calculated for liposomes containing 5% PG at $T<T_c$,, while in the 35-40°C range polarisation dramatically decreased (to 0.12). This pattern indicated the gel→fluid crystalline phase transition $(T_{c(DPPC)}\sim40.5°C)$ (Fig.7a). P values of DPH after incubation with buffer or polypeptide solutions were almost the same. Among polycationic compounds the highest - but moderate increase in polarisation ($\Delta P\sim0.04$) was observed in the presence of polylysine in the whole temperature range tested. This polypeptide initiated the highest increase (+1°C) of T_c, which is indicative for interaction.

For liposomes containing 20% negatively charged phospholipid $P\sim0.35$ values were observed at $T<T_c$ temperature. This finding can be explained by the increase of bilayer fluidity originated from the presence of higher amount of negatively charged lipids. The gel→liquid crystalline phase transition of this bilayer occurred at lower temperature (38.5°C).

Polylysine had the most pronounced rigidifying effect ($\Delta P\sim0.05$) (Fig.7b). Polycationic branched polypeptides (AK, SAK, OAK) had moderate ($\Delta P=0.01-0.04$), while amphoteric (EAK) and polyanionic (Ac-EAK) derivatives had no influence on polarisation. Presence of polycationic polypeptides increased the transition temperature ($\Delta T_c\sim+0.5-1.3°C$) and widened the temperature range of the phase transition, indicating that penetration of polymeric polypeptides into phospholipid bilayers resulted in increased distance of phospholipid molecules.

From these data we can conclude that polymeric polypeptides had no (AK, SAK, EAK and Ac-EAK) or moderate (polylysine > OAK) influence on the fluidity of hydrophobic alkyl chain region of bilayers. The effect of polymeric polypeptides on hydrophobic core was dependent on PG content of the liposome. The presence of higher amount of PG resulted in more pronounced interaction, what can be explained with the viscosity changes of the bilayer.

7.   CONCLUSION

The surface properties at the air/water interface and the interaction of polycationic (polylysine, AK, SAK, OAK), amphoteric (EAK) and polyanionic (Ac-EAK) polypeptides with mono- and bilayers composed of DPPC or DPPC/PG was investigated.

It has been demonstrated that branched chain polymers are capable to form stable monolayers at the air/water interface. This behaviour were dependent on the structure of polymers and can be described by SAK>AK>EAK>Ac-EAK>OAK>polylysine order.

In DPPC or DPPC/PG monolayer experiments changes in surface pressure (penetration kinetics) and area/phospholipid molecule values (compression isotherms) indicated similar differences in expansion of membranes. It can be concluded that the effect of polymeric polypeptides on phospholipid monolayers depends not only on the polymer charge (positive/negative, neutral), but also on charge density.

The effect of branched polypeptides on phospholipid bilayers saturated with fluorescent probes located either at the polar surface (ANS) or within the hydrophobic part (DPH) of the liposome indicate that only polymers with high positive charge density are capable to initiate significant changes.

Taken together, data obtained from mono- or bilayer experiments suggest that the interaction between branched polymers and phospholipid model membranes is highly dependent on the charge properties (Ser vs. Glu, Ac-Glu), on the identity (Ser vs. Ala, Orn) of side chain terminating amino acid and for polycationic polypeptides on type of amino group (α/ϵ). Under nearly physiological conditions membrane activity of the polymer can be controlled by the proper selection of the side chain terminating amino acid. Positioning amino acid residues with α-amino group (with pK_a 7.5 - 9) could result in polymers with less pronounced positive charge, while the presence of ϵ-amino group (with pK_a 10-11) at the same position create highly positive charged polymers.

Data presented in this communication could be important for the design of drug/epitope - polymer conjugates, since preliminary results with a limited set of epitope-branched polymer constructs indicate the importance of the carrier part in membrane activity.[43,44,45,46]

ACKNOWLEDGMENTS

Experimental work summarised in this paper was supported by grants from the Hungarian-Spanish Intergovernmental Programme (5/1998), from the Hungarian Research Fund (OTKA) No. T-014964, T-03838 and from the Hungarian Ministry of Welfare (ETT No. 115/1996).

REFERENCES

1. Seymour, L. W., 1992, *Crit. Rev the Drug. Syst.* 9: 135-162.
2. Johnson, K. P., Brooks, B. R., Cohen, J. A., Ford, C. C., Goldstein, J., Lisak, R. P., Myers, L. W., Panitch, H. S., Rose, J. W., Schiffer, R. B., Vollmer, T., Weiner, L. P. and Wolinsky, J. S., 1998, *Neurology* 50: 701-708.
3. Friend, D. R. and Pangburn, S., 1987, *Medicinal Research Reviews* 7: 53-106.
4. Duncan, R., 1992, *Anti-Cancer Drugs*, 3: 153-156.
5. Maeda, H. and Seymour, L.W., Miyamoto, Y., 1992, *Bioconjugate Chem.*, 3: 351-362.
6. Takakura, Y.and Hashida, M., 1995, *Critical Reviews in Oncology/Haematology* 18: 207-231.
7. Hershfield, M.S.and Mitchell, B.S., 1995, *The Metabolic and Molecular bases of Inherited Disease*, (Scriver, C.R., Beaduet, A.L., Sly, W.S.and Valle, D. eds.) McGraw-Hill, New York, pp. 1725-1768.
8. Nucci, M.L., Shorr, R. and Abuchowski, A., 1991, *Advanced Drug Delivery Reviews,* 6: 133-151.
9. Arnon, R., 1997, *BEHRING Inst. Mitt.* 98: 184-190.
10. Del Giudice, G., 1992, *Current Opinion in Immunology.* 4: 454-459.
11. Hudecz, F. and Tóth, G.K., 1994, *Synthetic peptides in the search for B-and T-cell epitopes* (Rajnavölgyi, É. ed.), R.G.Landes Company, Austin, pp 97-119.
12. Hudecz, F., Votavova, H., Gaál, D., Sponar, J., Kajtár, J., Blaha, K. and Szekerke, M., 1985, *Polymeric Materials in Medication* (Gebelein, Ch.G. and Carraher, Ch.E. eds.) Plenum Press, New York, pp 265-289.
13. Hudecz, F., 1995, *Anti-Cancer Drugs* 6; 171-193.
14. Hudecz, F., Gaál, D., Kurucz, I., Lányi, S., Kovács, A.L., Mező, G., Rajnavölgyi, É. and Szekerke, M., 1992, *J. Controlled Release* 19: 231-243.
15. Clegg, J.A., Hudecz, F., Mező, G., Pimm, M.V., Szekerke, M. and Baldwin, R.W., 1990, *Bioconjugate Chem.*, 2: 425-430.
16. Hudecz, F., Clegg, J.A., Kajtár, J., Embleton, M.J., Szekerke, M. and Baldwin, R.W., 1992, *Bioconjugate Chem.* 3: 49-57.
17. Vincze, B., Pályi, I., Daubner, D., Kálnay, A., Mező, G., Hudecz, F., Szekerke, M., Teplán, I., and Mező, I., 1994, *J. Cancer Res. Clin. Oncol.* 120: 578-584.
18. Mező, G., Mező, I., Seprődi, A., Teplán, I., Kovács, M., Vincze, B., Pályi, I., Kajtár, J., Szekerke, M. and Hudecz, F., 1996, *Bioconjugate Chem.* 7: 642-650.
19. Mező, G., Sármay, G., Hudecz, F., Kajtár, J., Nagy, Zs., Gergely, J., Szekerke, M., 1996, *J. Bioactive and Compatible Polymers* 11: 263-285.
20. Pimm, M.V., Gribben, S.J., Mező, G. and Hudecz, F., 1995, *J. Labelled Compounds and Radiopharmaceuticals* 36: 157-172.
21. Perkins, A.C., Frier, M., Pimm, M.V., Hudecz, F., 1998, *J. Labelled Compounds* 41: 631-638.
22. Gaál, D. and Hudecz, F., 1998, *Eur.J.Cancer* 34: 155-161.
23. Hudecz, F., 1995, *Biomed. Peptides, Proteins and Nucleic Acids* 1: 213-220.
24. Hudecz, F., Hilbert, Á., Mező, G., Mucsi, I., Kajtár, J. Bősze, Sz., Kurucz, I. Rajnavölgyi, É., 1994, *Innovation and Perspectives in Solid Phase Synthesis - Peptides, Polypeptides and Oligonucleotides - 1994* (Epton, R. ed.) Intercept, Andover, pp.315-320.

25. Hudecz, F., Hilbert, Á., Mező, G., Kajtár, J. and Rajnavölgyi, É., 1994, *Synthetic peptides in the search for B-and T-cell epitopes.* (Rajnavölgyi, É. ed.) R.G.Landes Company, Austin, pp. 157-169.

26. Wilkinson, K.A., Vordermeier, M.H., Wilkinson, R., Iványi, J. and Hudecz, F., 1998, *Bioconjugate Chemistry,* 9: 539-547.

27. Pimm, M.V., Gribben, S.J., Bogdán, K. and Hudecz, F., 1995, *J. Controlled Release* 37: 161-172.

28. Nagy, I.B., Haro, I., Alsina, A., Reig, F.and Hudecz, F.,1998, *Biopolymers* 46: 169-179.

29. Sela, M., 1980, *Molecules, cells, and parasites in immunology* (Larralde, C., Wills, K., Ortiz-Ortiz, L.and Seka, M., eds.) Acad Press, New York, pp. 215-228.

30. Hudecz, F. and Szekerke, M., 1980, *Coll. Czech. Chem. Commun.* 45: 933-940.

31. Mező, G., Kajtár, J., Nagy, I., Majer, Zs., Szekerke, M. and Hudecz, F., 1997, *Biopolymers* 42: 719-730.

32. Hudecz, F., Kovács, P., Kutassi-Kovács, S., Kajtár, J., 1984, *Colloid and Polymer Sci.* 262: 208-212.

33. Idei, M., Dibó, G., Bogdán, K., Mező, G., Horváth, A., Érchegyi, J., Mészáros, Gy., Teplán, I., Kéri, Gy. and Hudecz, F., 1996, *Electrophoresis* 17: 1357-1360.

34. Verger, R. and De Haas, G. H., 1973, *Chem. Phys. Lipids* 10: 127-135.

35. Gómez, V., Colomé, C, Reig, F., Rodriguez, L. and Alsina, M. A., 1994, *Anal. Chim. Acta* 294: 65-74.

36. Goupil, D. W. and Goodrich, F. C., 1977, *J. Coll.Interf. Sci.* 62:142-148.

37. Winterhalter, M., Bürner, H., Marzinka, S., Benz, R. and Kasianowicz, J. J., 1995, *Biophys. J.* 69: 1372-1381.

38. Reig, F., Busquets, M. A., Haro, I., Rabanal, F. and Alsina, M. A., 1992, *J. Pharm. Sci.* 81: 546-550.

39. Van Deenen, L. L. M., Demel, R. A., Geurtz van Kessel, W. S. H., Kamp, H. H., Roelfsen, B., Veerkelij, A. J., Wirtz, K. W. A. and Zwall, R. F. A., 1976, In *The structural basis of membrane function* (Hatefi, Y. and Djavadi-Ohaniance, L. eds.) Academic Press, New York, pp. 21-38.

40. Cajal, Y., Alsina, M. A., Rabanal, F. and Reig, F., 1996, *Biopolymers* 38: 607-618.

41. De Kruijff, B., 1994, *FEBS Letters* 346: 78-82.

42. Leenhouts J.M., van de Wijngaard, P.W., de Kroon, A.I.P.M and de Kruijff, B., 1995, *FEBS Letters* 370: 189-192.

43. Garcia, M., Nagy, I.B., Alsina, A., Mező, G., Reig, F., Hudecz, F. and Haro, I., 1998, *Langmuir* 14: 1861-1869.

44. Hudecz, F., Pimm, M.V., Rajnavölgyi, É., Mező, G., Fabra, A., Gaál, D., Kovács, A.L., Horváth, A. and Szekerke, M., 1999, *Bioconjugate Chemistry* 10: 781-790 (1999)

45. Sospedra, P., Nagy, I.B., Haro, I., Mestres, C., Hudecz, F., and Reig, F., 1999, *Langmuir* 15: 5111-5117.

46. Nagy, I.B., Alsina, M.A., Haro, I., Reig, F., and Hudecz, F., 2000, *Bioconjugate Chemistry*, 11: 30-38.

BLOCK COPOLYMER-BASED FORMULATIONS OF DOXORUBICIN EFFECTIVE AGAINST DRUG RESISTANT TUMOURS

[1]Valery Alakhov and [2]Alexander Kabanov
[1]*Supratek Pharma Inc., 531, Blvd. des Prairies, Bldg. 18, Laval, Quebec H7V 1B7, Canada*
[2]*Department of Pharmaceutical Sciences, 986025 Nebraska Medical Center, Omaha, Nebraska 68198-6025*

1. INTRODUCTION

Anthracyclines are amongst the most widely used anticancer agents in man, but are limited by toxicity considerations, as well as by inherent or induced drug resistance[1,2]. Several approaches were explored to improve therapeutic indexes of anthracyclines and to extend their utility by using various drug delivery systems, for example, liposomes[3,4], polymer and antibody-based conjugates[5,6]. The use of these systems significantly enhance the pharmacological properties of anticancer drugs. In particular, liposomal formulations of doxorubicin[3] and daunorubicin[7] demonstrated lowered toxicity and increased accumulation in tumours. Some of these products, for example, Doxil (liposomal doxorubicin) and Daunosone (liposomal daunorubicin) have been approved for clinical use[8]. Considerable improvement in safety profile was reported for some drug polymer conjugates, such as doxorubicin-HPMA (PK1) that is presently in phase 2 clinical trials[9] and doxorubicin-polyaspartate-polyethylene oxide[10]. Although many of these products demonstrated relatively low toxicity, in many cases, their efficacy was also significantly reduced[9,10] resulting in therapeutic indexes that did not differ much from those of the generic drugs. Moreover, drug resistance, which is one of the most significant obstacles that reduce the effectiveness of anthracycline-based chemotherapy, was not sufficiently addressed during the development of the above-mentioned products.

We have recently discovered that nonionic block copolymeric surfactants, *i.e.* hydrophobic polyethylene oxide polypropylene oxide block copolymers (Pluronics) can considerably reduce drug resistance of various tumour cells to anthracyclines and other cytotoxic drugs[11,12]. Following this finding, we have developed a Pluronic-based formulation of doxorubicin (SP1049C) that is thermodynamically stable, safe and provides doxorubicin with high efficacy against both drug resistant and drug sensitive tumours. In this report, we have summarised the results accumulated during the work performed to bring this product to clinical trials.

2.   SP1049 FORMULATION AND ITS *IN VITRO* PROPERTIES

2.1   Block Copolymer/Drug Formulation

Like many other amphiphilic surfactants, the Pluronic copolymers used in SP1049C formulation spontaneously form micelles at concentrations equal to or exceeding the critical micellar concentration (CMC). At polymer concentration above CMC, the drug added to the system is partitioned between the water phase and the micellar microphase. The equilibrium between the micelle-incorporated drug and the drug present in the water phase is described by the partitioning coefficient (P), which is equal to the ratio between the local concentration of the drug in the micellar microphase and its concentration in the water phase.

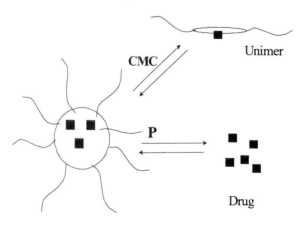

Figure 1. Thermodynamic equilibrium between unimers, micelles and drug.

The partitioning of the drug between the micellar microphase and the water phase depends on both the drug and polymer structural properties. At polymer concentrations exceeding CMC, the drug/polymer composition consists of three components: free drug, drug-micelle complexes and unimers (Fig.1).

2.2   Selective Effect of Copolymers on Cytotoxic Activity of Doxorubicin against Drug Resistant Tumour Cells

We have recently discovered that Pluronic copolymers can selectively potentiate the response of multidrug resistant cancer cells to cytotoxic drugs, such as anthracyclines, vinca alkaloids, epipodophilotoxins and taxanes[11,12]. It has been previously reported that some other nonionic surfactants enhance the membrane permeability of drug resistant cells, in this way increasing the cytotoxic response of these cells to anticancer drugs[13,14]. However, in the case of Pluronic copolymers, doxorubicin cytotoxicity is increased to the point where resistant cells become more susceptible to the drug than their sensitive counterparts (hypersensitisation effect) (Fig.2).

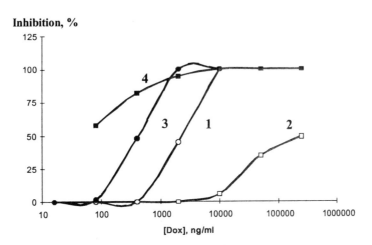

Figure 2. Cytotoxicity of doxorubicin (1 and 2) and doxorubicin in the presence of 0.01% w/v Pluronic L61 (3 and 4) with respect to drug sensitive (1 and 4) MCF-7 and drug resistant (2 and 3) MCF-7ADR cell sublines.

An analysis of the effect of various concentrations of copolymer on the response of drug resistant cells to the drug has revealed that the copolymer reduces IC_{50} of doxorubicin below CMC (Fig.3) suggesting that unimers rather than micelles are responsible for the hypersensitisation effect.

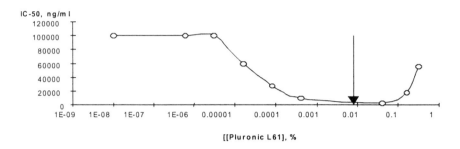

Figure 3. Dependence of doxorubicin cytotoxicity against drug resistant MCF-7ADR cells on concentration of Pluronic L61 in the medium. The vertical arrow shows CMC.

A comparison of effects of various Pluronic copolymers on the cytotoxic activity of doxorubicin against drug resistant cells has demonstrated that hydrophobic compounds in which polyethylene oxide block constitutes 50% or less of the total molecular mass have the highest modulating effect (Table 1). The dependence of the modulating effect of copolymer on its total molecular weight was less pronounced, however, copolymers with higher molecular masses were less effective (Table 1). This finding can be likely explained by lower CMC of these copolymers and, as a result, lower concentration of the unimers in the system. More detailed study on the panel of drug sensitive and drug resistant cell sublines was performed with doxorubicin formulated with Pluronic L61, which was found to be one of the most effective copolymers (Table 2). This study has demonstrated that in standard cytotoxicity assay, the copolymer formulation exhibits significantly higher efficacy than doxorubicin against a variety of drug resistant tumour cell lines.

Table 1. Comparison of effects of various Pluronic copolymers on cytotoxic activity of doxorubicin against drug resistant CH^RC5 cells

Pluronic	Average molecular mass[a]	Hydrophilic-Lipophilic Balance (HLB)	Doxorubicin IC_{50}, ng/ml in the presence of 0.1% w/v Pluronic
L61	2000	1-7	250±9
L81	2750	1-7	250±12
P84	4200	12-18	800±50
P85	4600	12-18	1100±120
F87	7700	>24	9900±670
L101	3800	1-7	290±15
L121	4400	1-7	490±35

[a]Average molecular masses and HLB are given as specified by the supplier

Table 2. Cytotoxicity of doxorubicin alone and of doxorubicin in the presence of 0.01% w/v Pluronic L61 against a panel of multidrug resistant cell sublines and their sensitive parental lines

Cell Lines	IC$_{50}$ Dox, ng/ml[a]	IC$_{50}$ Dox/L61, ng/ml[a]	Sensitivity Increase, -fold	Resistance Factor, Dox[b]	Resistance Factor, Dox/L61[b]
MCF-7	2,000±58	2,000±65	1		
MCF-7ADR	222,000±12,000	300±15	740	111	0.15
Aux-B1	1,000±29	700±21	1.3		
CH^{R}C5	70,000±2,470	250±9	280	70	0.35
SKV03	4,300±390	3,8000±211	1.1		
SKVLB	7,345+/-670	7.2±0.5	1,020	1.7	0.002
KB	1,560+/-125	980±125	1.6		
KBV	15,350±890	260±50	60	9.8	0.26
B16BL6	880±44	440±45	2.0		
B16F10	260,220±11,000	125±10	2,082	296	0.28
SP2/0	9,890±85	6,660±555	1.48		
SP2/0Dox	125,660±1,300	620±21	202.7	12.7	0.09
P388	55±2.2	62±4.5	0.89		
P388-Dox	1,250±140	8.4±0.6	149	22.7	0.14
LoVo	85±5.5	15±0.8	5.7		
LoVo/Dox	1,400±55	0.4±0.05	3,500	16.4	0.03

[a] IC$_{50}$ values were determined by standard XTT assay; each value is the mean ± SEM of triplicate assays[12].

[b] Resistance factor was calculated as IC$_{50}$ of each multidrug resistant cell line divided by IC$_{50}$ of the drug sensitive parent line.

The combination of the two aforementioned properties of Pluronic block copolymers, *i.e.* their flexibility in pharmaceutical formulations and their ability to produce strong and selective effects on drug resistant tumour cells, has suggested that clinically effective anticancer formulations may be developed on their basis.

2.3 Effect of Copolymers on the Drug Transport and its Intracellular Distribution in Drug Resistant and Drug Sensitive Tumour Cells

To assess the mechanism by which Pluronic copolymers hypersensitise drug resistant tumour cells, we have analysed the effect of these compounds on the drug transport into and within the cells, as well as the drug interaction with DNA, which is known to be the target for cytotoxic activity of doxorubicin[1].

2.3.1 Drug uptake

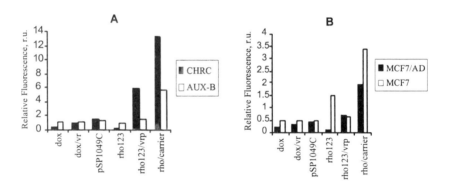

Figure 4. Intracellular accumulation of 10 µg/ml doxorubicin and 0.5 µM Rho123; Rho123 in the presence of 50 µM Vrp; and Rho123 in the presence of 0.1% w/v Pluronic L61 in drug-resistant MCF-7ADR (A) and CH^{R}C5 (B), drug sensitive (MCF-7 (A) and Aux-B1 (B) cells[12].

Doxorubicin uptake was analysed by flow cytometry in Chinese hamster ovary Aux-B1 (sensitive) and CH^{R}C5 (resistant) cells. The latter demonstrated about a 100-fold higher resistance to doxorubicin than the parental sensitive cells due to overexpression of P-glycoprotein (P-gp), which is known to be one of the major ATP-dependent transporters contributing to multidrug resistance of tumours[15]. This study revealed that Pluronic L61 induced a 7.2-fold higher increase in the drug accumulation in the drug resistant subline than free doxorubicin. A smaller but significant rise (1.6-fold) in the drug uptake was also observed in the sensitive cells. Similar results were obtained on human breast carcinoma MCF-7 (sensitive) and MCF-7ADR (resistant) cell sublines. In this case, the drug uptake increased by 2.4-fold and by 1.1-fold in MCF-7ADR and MCF-7 cells, respectively (Fig.4A)[12]. Doxorubicin uptake could not be assessed precisely by fluorescence analysis because of quenching of the drug fluorescence during its intercalation with DNA. To avoid the quenching effect, the experiment was repeated with Rhodamine 123 (Rho123), which is commonly used as a fluorescent probe for transport studies in multidrug resistant cells[12]. This compound accumulates in cells without any substantial loss in its fluorescence and interacts with copolymeric carriers similarly to doxorubicin. The copolymer significantly enhanced (124-fold) the accumulation of Rho123 in CH^RC5 cells and to a lesser extent (23-fold) - in Aux-B1 cells. In the case of MCF-7ADR and MCF-7 cells, the copolymer increased Rho123 uptake by 30- and 2.8-fold, respectively (Fig.4B). Verapamil (Vrp), which reverses multidrug resistance by inhibiting P-gp in a competitive manner[16], affected the drug uptake to a much lesser degree than Pluronic L61 suggesting that a substantial amount of the drug gets into resistant cells via a P-gp-independent pathway.

2.3.2 Drug efflux

To estimate the effect of the carrier on the P-gp function, the kinetics of Rho123 efflux in CH^RC5 cells was studied. As expected, the fast component in the efflux kinetic rate was observed in the case of free Rho123, which is characteristic of a P-gp-mediated process[15]. The drug efflux was completely blocked in the presence of Vrp due to competitive inhibition of P-gp activity by the latter. The fast component in the Rho123 efflux was missing when the Pluronic L61/Rho123 formulation was used. Instead, only low-rate efflux was registered (Fig.5) suggesting that the carrier inhibited the rapid drug efflux, which is usually attributed to the P-gp function.

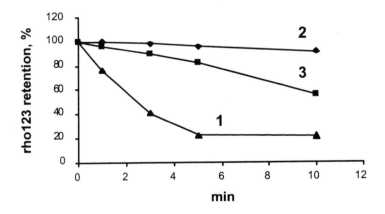

Figure 5. Rho123 efflux rate in resistant CH^RC5 cells: (1) 0.5 μM Rho123; (2) 0.5 μM Rho123 in the presence of 50 μM Vrp; (3) 0.5 μM Rho123 in the presence of 0.1% w/v Pluronic L61. The analysis was performed by using flow cytometry as described[12].

2.3.3   Drug intracellular distribution

Multidrug resistant cells sequester doxorubicin in cytoplasmic vesicles thereby diminishing the amount of the drug in the nucleus[17]. The effect of the copolymer and Vrp on the intracellular distribution of doxorubicin in resistant CH^RC5 cells was studied by fluorescent microscopy[12]. Figure 6 shows that after 1-h incubation of the cells with doxorubicin, the drug was concentrated in the cytoplasmic vesicles, and a very low level of fluorescence was associated with the nucleus. In the case of doxorubicin combined with Vrp, an increase in fluorescence was observed in the nucleus, while the drug level in the cytoplasmic vesicles remained significant. In the case of Pluronic L61/doxorubicin formulation, fluorescence appeared mainly in the nucleus and none was visible in the vesicles. Sequestering of doxorubicin and other positively charged compounds in acidic cytoplasmic vesicles is attributed to high pH gradients between the vesicles and the cytosol in multidrug resistant cells[17].

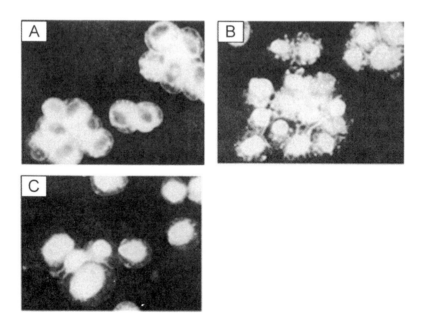

Figure 6. Intracellular distribution of 50 µg/ml doxorubicin in resistant CH^RC5 cells visualised by fluorescent microscopy (x500) after 60-min incubation. (A) doxorubicin alone; (B) doxorubicin in the presence of 50 µM Vrp; (C) doxorubicin in the presence of 0.1% w/v Pluronic L61.

High retention of the drug in these compartments can be explained by protonation of the drug molecules in the acidic compartments, which reduces their ability to diffuse across vesicular membranes. The fact that ionophoric compounds, such as nigoricin and monensin, reverse multidrug resistance by reducing pH gradients in the cytoplasm[17] supports this hypothesis.

The decrease in accumulation of doxorubicin in the cytoplasmic vesicles leads to an increase in the drug availability to DNA, which is the target for its cytotoxic action[1]. As a result, about a 10-fold increase in the drug binding with DNA was observed in SKVLB cells treated with Pluronic L61/doxorubicin as compared to doxorubicin alone (Fig.7)[11].

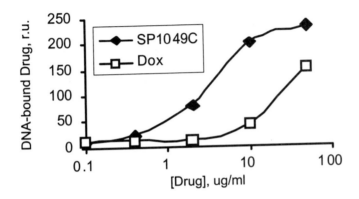

Figure 7. Binding of doxorubicin and SP1049C with DNA in drug resistant SKVLB cells. The analysis was performed by using fluorescent spectroscopy as described[11].

The ability of doxorubicin formulated with the copolymer to avoid accumulation in acidic cytoplasmic vesicles is probably the most important contributor to its mechanism of action. Resistance-modulating agents and some polymer conjugates can partially reduce the drug resistance mediated by ATP-dependent transporters by either directly inhibiting these transporters[18], or by switching the drug transport from passive diffusion to endocytosis[19]. At the same time, they cannot overcome the endosomal barrier that represents the second level of resistance in drug resistant cells. The fact that Pluronic L61/doxorubicin can effectively penetrate the plasma membrane of resistant cells and, at the same time, avoid sequestration in the vesicles suggests that this product may be more clinically effective than other doxorubicin-based products.

2.3.4 Combining Hydrophobic and Hydrophilic Pluronic Copolymers to Prevent Micellar Aggregation in the System

The cell screen studies have demonstrated that hydrophobic Pluronic copolymers with small to medium molecular masses, Pluronic L61 in particular, produce the highest modulating effect on doxorubicin activity against drug resistant cells (Table 1). However, the analysis of the micellar size by dynamic light scattering revealed that the copolymers of this group have a high tendency to aggregation in aqueous solution followed by liquid phase separation at relatively low copolymer concentrations. In fact, in the case of Pluronic L61, phase separation was observed at the concentration as low as 0.1% w/v at 37°C (Table 3), while in the preliminary *in vivo* studies, the most effective concentration of this copolymer in the dosing solution was established to be 0.25% w/v[20]. Therefore, it was necessary to improve physico-chemical properties of the formulation to avoid toxicity observed in the preliminary toxicity study (data not shown) related to microembolic effects caused by filtering of large copolymer aggregates in the lung, liver and kidney. To this end, we have combined Pluronic L61 with hydrophilic and higher molecular weight Pluronic F127.

Table 3. Effective diameter of micelles formed by Pluronic L61 and combinations of Pluronic L61 and Pluronic F127 in PBS at 37^{0}C in the presence and in the absence of 2.5% BSA

Pluronic L61, %w/v	Pluronic F127, % w/v	Effective diameter, nm; no BSA	Effective diameter, nm; with BSA
0.1	-	LPS[a]	N/D[b]
0.25	-	LPS	N/D
0.1	0.5	26.7	23.8
0.1	1.0	23.6	12.9
0.1	2.0	21.6	12.8
0.25	0.5	T[c]	N/D
0.25	1.0	T	N/D
0.25	2.0	22.4	15.0

[a]Liquid phase separation ; [b]Not determined; [c]High turbidity

The results presented in Table 3 show that the latter produced a significant stabilising effect on Pluronic L61 by preventing liquid phase separation and by preserving the effective diameter of the micelles below 30 nm, which is highly desirable for pharmaceutical formulations[21]. At the same time, the addition of Pluronic F127 did not significantly alter the cytotoxic activity of the Pluronic L61/doxorubicin formulation (Table 4). Therefore, doxorubicin formulated with the combination of Pluronic L61 and Pluronic F127 at the ratio 1:8 w/w (SP1049C formulation) was selected for further development of the product.

Table 4. Cytotoxicity of doxorubicin against various drug resistant cell lines in the presence of 0.02% w/v Pluronic L61; 0.025 w/v Pluronic L61 and 0.2% w/v Pluronic F127

Cell line	IC_{50} ng/ml (Pluronic L61)	IC_{50} ng/ml (Pluronic L61/ Pluronic F127)
MCF-7ADR	290±18	320±33
KBV	220±30	260±50
LoVo/Dox	0.6±0.05	0.8±0.05

3. EFFICACY COMPARISON OF DOXORUBICIN AND SP1049C ON THE *IN VIVO* TUMOUR MODELS

Table 5. Tumour panel efficacy studies

Tumour Type	Tumour Site	SP1049C TI, % [a]	Dox TI, %	NCI Criteria for TI, %	SP1049C ILC, % [b]	Dox ILC, %	NCI Criteria for ILC, %
Murine Leukaemia, P388	i.p.	-	-	-	176	96	120
Murine Leukaemia, P388-Dox	i.p.	-	-	-	175	92	120
Murine Myeloma, Sp2/0	s.c.	96.7	58.2	58	182	147	145
Murine Myeloma, Sp2/0-Drn	s.c.	98.2	28.4	58	177	122	145
Lewis Lung Carcinoma, 3LL-M27	i.v.	-	-	-	222	118	140
Lewis Lung Carcinoma, 3LL-M27	s.c.	68.4	18.2	58	220	115	140
Human Breast Carcinoma, MCF-7	s.c.	71.2	48.5	58	203	134	140
Human Breast Carcinoma, MCF-7ADR	s.c.	69.4	9.3	58	210	103	140
Human Head and Neck Carcinoma, KBV	s.c.	90.7	68.7	58	-	-	-
Responses	Dox	2/9					
	SO1049C	9/9					

[a]Tumour inhibition; [b]Increased life-span

To design this study and to evaluate the results obtained, we have used recommendations by National Cancer Institute, NIH, for evaluating the *in vivo* activity of anticancer drugs[22,23]. These recommendations and evaluation criteria were established on the basis of retrospective studies of a large number of anticancer drugs that are being used in clinic. The results of the *in vivo* screen tumour panel study shown in Table 5 demonstrate that SP1049C met the efficacy criteria in all models (9 out of 9), whereas doxorubicin was effective only in 2 out of 9 models.

4.   COMPARISON OF PLASMA PHARMACO-KINETICS AND BIODISTRIBUTION OF SP1049C AND DOXORUBICIN IN NORMAL AND TUMOUR-BEARING MICE

Female 6- to 7-week-old C57Bl/6 mice were used to analyse pharmacokinetics and biodistribution of doxorubicin and SP1049C. In the tumour-bearing group, the animals were implanted *s.c.* with Lewis lung carcinoma 3LL M-27 cells. To analyse plasma pharmacokinetics, the experimental data from both doxorubicin and SP1049C series were fitted into tri-exponential function. The plasma pharmacokinetic parameters calculated from the fitting results are shown in Table 6.

Table 6. The plasma pharmacokinetic parameters of doxorubicin and SP1049C in normal and tumour-bearing mice

Parameter	Dox	SP1049C
NORMAL MICE		
AUC, μg x h / ml	7.1	14.6
$t_{1/2}$, h	15.4	21.2
MRT, h	21.6	29.9
CL, L / kg x h	1.41	0.68
V_{ss}, L / kg	30.4	20.5
TUMOUR-BEARING MICE		
AUC, μg x h / ml	7.1	8.5
$t_{1/2}$, h	15.1	15.8
MRT, h	18.6	16.1
CL, L / kg x h	1.4	1.2
V_{ss}, L / kg	26.1	18.9

The AUC values for the organs were calculated by using the trapezoidal method. The C_{max} and AUC values determined for single 10 mg/kg injections of SP1049C and doxorubicin in normal and tumour-bearing mice are represented in Table 7.

Table 7. Biodistribution of doxorubicin and SP1049C in normal and tumour-bearing mice

Organ	Dox C_{max} µg/g tissue	SP1049C C_{max} µg/g tissue	C_{max} Ratio SP1049C/ Dox	Dox AUC µg x h/g	SP1049C AUC µg x h/g	AUC Ratio SP1049C/ Dox
NORMAL MICE						
Plasma	2.77	4.47	1.6	7.1	14.6	2.1
Brain	0.36	1.01	2.8	9.0	26.0	2.9
Heart	12.50	17.50	1.4	111.5	139.8	1.3
Kidney	26.1	36.5	1.4	271.5	312.1	1.1
Liver	20.70	33.80	1.6	147.1	192.3	1.3
Lung	10.10	12.00	1.2	307.5	282.2	0.9
TUMOUR-BEARING MICE						
Plasma	2.94	3.08	1.0	7.1	8.5	1.2
Tumour	1.00	0.86	0.9	30.1	50.8	1.7
Brain	0.18	0.30	1.7	5.6	9.2	1.6
Heart	9.17	10.63	1.2	156.2	177.2	1.1
Kidney	11.84	17.25	1.5	263.7	270.9	1.0
Liver	14.00	10.50	0.8	154.2	173.5	1.1
Lung	15.30	17.83	1.2	207.5	268.2	1.3

SP1049C and doxorubicin exhibited similar C_{max} and AUC in liver, kidney, heart and lung of both normal and tumour-bearing mice. Increased AUC was detected in the tumour and brain of SP1049C-treated animals. The pharmacokinetic curves (Fig.8) indicate that in the tumour, the peak levels of SP1049C are delayed and the drug residence time is prolonged in comparison to those observed when an equivalent dose of doxorubicin was used. The enhanced distribution of the drug in the tumour tissue may also explain the fact that the levels of SP1049C in plasma are typically higher in normal mice than in tumour-bearing animals.

The potentiating effects of polymers on drug accumulation in tumours have been previously reported for several polymer-based compositions[24]. This phenomenon, also known as Enhanced Permeability and Retention effect (EPR), was described for drug/polymer conjugates (PK-1, PEG-polyaspartate-doxorubicin) and some other supramolecular complexes, such as liposomes. Moreover, the ability to facilitate accumulation of various compounds in tumour tissues was specifically attributed to Pluronic copolymers[25-28]. Therefore, it can be concluded that the increased drug level and residence time in tumour tissue observed in the SP1049C-treated group of animals are likely related to EPR effect.

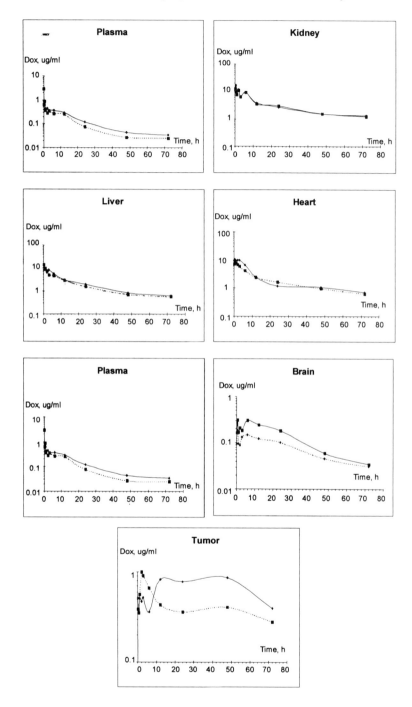

Figure 8. Pharmacokinetics of doxorubicin (dashed lines) and SP1049C (solid lines) in plasma and major organs of C57Bl/6 mice bearing *s.c.* 3LL-M27 tumours.

5.   TOXICITY OF SP1049C

To establish the safety of this formulation prior to its use in human clinical trials, SP1049C was subjected to detailed comparative (*vs.* doxorubicin) *i.v.* toxicity studies in rodents and a vascular irritation study in rabbits. In the dose range finding studies in mouse and rat, the maximum tolerated single doses (MTD) were analysed for SP1049C and doxorubicin. Within the species, MTDs of both products were the same, being established as 15 mg/kg in mouse and 7.5 mg/kg in rat. In the detailed toxicity study, three groups (20 male rats each) were dosed once with SP1049C at MTD, circa 1/2MTD and 1/3MTD, while the fourth received the carrier only. Similar four groups were treated with doxorubicin or its vehicle (saline). Three days after, half of the animals was sacrificed and subjected to gross pathological evaluation, and a full range of organs and tissues was collected into fixative. For control and high-dose-treated groups, the major tissues were then subjected to histopathological evaluation. The remaining animals were sacrificed 30 days after dosing to allow the assessment of the progression or reversibility of any treatment-related toxic effects. For both doxorubicin and SP1049C, in-life clinical findings were restricted to fur thinning and submandibular/abdominal edema in high-dose rats. Impairment of body weight growth and consumption occurred, particularly, at the two highest dose levels, although, the effects were less pronounced in rats treated with SP1049C. Blood evaluation showed trombocytopenia, reduction in red blood cell parameters, increases in cholesterol, triglicerides, ALT, bilirubin and blood urea nitrogen, and decreases in AST, AP, glucose and albumin. The gross and histopathological findings obtained showed changes expected subsequent to the administration of doxorubicin (29-31). By day 3, these included changes in organ weights, in particular significant reduction in the spleen and thymus weights and some increase in the liver weight, especially at high dose level. Doxorubicin-related toxic effects were observed histologically both in doxorubicin- and SP1049C-treated groups in numerous organs, particularly, in mitotically active tissues, such as testes, gastrointestinal tract, mammary gland, skin and the hematopoietic system. Early indications of glomerulonephropathy were evident in kidneys. By day 30, the organ weight changes had extended to reductions in testicular weight and to increases in kidney weight, both changes being dose-dependent. In high-dose animals, the expected cardiotoxicity was evident. The aforementioned gross and histopathological changes were equivalent for both doxorubicin- and SP1049C-treated animals, except for histopathological changes in skin, thymus and testes that were less severe in rats dosed with SP1049C.

To evaluate any difference in the vascular irritation properties between doxorubicin and SP1049C, the study using New Zealand white rabbits was performed. Single injections of SP1049C and of the carrier were made into the marginal vein of the right and left ears, respectively, in six animals. Six more rabbits were similarly treated with doxorubicin and saline. Half of the animals were sacrificed seven days after injection and the remainder - 28 days later. Gross and histopathological evaluation of the right and left marginal ear vein of all rabbits was then performed. The results obtained revealed no difference in the irritation potential of doxorubicin and SP1049C.

6.   CONCLUSION

SP1049C has a novel mechanism of action based on the altered transport of the drug and its intracellular distribution. The increased efficacy of this product compared to doxorubicin, particularly in drug resistant settings, is due to: (i) increase in the drug influx; (ii) inhibition of the energy-dependent efflux; and (iii) changes in intracellular trafficking, as discussed above. As a result, a SP1049C binds with DNA 10 times more actively than doxorubicin. The ability of SP1049C to kill multidrug resistant cells more effectively than their drug sensitive counterparts distinguishes this product from other doxorubicin-based formulations. This specific property of the product suggests that SP1049C may have the potential to be used not only for treatment of primary and relapsed tumours, but also for prevention of multidrug resistance development.

ACKNOWLEDGMENTS

We thank Dr. Brian Leyland-Jones (McGill University, Montreal, QC) for his valuable critical comments and scientific support of the work; Dr. Barry Osborne (STS Inc., Montreal, QC) for monitoring and Dr. Bryan Proctor (ITR Laboratories Canada Inc., Montreal, QC) for supervision of the toxicity studies; and Mrs. Irina Deinekina for assistance in the preparation of the manuscript.

The work was supported in part by Industrial Research Assistance Program of the National Research Council of Canada (grants # 14113Q and 14835Q), as well as by Nebraska Research Initiative.

REFERENCES

1. Lehnert, M., 1996, *Eur. J. Cancer* 32a: 912-918.

2. Pastan, M.M., and Gottesman, N., 1987, *Engl. J. Med.* 316: 1388-1392.

3. Gabizon, A., Goren, D., Cohen, R., and Barentidz, Y., 1998, *J Controlled Release* 53: 275-278.

4. Lopes de Meneres, D.E., 1998, *Cancer Res.* 58: 3320-3330.

5. Omelyanenko, V., Kopeckova, P., Gentry, C., and Kopecek, J., 1998, *J. Controlled Release* 53: 25-33.

6. Jelinkova, M., Strohalm, J., Plocova, D., Subr, V., St'astny, M., Ulbrich, K., and Rihova, B., 1998, *J. Controlled Release* 52: 253-259.

7. Richardson, D.S., and Johnson, S.A., 1997, *Blood Rev.* 4: 201-223.

8. Bennett, C.L., Golub, R.M., Stinson, T.J., Aboulafia, D.M., Von Roenn, J., Bogner, J., Goebel, F.D., and Stewart, S., 1998, *J. Acquir. Imm. Defic. Syndr. Hum. Retrov.* 18: 460-472.

9. Duncan, R., Coatsworth, J.K., and Burtles, S., 1998, *Hum. Exp. Toxicol.* 2: 93-99.

10. Yokoyama, M., Kwon, G.S., Okano, T., Sakurai, Y., Seto, T., and Kataoka, K., 1992, *Bioconjug Chem* 4: 295-302.

11. Alakhov, V.Y., Moskaleva, E.Y., Batrakova, E.V., and Kabanov, A.V., 1996, *Bioconjug. Chem.* 7: 209-216.

12. Venne, A., Li, S., Mandeville, R., Kabanov, A., and Alakhov, V., 1996, *Cancer Res.* 56: 3626-3629.

13. Buckingham, L.E., Balasubramanian, M., Emanuele, R.M., Clodfelter, K.E., and Coon, J.S., 1995, *Int. J. Cancer* 62: 436-447.

14. Buckingham, L.E., Balasubramanian, M., Safa, A.R., Shah, H., Komarov, P., Emanuele, R.M., and Coon, J.S., 1996, *Int. J. Cancer* 65: 74-85.

15. Pereira, E., Tarasiuk, J., and Garnier-Suillerot, A., 1998, *Chem. Biol. Interact.* 114: 61-69.

16. Naito, S., Kotoh, S., Omoto, T., Osada, Y., Sagiyama, K., Iguchi, A., Ariyoshi, A., Hiratsuka, Y., and Kumazawa, J., 1998, *Cancer Chemother. Pharmacol.* 42: 367-375.

17. Benderra, Z., Morjani, H., Trussardi, A., and Manfait, M., 1998, *Leukemia* 10: 1539-1549.

18. Sikic, B.I., Fisher, G.A., Lum, B.L., Halsey, J., Beketic-Oreskovic, L., and Chen, G., 1997, *Cancer Chemother. Pharmacol.* 40: S13.

19. Reynolds, T., 1995, *J. Natl. Cancer Inst.* 87: 1582-1591.

20. Batrakova, E.V., Dorodnych, T.Y., Klinski, E.Y., Kliushnenkova, E.N., Shemchukova, O.B., Goncharova, O.N., Arjakov, S.A., Alakhov, V.Y., and Kabanov, A.V., 1996, *Br. J. Cancer* 74: 1545-1549.

21. Alakhov, V., and Kabanov, A., 1998, *Exp. Opin. Invest. Drugs* 7: 1453-1473.

22. Goldin, A.L., Sandri-Goldin, R.M., Glorioso, J.C., and Levine, M., 1981, *Eur. J. Cancer* 17: 129-133.

23. Dexter, D.L., Hesson, D.P., Ardecky, R.J., Rao, G.V., Tippett, D.L., Dusak, B.A., Paull, K.D., Plowman, J., DeLarco, B.M., and Narayanan, V.L., 1985, *Cancer Res.* 45: 5563-5568.

24. Meada, H., and Seymur, L., 1992, *Bioconjugate Chem.* 3: 351-356.

25. Cassel, D.M., Young, S.W., Brody, W.P., Muller, H.H., and Hall, A.L., 1982, *J. Comput. Assist. Tomogr.* 6: 141-148.

26. Grover, F.L., Kahn, R.S., Heron, M.W., and Paton, B.C., 1973, *Arch. Surg.* 106: 307-312.

27. Moore, A.R., Paton, B.C., Eiseman, B., 1968, *J. Surgical Res.* 8: 563-571.

28. Silk, M., and Sigman, E., 1072, *Cancer* 25: 171-181.

29. Comereski, C.R., Peden, W.M., Davidson, T.J., Warner, G.L., Hirth, R.S., and Frantz, J.D., 1994, *Toxic. Pathol.* 22: 473-484.

30. *Toxicology: The Basic Science of Poisons*, 3rd Edit., 1986, Casarett and Doul, p. 403.

31. Yeung, T.K., Jaenke, R.S., Wilding, D., Creighton, A.M., and Hopewell, J.W., 1992, *Cancer Chemother. Pharmacol.* 30: 58.

PREPARATION OF POLY(LACTIC ACID)-GRAFTED POLYSACCHARIDES AS BIODEGRADABLE AMPHIPHILIC MATERIALS

Yuichi Ohya, Shotaro Maruhashi, Tetsuya Hirano, and Tatsuro Ouchi
[1]Department of Applied Chemistry, Faculty of Engineering and High Technology Research Center, Kansai University, Suita, Osaka 564-8680, Japan

1.   INTRODUCTION

Biodegradable and biocompatible polymers have become of interest from the standpoints of biomedical and pharmaceutical applications. Based on their biodegradation properties, biocompatibility, high mechanical strength, and excellent shaping and moulding properties, poly(lactic acid)s have been frequently applied as implantable carriers for drug delivery systems as well as surgical repair materials[1,2,3,4]. However, the high crystallinity of the polymers interfere with the controlled degradation, cause decrease in compatibility with soft tissues as biomedical materials, and was an obstacle to applications as biodegradable soft plastics. Possible promising approaches to overcome these problems are introduction of hydrophilic segments and branched structure in poly(lactic acid)s. Many approaches, for example, synthesis of block copolymer with polyethers, were carried out to control the degradation rate by varying the crystallinity[5,6,7,8]. On the other hand, polysaccharides, such as amylose and pullulan, are typical examples of natural biodegradable hydrophilic polymers, which show enzymatic degradation behavior and relatively good biocompatibility. However, polysaccharides are insoluble in common organic solvents.

Saccharides and polysaccharides have many hydroxyl groups and have been used as hydrophilic and bioactive segments in some hybrid-type

Biomedical Polymers and Polymer Therapeutics
Edited by Chiellini *et al.*, Kluwer Academic/Plenum Publishers, New York, 2001

biomaterials. Pullulans carrying hydrophobic cholesterol or alkyl groups formed aggregates in aqueous solution and entrap hydrophobic or hydrophilic compounds[9]. Many kinds of glycopolymers having saccharide moieties have been synthesized and studied as biofunctional materials. Biodegradable polymers having both hydrophobic aliphatic groups and hydrophilic saccharide units in their main chains were synthesized using mono- or oligosaccharides[10].

Poly(lactic acid)s are commonly synthesized by ring-opening polymerization of lactic acid dimer (lactide). This ring-opening polymerization reaction can proceed in the presence of an alkali metal alkoxide to give poly(lactic acid) containing the alkoxy group as its terminal group[8]. Our group have tried to obtain graft copolymers of poly(lactic acid) and polysaccharides using the hydroxyl groups of the polysaccharides as initiating groups. However, these attempts were unsuccessful, because of the poor solubility of polysaccharides in organic solvents to give a heterogeneous reaction. Although the ring-opening reaction actually occurred, the degree of polymerization was quite low, because polysaccharides have too many hydroxyl groups as initiating groups.

Very recently, we reported a new method to achieve graft polymerization of lactide on polysaccharide using mostly trimethylsilyl (TMS) protected polysaccharides[11,12]. Using the protection technique via TMS groups, amylose and pullulan became soluble in organic solvents and the number of initiating groups, which means number of grafted chains, could be controlled. As a result, poly(lactic acid)-grafted amylose and poly(lactic acid)-grafted pullulan could be obtained by homogeneous graft polymerization of L-lactide on mostly TMS protected polysaccharides in tetrahydrofuran (THF) by using potassium *t*-butoxide (*t*-BuOK) and consequent deprotection of TMS groups. Initial experiment for the evaluation of the graft copolymers as biomaterials, the biodegradability of the films of poly(lactic acid)-grafted polysaccharides obtained was also investigated.

2.   EXPERIMENTAL

2.1   Materials

Low-molecular-weight amylose having narrow polydispersity (GPC standard grade, AS-10 lot. No. 68, $M_n = 1.0 \times 10^4$, $M_w/M_n = 1.05$),

L-lactide, anhydrous THF, α-amylase and 4% osmic acid aqueous solution were purchased from Wako Pure Chemical (Tokyo, Japan). Pullulan (Mn = 1.5 x 10^5, Mw/Mn = 1.5) was obtained from Hayashibara Co. (Japan). High-molecular-weight amylose having wider polydispersity (from potato, lot. No.A-9262, Mn = 1.5 x 10^5, Mw/Mn = 1.5) were purchased from Sigma. Chlorotrimethylsilane (TMS-Cl) and *t*-BuOK were purchased from Kanto Chemical (Tokyo, Japan). L-Lactide was recrystallized twice from ethyl acetate before using. Amyloses and pullulan were dried at 60°C under vacuum before using. Anhydrous THF was used without purification. Pyridine, n-hexane and other organic solvents were purified by usual distillation method. Other reagents were commercial grades and used without purification.

2.2 Preparation of trimethylsilylated polysaccharides[11,12]

Mostly TMS-protected polysaccharides were prepared according to Scheme 1. Typical example of the procedures using amylose is described as follows. Low-molecular-weight amylose (381 mg, 2.35 mmol glucose unit) was suspended in pyridine (8.4 mL). TMS-Cl (1.76 mL, 14.1 mmol) dissolved in n-hexane (6.3 mL) was added to the solution. After 3 hours stirring, the reaction mixture was washed with sat. NaCl aqueous solution to remove pyridine hydrochloride. After reprecipitation with n-hexane and evaporation under reduced pressure, mostly trimethylsilylated amylose (TMSAm-1) was obtained. The introduction of TMS groups was confirmed by the methyl proton signal at δ = 0.10 ppm besides the broad methyne and methylene proton signals of amylose at δ = 3.5 ~ 5.5 ppm on ^{1}H-NMR spectra in CDCl$_3$. The molecular weight was estimated by GPC (Column: Toso TSKgel G4000HXL + G2500HXL, eluent: THF, detector: refractive index (RI), standard: polystyrene) to be Mn = 1.5 x 10^4 (Mw/Mn = 1.07). The degree of trimethylsilylation (DTMS) was estimated to be 93.1 mol%/hydroxyl group by ^{1}H-NMR. Yield 777 mg (90.5%). High-molecular-weight amylose and pullulan were used after degradation in acidic condition (1M-HCl aqueous solution at 95°C for 5 min) to enlarge the solubility and reactivity. The obtained partially degraded amylose (Mn = 7.0 x 10^3, Mw/Mn = 2.0) and pullulan (Mn = 4.4 x 10^3, Mw/Mn = 1.5) were trimethylsilylated to give TMSAm-2 (Mn = 1.4 x 10^4, Mw/Mn = 2.1, DTMS = 96.2 mol%/hydroxyl group) and TMSP (Mn = 6.2 x 10^3, Mw/Mn = 1.5, DTMS = 90.4 mol%/hydroxyl group) according to the same procedures described above.

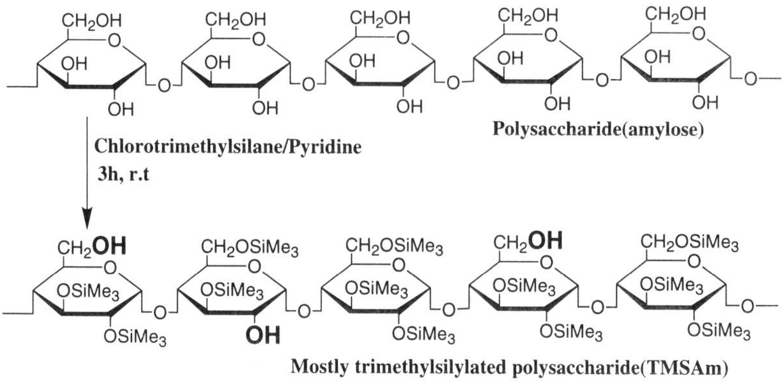

Scheme 1. Preparation of trimetylsilyated polysaccharides

2.3   Preparation of graft copolymers [11,12]

Graft polymerization of L-lactide on mostly trimethylsilylated polysaccharides were carried out according to Scheme 2. Following procedures show the typical example using TMSAm-1. The procedures were basically carried out in a dry glove box under Ar atmosphere. TMSAm-1 (36.6 mg) was dissolved in dry THF. *t*-BuOK (2.2 mg, 0.02 mmol) dissolved in dry THF (0.5 mL) was added to the solution. After stirring for 1 hour, L-lactide (228.3 mg, 2.0 mmol) in THF (0.75 mL) was added. The polymerization was terminated when the reaction proceeded for 5, 10, 30 or 90 min, by the addition of excess amount of acetic acid in THF to the reaction mixture. The obtained products were precipitated with ethyl ether. Characterization of the TMS protected graft copolymer was performed by ^{1}H-NMR.

^{1}H-NMR (CDCl$_3$): δ = 0.10 (s; Si(CH$_3$)$_3$), 1.55 (d; -CH(C$\underline{H}_3$)-), 3.5 ~ 5.5 (broad and very weak; amylose), 4.10 (q; -C$\underline{H}$(CH$_3$)OH), 5.15 (q; -C$\underline{H}$(CH$_3$)O-CO-)

The obtained TMS protected graft copolymer was dissolved in 50mL of CHCl$_3$, poured into 250 mL of methanol, and stirred for 1 ~ 2 days to deprotect TMS groups. The deprotection of TMS groups was confirmed by the disappearance of the methyl proton signal at δ = 0.10 ppm by ^{1}H-NMR. The broad and very weak methyne and methylene proton signals of amylose existed at δ = 3.5 ~ 5.5 ppm. The graft polymerizations of L-lactide on TMSAm-2 and TMSP were carried out in the same manner described above.

^{1}H-NMR (CDCl$_3$): δ = 1.55 (d; -CH(C$\underline{H}_3$)-), 3.5 ~ 5.5 (broad and very weak; amylose), 4.10 (q; -C$\underline{H}$(CH$_3$)OH), 5.15 (q; -C$\underline{H}$(CH$_3$)O-CO-)

IR (KBr disk): 3000-3800 (-OH), 1730 cm^{-1} (C=O).

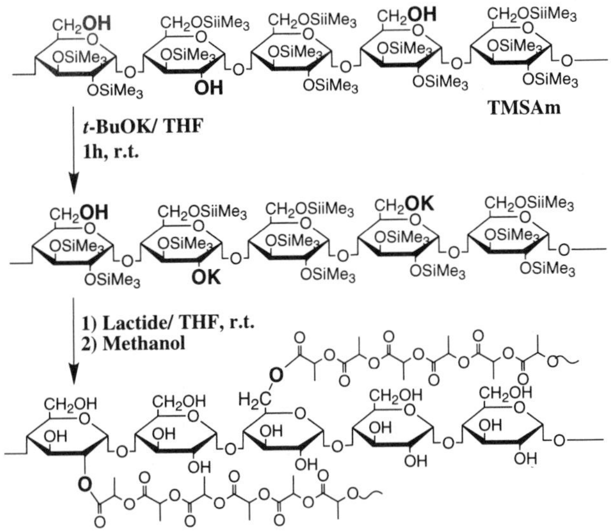

Scheme 2. Preparation of trimethylsilylated polysaccharides

2.4   Degradation behavior

Poly(lactic acid) ($Mn = 5.7 \times 10^3$) was obtained by the similar procedure described above without trimethylsilylated polysaccharides. The polymers (30 mg) were casted from chloroform solution (4 wt%) to give colorless films having ca. 0.3 mm thickness. The films were incubated in 1/15 M-KH_2PO_4/NaHPO₄ buffer (pH = 7.0) at 37°C in the presence or absence of α-amylase (1.2 unit/mL) for 5 days. After certain periods, the films were washed with distilled water and dried in vacuo and then dissolved in THF. The molecular weights of the polymers were measured by GPC (Column: Toso TSKgel G4000HXL + G2500HXL, eluent: THF, detector: RI, standard: polystyrene). The hydrolysis rates were estimated by the molecular weight reduction (%) calculated with the following equation:

Molecular weight reduction (%) = 100 x (M_0 - M_t) / M_0
M_0: initial molecular weight
M_t: molecular weight after degradation

3.   RESULTS AND DISCUSSION

Most of hydroxyl groups of polysaccharides were protected by TMS groups to achieve solubility in organic solvents and to control the number of reaction sites with the alkali metal initiator. Preparation of mostly trimethylsilylated amyloses and pullulan (TMSAm-1, 2 and TMSP) was carried out by the procedure shown in Scheme 1. The degrees of trimethylsilylation (DTMS) of TMSAm-1 (Mn = 1.5 x 10^4, Mw/Mn = 1.07), TMSAm-2 (Mn = 1.4 x 10^4, Mw/Mn = 2.1) and TMSP (Mn = 6.2 x 10^3, Mw/Mn = 1.5) were determined using ^{1}H-NMR spectroscopy based on the area ratio of signals from methyl protons of TMS groups at δ = 0.10 ppm and protons of amylose at 3.5 ~ 5.5 ppm in $CDCl_3$ to be 93.1 mol%, 96.2 mol% and 90.4 mol% per hydroxyl group, respectively. These value means that one free hydroxyl group exists per 4.8, 8.8 and 3.5 glucose residues, respectively. The obtained TMSAm-1, 2 and TMSP were soluble in THF, $CHCl_3$ and other typical organic solvents.

Preparation of poly(lactic acid)-grafted polysaccharides was carried out according to the procedure shown in Scheme 2. Polymerization reaction could be carried out homogeneously in THF with *t*-BuOK. Unprotected hydroxyl groups of TMS-polysaccharides were reacted with *t*-BuOK to be converted into corresponding alkoxides in THF. The ring-opening polymerization of L-lactide was carried out in THF at room temperature for various periods using the polymer alkoxides as initiators. The deprotection of TMS groups of the obtained graft copolymers was carried out by incubation in methanol and confirmed by the disappearance of ^{1}H-NMR signals of methyl protons from TMS groups at δ = 0.10 ppm. All of the graft copolymers obtained were soluble in THF, chloroform, dimethylformamide and other organic solvents, but not soluble in water.

Table 1 summarizes the reaction conditions and the results of graft copolymerizations using TMSAm-1. The initial concentration of L-lactide in feed was adjusted at 1.0 mol/L. The molar ratios of L-lactide to *t*-BuOK in the feed were adjusted to 100. The degree of polymerization (Pn) of lactic acid was calculated using ^{1}H-NMR spectroscopy based on the area ratio of the terminal methyne proton signal at δ = 4.10 ppm to internal methyne proton signal at 5.15 ppm. The contents of sugar unit (wt%) in the graft copolymers were calculated by the equation shown in the Table, and tended to decrease with increasing polymerization time. Molecular weight of the graft copolymer obtained, yield and Pn of lactic acid tended to increase with increasing polymerization time. Table 2 summarizes the reaction conditions and the results of graft copolymerizations using TMSAm-2 having wider polydispersity in larger scales. Small amount of *t*-BuOK in feeding ratio to hydroxy group were used to obtain graft polymers having higher contents of sugar unit. In these experiments with TMSAm-2, the yields were not high and the degrees of polymerization were low compared with the results using TMSAm-1, because of the difficulty of dehydration, low stirring efficiency in larger scales and the polydispersity of the amylose used. As results, the graft copolymers having higher contents of sugar unit, 40 wt% and 25 wt%, were obtained.

Table 1. Results of the graft-polymerization of L-lactide onto TMSAm-1[a) 11]

Run	Reactiuon time (min)	Yield (%)	$Mn^{b)} \times 10^{-4}$ $(Mw/Mn)^{b)}$	Degree of polymerization of lactic acid[c)]	Content of sugar unit[d)] (wt%)
1	5	40.0	8.5(1.11)	66.7	7.9
2	10	67.2	12.0(1.10)	94.3	5.6
3	30	76.8	15.1(1.10)	115.0	4.4
4	90	84.7	15.8(1.08)	118.0	4.2

a) Polymerization was carried out with *t*-BuOK in THF at room temperture. Initial concentration of L-lactide = 1.0mol/l; molar ratio of total OH group to *t*-BuOK = 21; molar ratio of L-lactide to *t*-BuOK = 100.

b) Mn: number-average molecular weight; Mw: weight-average molecular weight; determined by GPC.

c) Number-aveage; calculated from ¹H-NMR.

d) Content of sugar unit (wt%)

$$= \frac{\text{glucose unit(g)}}{\text{poly(lactic acid)-grafted amylose(g)}} \times 100 = \frac{\text{Mn of initial amylose}}{\text{Mn of poly(lactic acid)-grafted amylose}} \times 100$$

Table 2. Results of the graft-polymerization of L-lactide onto TMSAm-2[a) 11]

Run	Total OH group/ t-BuOK in feed (molar ratio)	Yield (%)	$Mn^{b)} \times 10^{-4}$ $(Mw/Mn)^{b)}$	Degree of polymerization of lactic acid[c)]	Content of sugar unit[d)] (wt%)
1	140	27.4	1.6(1.9)	11.1	40
2	70	35.5	2.5(2.0)	8.3	25

a) Polymerization was carried out with *t*-BuOK in THF for 10 min at room temperture. Initial concentration of L-lactide = 1.0mol/l; molar ratio of L-lactide to *t*-BuOK = 100.

b) Mn: number-average molecular weight; Mw: weight-average molecular weight; determined by GPC.

c) Number-aveage; calculated from ¹H-NMR.

d) Content of sugar unit (wt%)

$$= \frac{\text{glucose unit(g)}}{\text{poly(lactic acid)-grafted amylose(g)}} \times 100 = \frac{\text{Mn of initial amylose}}{\text{Mn of poly(lactic acid)-grafted amylose}} \times 100$$

Table 3 shows the reaction conditions and the results of graft copolymerizations using TMSP. The molar ratios of L-lactide to *t*-BuOK in the feed were adjusted to 100. Molecular weight of the graft copolymer obtained, yield and Pn of lactic acid tended to increase with increasing polymerization time, and the contents of sugar unit (wt%) in the graft copolymer tended to decrease with increasing polymerization time, when the reaction time was no less than 30 min. However, when the reaction time was

300 min, Pn of lactic acid was lower than the other results. This is probably because of degradation of poly(lactic acid) segment under a basic condition. On the basis of the values in Table 3, the numbers of graft chains on each pullulan were calculated to be 7.3-7.9 for the graft copolymers obtained, and were almost constant. These results mean the number of graft chain was depend on the feeding ratio of lactide to *t*-BuOK but not on the reaction time, and can be controlled.

Table 3. Results of the graft-polymerization of L-lactide on TMSP[a) 12]

Run	Reaction time (min)	Yield (%)	$Mn^{b)} \times 10^{-4}$ $(Mw/Mn)^{b)}$	Degree of polymerization of lactic acid[c)]	Content of sugar unit[d)] (wt%)
1	5	43.9	8.3(1.60)	74.0	3.4
2	15	44.0	10.5(1.57)	96.4	2.7
3	30	46.0	15.1(1.52)	140.9	1.8
4	300	43.6	8.3(1.59)	70.4	3.4

a) Polymerization was carried out with *t*-BuOK in THF at room temperture. Initial concentration of L-lactide = 1.0mol/l; molar ratio of total OH group to *t*-BuOK = 12; molar ratio of L-lactide to *t*-BuOK = 100.
b) Determined by GPC.
c) Number-aveage; calculated from ^{1}H-NMR.
d) Content of sugar unit (wt%) descrived in Table 1.

The hydrolysis behavior of the poly(lactic acid)-grafted amyloses was investigated in 1/15 M-KH_2PO_4/$NaHPO_4$ buffer (pH = 7.0) at 37°C in the presence or absence of α-amylase and compared with poly(lactic acid). Figure 1 shows the in vitro degradation profiles of the films of the graft copolymers (content of sugar unit in graft copolymer = 40 or 25 wt%) and poly(lactic acid) in the presence or absence of α-amylase. The degradation rates of the graft copolymers were significantly larger than that of poly(lactic acid). The graft copolymer having a higher content of sugar unit (40%) showed a larger degradation rate than the graft copolymer having a lower content of sugar unit (25%). The molecular weight reduction of the graft copolymer having a higher content of sugar unit (40%) after day 3 could not be estimated because the hydrolyzed products became insoluble in both THF and water as GPC eluents. The degradation rates of the graft copolymers with α -amylase were slightly larger than those without enzyme, but not significant. The amylose unit of the graft copolymer in solid state was hardly recognized by α -amylase.

The hydrolysis behavior of the poly(lactic acid)-grafted pullulans (content of sugar unit in graft copolymer = 1.8 or 3.4 wt%) was also investigated (Fig.2). The degradation rates of the graft copolymers were significantly larger than that of poly(lactic acid). The graft copolymer having a higher content of sugar unit (3.4%) showed a larger degradation rate than the graft copolymer having a lower content of sugar unit (1.8%).

The degradation rate of the poly(lactic acid)-grafted amylose (content of sugar unit = 25%) and poly(lactic acid)-grafted pullulans became slow after day 2-4. These results suggest that the initial rapid degradation occurred at the low crystallinity region of the graft copolymer, however, the degradation rate of the high crystallinity region of the graft copolymer was slow and should be similar to poly(lactic acid) homopolymer. These all results suggest that the introduction of hydrophilic polysaccharide unit decreases the crystallinity of poly(lactic acid) and increases the degradation rate, and the hydrolytic degradation was occur at poly(lactic acid) segment.

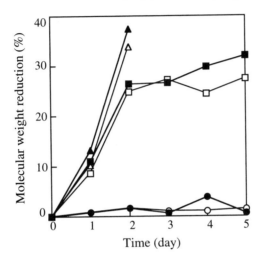

Figure 1. Degradation behavior of poly(lactic acid)-grafted amyloses and poly(lactic acid) in 1/15M-KH$_2$PO4-Na$_2$HPO$_4$ (pH = 7.0) at 37°C with or without α-amylase in vitro[11]. ▲ and △: poly(lactic acid)-grafted amylose (content of sugar unit = 25 wt%), ■ and □: poly(lactic acid)-grafted amylose (content of sugar unit = 25 wt%), ● and O: Poly(lactic acid), ▲, ■ and ●: with α-amylase, △, □ and ■: without a-amylase.

4.   CONCLUSION

We could synthesize new amphiphilic degradable graft copolymers composed of poly(lactic acid) and polysaccharides (amylose and pullulan) through a TMS protection technique. The obtained graft copolymers showed higher degradability compared with poly(lactic acid) by introduction of hydrophilic segments and branched structures. This TMS protection / graft-polymerization method can be applied to various combinations of polysaccharides and aliphatic polyesters, poly(lactide, glycolide or lactone). Such hybrid polymers are expected to be applied as novel biodegradable materials.

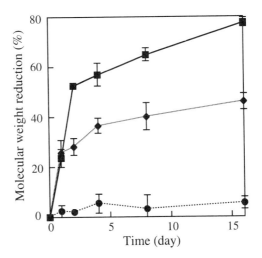

Figure 2. Degradation behavior of poly(lactic acid)-grafted pullulans and poly(lactic acid) in 1/15 M-KH$_2$PO$_4$/Na$_2$HPO$_4$ (pH = 7.0) at 37°C in vitro[12]. ◆: gpoly(lactic acid)-grafted pullulan (content of sugar unit = 1.8 wt%), ■: poly(lactic acid)-grafted pullulan (content of sugar unit = 3.4 wt%), ●: Poly(lactic acid).

ACKNOWLEDGMENTS

This research was financially supported by a Grant-in-Aid for Scientific Research (11780635) from the Ministry of Education, Science, Culture and Sports, Japan.

REFERENCES

1. Frazza, E. J. and Schmitt, E. E., 1971, *J. Biomed. Mater. Res. Symp.* 1: 43-49.
2. Ogawa, M., Yamamoto, M., Okada, H., Yashiki, T. and Shimamoto, T., 1988, *Chem. Pharm. Bull.* 36: 1095-1103.
3. Wise, D. L., Fellmann, T, D., Sanderson, J. E. and Wentworth, R. L., 1979, In *Drug carrier in biology and medicine.* (G. Gregoriadis ed.), Academic Press, pp.237-270
4. Echeverria, E. A. and Jimenez, J., 1970, *Surgery* 131: 1-11.
5. Kimura, Y., Matsuzaki, Y., Yamane, H. and Kitao, T., 1989, *Polymer,* 30: 1342-1348.
6. Hu, D. S. G., Liu, H. J. and Pan, I. L., 1993, *J. Appl. Polym. Sci.* 50: 1391-1396.
7. Kissel, T., Li, Y. X., Volland, C., Gorich, S. and Koneberg, R., 1996, *J. Contr. Rel.* 39: 315-326.
8. Kricheldolf, H. R. and Boettcher, C., 1993, *Makromol. Chem. Macromol. Symp.* 73: 47-64.
9. Akiyoshi, K. and Sunamoto, J., 1996, *Supramol. Sci.* 3: 157-163.
10. Kurita, K., Hirakawa, H. and Iwakura, Y., 1980, *Makromol. Chem.* 181: 1861-1870.
11. Ohya, Y., Maruhashi, S. and Ouchi, T., 1998, *Macromolecules* 31: 4662-4665.
12. Ohya, Y., Maruhashi, S. and Ouchi, T., 1998, *Macromol. Chem. Phys.* 199: 2017-2022.

PHOSPHOLIPID POLYMERS FOR REGULATION OF BIOREACTIONS ON MEDICAL DEVICES

Kazuhiko Ishihara
Department of Materials Science, Graduate School of Engineering, The University of Tokyo, Tokyo, Japan

1.   INTRODUCTION

Various polymeric materials have been used as for manufacturing medical devices including artificial organs in contact with blood[1]. However, the only polymers presently used are conventional materials, such as poly(vinyl chloride), polyethylene, segmented polyetherurethane (SPU), cellulose, and polysulfone (PSf). These materials do not have enough blood compatibility and anti-thrombogenicity; therefore, infusion of an anticoagulant is required during clinical treatments when using these medical devices, to avoid clot formation. To improve their blood compatibility and prevent clotting, some surface modification methods using newly designed polymeric materials have been studied. One of the most effective polymers for this purpose is a phospholipid polymer (MPC polymer) which can create a mimetic biomembrane surface[2]. In this review, molecular design and fundamental performance of MPC polymers are described. Moreover, improvement of the bio/blood compatibility of the medical devices using the MPC polymers to suppress unfavourable bioreactions at the interface between the medical devices and living organisms are explained.

Biomedical Polymers and Polymer Therapeutics
Edited by Chiellini *et al.*, Kluwer Academic/Plenum Publishers, New York, 2001

149

2. MOLECULAR DESIGN OF THE MPC POLYMERS

A biomembrane is a hybrid mainly consisting of two chemical classes, phospholipids and proteins. There is no covalent bonding to bind each molecule; thus, the surface of a biomembrane is heterogeneous and dynamic. The development of new biomaterials was proposed based on mimicing a simple component present on the extracelluar surfaces of the phospholipid bilayer that forms the matrix of the cell membranes of blood cells, namely, the phosphorylcholine group of phosphatidylcholine and sphingomyelin. Phosphorylcholine is an electrically neutral, zwitterionic head group of the phospholipid headgroups present on the external surface of blood cells. It is inert in coagulation assays. The blood compatibility of a polymer surface coated with phospholipids was confirmed by several researchers[3-5]. For example, the interactions between polyamide microcapsules coated with a phospholipid bilayer membrane and platelets were investigated. It was found that platelet adhesion onto the microcapsules was significantly suppressed by the phospholipid coating.

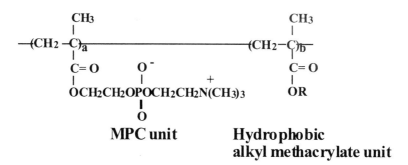

Figure. 1. Chemical structure of the MPC polymer.

New concepts for making blood compatible polymer materials have been proposed based on the characteristics of natural phospholipid molecules in plasma. It was considered that if a polymer surface possesses a phospholipid-like structure, a large amount of natural phospholipids in plasma can be adsorbed on the surface by their self-assembling character[6]. Based on this idea, a methacrylate monomer with a phospholipid polar group, 2-methacryloyloxy ethyl phosphorylcholine (MPC), was designed and synthesized [7,8]. The polymers, composed of MPC and various alkyl methacrylates (as shown in Fig.1) or styrene derivatives were prepared and their blood compatibility carefully evaluated[9-11]. Platelet adhesion and activation were significantly suppressed on the surface of the MPC polymers when the MPC composition was above 30 mol%[12]. These excellent nonthrombogenic properties appeared when the MPC polymers came in contact with human whole blood even in the absence of anticoagulant[13]. Figure 2 shows SEM pictures of SPU surfaces after various treatments after contact with rabbit platelet-rich plasma (PRP) for 3 h. Clearly the original SPU and heparin-immobilized SPU induced platelet adhesion and activation. On the other hand, the MPC polymer coating completely suppressed platelet adhesion. That is, the MPC polymer coating effectively inhibited platelet adhesion.

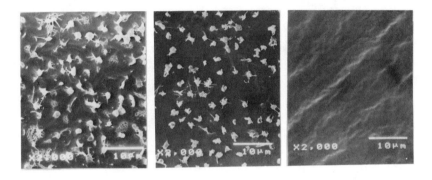

Figure 2. SEM pictures of the SPU with various surface treatments after contact with PRP for 3 h.

The amount of phospholipids adsorbed from plasma on the MPC polymers increased with increasing amounts of MPC[13]. In the case of hydrophobic poly(*n*-butyl methacrylate (BMA)) and hydrophilic poly(2-hydroxyethyl methacrylate (HEMA)), the amount of adsorbed phospholipids was the same level (about 0.5 μg/cm^2). This means that the MPC moiety in the poly(MPC-*co*-BMA) played an important role to increase adsorption . To clarify the state of phospholipids adsorbed on the polymer surface, the phospholipid liposomal suspension was put in contact with these polymers. Phospholipids were adsorbed on every polymer surface, however, the adsorption state of the phospholipids was different on each polymer. On the surface of the MPC polymer, the adsorbed phospholipids maintained a liposomal structure, like a biomembrane; this was confirmed by a differential scanning calorimetry, X-ray photoelectron spectroscopy (XPS), quarts crystal microbalance, and atomic force microscope. The phospholipids adsorbed on the poly(BMA) and poly(HEMA) did have any organized form[14-16]. It was concluded that the MPC polymers stabilized the adsorption layer of phospholipids on the surface. Small amounts of plasma proteins were adsorbed on the MPC polymer surface pretreated with phospholipid liposomal suspension to prepare a model biomembrane surface. On the other hand, on the poly(BMA) and poly(HEMA) surfaces pretreated with the liposomal suspension were not as effective in suppressing protein adsorption[17]. Thus, the protein adsorption on these polymer surfaces strongly depended on the adsorption state of the phospholipids that covered the polymer surface.

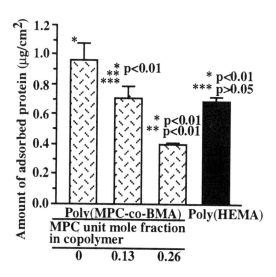

Figure 3. Amount of plasma proteins adsorbed on the MPC polymers after 60-min contact.

The amount of plasma protein adsorbed on the MPC polymer from human plasma was smaller than that on poly(BMA) and poly(HEMA) as shown in Figure 3[13]. Moreover, radioimmunoassay and immunogold-labeling of the adsorbed proteins on the polymer surface indicated that the MPC polymer could reduce adsorption of every protein tested, that is, albumin, γ-globulins, fibrinogen, coagulation factors (factor XII and VIII), high-molecular-weight kininogen, complement components, and fibronectin[18].

3.   MECHANISM OF THE PROTEIN ADSORPTION RESISTANCE BY THE MPC POLYMERS

Protein adsorption is one of the most important phenomena to determine the biocompatibility of materials[19]. In general, proteins adsorb on a material surface within a few minutes after being in contact with body fluids such as, blood, plasma and tears. Some methods to reduce protein adsorption have been proposed. Among them, the construction of a hydrophilic surface which is believed to be effective. Therefore, surface modification of materials with hydrophilic or a water-soluble polymer chains was investigated. However, it was wondered whether some modifications with hydrophilic polymers would be effective not only to obtain blood compatibility but also to reduce protein adsorption.

The environment surrounding the biomolecules and polymeric materials used in our body is aqueous. Therefore the characteristics of the water in the material or on the surface of the material are important to recognize the interactions. In particular, it is considered that the structure of water surrounding proteins and polymer surfaces influences the protein adsorption behaviour.

When the water structure in the hydrated polymers was examined using the differential scanning calorimetric technique, it was found that a large amount of free water existed in the MPC polymers [poly(MPC-co-BMA) and poly(MPC-co-*n*-dodecyl methacrylate (DMA) with 30 unit mol% of MPC] compared with that in poly(HEMA), poly(acrylamide-co-BMA) (PAM), poly(sodium 2-methyl-2-acrylamide propane sulfonate-co-BMA) (PAMB), and poly(N-vinyl pyrrolidone-co-BMA) (PVB)[20].

The MPC polymers suppressed protein adsorption well. It was reported that the theoretical amounts of BSA and BPF adsorbed on the surface in a monolayered state are 0.9 µg/cm^2 and 1.7 µg/cm^2, respectively. On the surface of PMB30, the amount of adsorbed proteins was less than these theoretical values. Thus, the phosphorylcholine group is considered to reduce protein adsorption effectively. The effectiveness of the phosphorylcholine group, particularly the MPC unit, for reducing protein adsorption has been reported by other groups[21-23].

Water molecules bind at the hydrophobic part of the polymer through van der Waals force, which is the so-called "hydrophobic hydration". These bound water molecules cause protein adsorption by the hydrophobic interaction. When a protein molecule is adsorbed on a polymer surface, water molecules between the protein and the polymer need to be replaced[24]. The protein adsorbed on the surface looses bound water at the surface-contacting portion. This phenomenon induces a conformational change in the proteins, that is, the exposed hydrophobic part of the protein in contact with the polymer surface directly. If the state of water at the surface is similar in the aqueous solution, the protein does not need to release the bound water molecules even if the protein molecules contact the surface. This means that the hydrophobic interaction does not occur between the proteins and the polymer surface. Moreover, the conformational change during the protein adsorption on or contact with the surface is also suppressed.

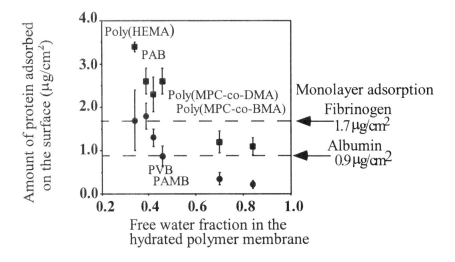

Figure 4. Relationship between free water fraction in the hydrated polymer membrane and amount of plasma proteins adsorbed on the polymer at 37 °C.

The relationship between free water fraction in the hydrated polymers and amount of plasma proteins adsorbed on the polymer surfaces[25] is shown in Figure 4. Clearly the MPC polymers have an extremely high free water fraction and suppress protein adsorption. In this case, though we applied a single protein solution with a 10 % plasma concentration as the initial solution, the amount of proteins adsorbed on the MPC polymer was below the theoretical value for monolayered adsorption. This means that the proteins attached on the surface could be very easily detached by rinsing. It was observed that the conformational change in the adsorbed protein did not occur on the MPC polymer surface[26]. These results clearly showed that the interactions between plasma protein and the MPC polymer are weak compared with that observed on the other polymer surfaces.

4.   APPLICATIONS OF THE MPC POLYMERS

4.1   Blood circulating devices

The surface modification of SPU to improve the blood compatibility using the MPC polymer was performed. For in vivo evaluation, an artificial heart made of SPU coated with the MPC polymer was implanted in a sheep[27]. There was no clot formation on the surface after 1-month implantation.

To obtain a more reliable modification of SPU with MPC, blending the MPC polymer as a polymeric additive was considered[28]. The SPU or SPU/MPC polymer blend was coated onto a polyester prosthesis with a 2 mm internal diameter and grafted to a rabbit carotid artery[29]. The coating process was very easy using a solvent evaporation method. Even when the composition of the MPC polymer was 10 wt% against that of SPU, the MPC units were located near the surface, (revealed with XPS). The mechanical properties of the SPU did not change after blending with the MPC polymer. The prosthesis coated with SPU was occluded, due to adherent and activated blood cells, within 90 min after implantation. On the other hand, the prosthesis coated with SPU/MPC polymer was functional for more than 8 months. The surface of the prosthesis covered with the SPU/MPC polymer blend membrane was clean and no protein deposition or clot formation was observed. These results suggested that the stability of the MPC polymer was sufficient even when the MPC polymer was in contact with blood continuously for 8 months.

4.2 Blood purification devices

Hemodialysis using a cellulose membrane is one of the most useful methods to purify blood. Although the cellulose membrane has both good permeability and mechanical strength, its blood compatibility must be improved. Therefore, if the MPC polymer chains are immobilized on the surface of the cellulose membrane, the blood compatibility can be improved[30]. For efficient grafting with MPC polymer, a methyl cellulose grafted with MPC (MGMC) was synthesized. Coating on the cellulose membrane with the MGMC made it blood compatible, and prevented both protein adsorption and clot formation even when the membrane is in contact with blood without anticoagulant.

Recently, MPC polymer was blended with PSf to make asymmetric porous membranes and hollow fibber membranes[31]. These membranes were made by a wet-membrane processing method. The permeability of these porous membranes was as the same level as that of commercially available PSf membranes. Protein adsorption and platelet adhesion was suppressed by blending with the MPC polymer.

4.3 Blood glucose level regulating devices

To prepare a stable, long-life glucose sensor, the sensor surface was covered with a MPC polymer membrane[32]. The glucose sensor showed excellent stability compared with a poly(vinyl alcohol) membrane, that is, the output current after 7 days was 94% of the initial value and 74 % even after a 14-day continuous insertion subcutaneously. The sensor applied to human volunteers with satisfactory performance with continuous monitoring for 14 days.

When the glucose sensor covered with the MPC polymer membrane, was combined with the insulin infusion pump system, control of the glucose levels using the system was achieved for more than 8 days[33]. Thus, the MPC polymer is useful for prolonging biosensor applications, particularly, implantable sensors.

4.4   Drug delivery devices

Biodegradable poly(L-lactic acid) (PLA) nanoparticles covered with MPC polymer were prepared by a solvent evaporation technique with the amphiphilic water-soluble MPC polymer as the emulsifier[34]. The diameter of the nanoparticles was approximately 200 nm which was determined by atomic force microscopy and dynamic light scattering. The XPS analysis indicated that the surface of the nanoparticles was fully covered with MPC polymer. The amount of plasma protein, BSA adsorbed on the nanoparticles coated with the MPC polymer was significantly smaller compared with that on the polystyrene nanoparticles. The PLA nanoparticles coated with the MPC polymer is suggested to be a safer drug carrier in the bloodstream. One of the anticancer drugs, adriamycin (ADR), can be adsorbed on the nanoparticles through a hydrophobic interaction. Even when the nanoparticles were stored in phosphate-buffered solution for 5 days, 40 % of the ADR remained on the surface. We concluded that nanoparticles coated with MPC polymer could be useful as a adsorption-type drug carrier which could be applied through the bloodstream.

5.   CONCLUSION

MPC polymers are useful not only for artificial organs but also medical devices due to their excellent properties, that is, resistance to protein adsorption and cell adhesion. MPC polymer may be the trigger to open surface modifications of substrates with phospholipid derivatives. These surfaces demonstrate improved protein adsorption and cell adhesion resistance[35-38]. Therefore, the construction of biomembrane-like surfaces may become a general concept to obtain not only blood-contacting medical devices, which can be safely and clinically applied for longer periods, but also devices or equipment for bioengineering and tissue engineering. We are convinced that MPC polymers will become an important material not only in the biomedical and pharmaceutical fields, but in every biotechnology field.

ACKNOWLEDGMENTS

This study was partially supported by the Grants-in-Aid for Scientific Research from the Ministry of Education, Japan (09480250).

REFERENCES

1. Ishihara, K., 1993, Blood compatible polymers, In *Biomedical applications of polymeric materials* (T.Tsuruta, K.Kataoka, T.Hayashi, K.Ishihara, and Y.Kimura, eds.), CRC, Boca Raton, pp. 89-115.
2. Nakabayashi, N. and Ishihara, K., 1996, *Macromol. Symp.*, 101: 405-412.
3. Hayward JA, Chapman D., 1984, *Biomaterials*, 15: 135-142.
4. Coyle. L.C., Danilov, Y.N., Juliano, R.L., and Regen, S.L., 1989, *Chem. Mater.*, 1: 606-611.
5. Kono, K., Ito, Y., Kimura, S., and Imanishi, Y., 1989, *Biomaterials*, 10, 455-461.
6. Ishihara, K. , 1997, *Polymer Science*, 5: 401-407.
7. Kadoma, Y., Nakabayashi, N., Masuhara, E., and Yamauchi, J., 1978, *Kobunshi Ronbunshu*, 35: 423-427.
8. Ishihara, K., Ueda, T., and Nakabayashi, N., 1990, *Polym. J.*, 23: 355-360.
9. Kojima, M., Ishihara, K., Watanabe, A., and Nakabayashi, N., 1991, *Biomaterials*, 12, 121-124.
10. Ueda, T., Oshida, H., Kurita, K., Ishihara, K., and Nakabayashi, N., 1992, *Polym. J.*, 24: 1259-1269.
11. Ishihara, K., Hanyuda, H., and Nakabayashi, N., 1995*Biomaterials*, 16: 873-879.
12. Ishihara, K., Aragaki, R., Ueda, T., Watanabe, A., and Nakabayashi, N., 1990, *J. Biomed.Mater. Res.*, 24: 1069-1077.
13. Ishihara, K., Oshida, H., Ueda, T., Endo, Y., Watanabe, A., and Nakabayashi, N., 1992, *J. Biomed.Mater. Res* ., 26: 1543-1552.
14. Ueda, T., Watanabe, A., Ishihara, K., and Nakabayashi, N., 1991, *J. Biomater.Sci. Polym. Edn*, 3: 185-194.
15. Tanaka, S., Iwasaki, Y., Ishihara, K., and Nakabayashi, N., 1994, *Macromol. Rapid Commun.*, 15: 319-326.
16. Iwasaki, Y., Tanaka, S., Hara, M., Ishihara, K., and Nakabayashi, N., 1997, *J.Colloid Interface Sci.*, 192: 432-439.
17. Iwasaki, Y., Nakabayashi, N., Nakatani, M., Mihara, T., Kurita, K., and Ishihara, K., 1999, *J.Biomater.Sci. Polymer Edn* , 10: 513-529.
18. Ishihara, K., Ziats,.N.P., Tierney , B.P., *J. Biomed. Mater. Res.*, 25: 1397-1407.
19. Horbett, T.A. and Brash, J.L, (eds) 1995, *Proteins at interfaces II; Fundamentals and applications*, American Chemical Society, Washington D.C.
20 Ishihara, K., Nomura, H., Mihara, T., Kurita, K., Iwasaki, Y., and Nakabayashi, N., 1998, *J.Biomed.Mater.Res.*, 39: 323-330.
21. Sugiyama, K., Aoki, H. , 1994, *Polym. J.,* 26: 561-569.
22. Campbell, E.J., Byrne, V.O., Stratford, P.W., Quirk, I., Vick, T.A., Wiles, M.C., Yianni , Y.P., 1994, *ASAIO J.*, 40: 853-857.
23. Murphy, E.F., Keddie, J.L., Lu, J.R., Brewer, J., and Russell, J., 1999, *Biomaterials*, 20: 1501-1511.
24. Lu, D.R., Lee, S.J., and Park, K., 1991, *J. Biomater. Sci. Polymer Edn*, 3: 127-147.

25.Ishihara, K., Ishikawa, E., Iwasaki, Y., and Nakabayashi, N., 1999, *J.Biomater.Sci., Polymer Edn.,* in press.

26.Sakaki, S., Nakabayashi, N., and Ishihara, K., 1999, *J.Biomed.Mater.Res.*, 47: in press.

27. Kido, T., Nojiri, C., Kijima, T., Horiuchi, K., Mori, T., Tanaka, T., Maekawa, J., Sugiyama, T., Sugiura, N., Asada, T., Ishihara, K., Akutsu, T., 1999, *Jpn.J.Artif.Organs*, 28:, 196-199.

28.Ishihara, K., Tanaka, S., Furukawa, N., Kurita, K., Nakabayashi, N., 1996, *J.Biomed. Mater. Res.*, 32: 391-399.

29.Yoneyama,T., Ishihara, K., Nakabayashi, N., Ito, M., Mishima, M., 1998, *J. Biomed. Mater. Res. Appl. Biomat*, 41: 15-20.

30.Ishihara, K., Fukumoto, K., Miyazaki, H., and Nakabayashi, N., 1994, *Artifi.Organs*, 18: 559-564.

31. Ishihara, K., Fukumoto, K., Iwasaki, Y., and Nakabayashi, N., 1999, *Biomaterials*, 20: in press.

32. Ishihara, K., Ohta, S., Yoshikawa, T., and Nakabayashi, N.,1992, *J.Polym.Sci., Part A: Polym.Chem.*30: 929-932.

33.Nishida, K., Sakakida, M., Ichinose, M., Umeda, T., Uehara, M., Kajiwara, K., Miyata, T., Shichiri, M., Ishihara, K., and Nakabayashi, N., 1995, *Med. Prog. through Techn.*, 21, 91-103.

34.Ishihara, K., Konno, T., Kurita, K., Iwasaki, Y., and Nakabayashi, N., 1999, *Polymer Preprints*, 40: 283-284.

35.Marra, K.G., Winger, T.M., Hanson, S.R., and Chaikof, E.L., 1997, *Macromolecules*, 30: 6483 –6489.

36 Kohler, A.S., Parks, P.J., Mooradian, D.L., Rao, G.H. and Furcht, L.T., 1996, *J. Biomed. Mater. Res.*, 32: 237-242.

37.van der Heiden, A.P., Goebbels, D., Pijpers, A.P., and Koole, L.H., 1997, *J. Biomed. Mater. Res.*, 37: 282-287.

38.Baumgartner, J.N., Yang, C.Z., and Cooper, S.L., 1997, *Biomaterials*, 18: 831-837

RELATIVE POLYCATION INTERACTIONS WITH WHOLE BLOOD AND MODEL MEDIA

[1]Dominique Domurado, [2]Élisabeth Moreau, [2]Roxana Chotard-Ghodsnia, [2]Isabelle Ferrari, [1]Pascal Chapon, and [1]Michel Vert
[1]*Centre de Recherche sur les Biopolymères Artificiels, URA 5473 CNRS, Faculty of Pharmacy, University of Montpellier 1, 34060 Montpellier cedex 2, France: [2]Biomécanique et Génie Biomédical, UMR 6600 CNRS, Université de Technologie de Compiègne, BP 20529, 60205 Compiègne cedex, France*

1. INTRODUCTION

Many carrier systems have been proposed to solubilize hydrophobic pharmacologically active compounds in aqueous media, to transport drugs and to overcome cell defenses (liposomes, particles, red blood cell (RBC) ghosts, ...). Among them, polycationic systems or surfaces received little attention as compared to polyanionic ones, mostly because of their relatively high inherent toxicity[1,2], although to a variable degree[3]. Nowadays the situation is changing rapidly. Polycations are regarded as attractive carriers, especially because of their ability to condense linear DNA or plasmids[4]. Polynucleotides do not enter easily into cells mostly because of their inherent hydrophily and high negative electrostatic charge, which preclude cell membrane crossing.

Because they neutralize the negative charges of polynucleotides and could therefore alleviate their consequences, namely lipophoby and electrostatic repulsion by the surface of cells, polycations promote to some extent cell penetration and are worldwide studied as gene transfecting agents. Although it is generally not taken into consideration as a major drawback, the toxicity of polycations could be a limiting factor to the effective use of gene transfer in human therapy.

Biomedical Polymers and Polymer Therapeutics
Edited by Chiellini *et al.*, Kluwer Academic/Plenum Publishers, New York, 2001

On the other hand polyanions modulate the behavior of polycations through formation of polyanion-polycation complexes. For instance, heparin reversed the anatomo-pathological abnormalities induced by poly(L-lysine) (PLL)[1]. In contrast, it had little effect on the in vivo toxicity of a polybase of the partially quaternized poly(tertiary amine)-type, namely poly[thio-1-(*N,N*-diethyl-aminomethyl)-1-ethylene], (Q-P(TDAE)$_x$ with x = percentage of quaternized subunits)[5]. It was thus observed that a same polyanion, namely heparin, can alleviate harmful effects of one polycation and have little effect on the toxicity of another one. The latter author[5] even noticed that a polyanion-polycation complex can unexpectedly be more toxic than the isolated polycation in the case of a poly(ethyleneglycol)-block-poly(aspartic acid) and Q-P(TDAE)$_x$.

The mechanism of RBC agglutination by polycations has been studied since 1951[6], particularly by Katchalsky and coworkers[7-9], who pointed out the occurrence of hemolysis in the presence of PLL[9]. In all these instances, authors used washed RBC in agglutination studies thus precluding the observation of any effect due to plasma proteins. Under physiological conditions, blood proteins have generally a net negative charge and can therefore form polyelectrolyte-type complexes with polycations upon intravenous administration.

This is the reason why we decided to investigate the interactions between blood elements, namely RBC and plasma proteins, and six polycations: poly(L-lysine) (PLL), poly(ε-lysine) (PEL), diethylaminoethyl-dextran (DEAE-dextran), poly(dimethyldiallylammonium) chloride (PDDAC), poly[2-(dimethylamino)ethyl methacrylate] (P(DMAEMA)) and Q-P(TDAE)$_{11}$.

PLL was chosen as a reference because it is one of the most investigated polycations in gene transfection. Two different molar masses, 19 and 124 kDa, were compared since the influence of molar mass was previously pointed out[6,7]. PEL was selected to study the structural effect since these two polymers bear similar primary amino groups but the four methylene groups present in the PLL lateral chains are included in the main chain of PEL.

DEAE-dextran was chosen because it has been tested as a DNA carrier[10] and because it is one of the less toxic polycations in cells cultures[3]. This polymer bears a quaternary ammonium and two tertiary amino groups of pK$_a$ 9.2 and 5.5, all of them being substituted by ethyl groups.

P(DMAEMA) bears tertiary amino groups too, but pendent nitrogen atoms are substituted by methyl groups, and it has also been tested as a DNA carrier[11]. PDDAC is a quaternary ammonium polybase.

Q-P(TDAE)$_x$ is a bifunctional polybase bearing quaternary ammonium groups that are always ionized, and tertiary amine ones. At acidic pH, the latter are protonated, and thus ionized and hydrophilic. Q-P(TDAE)$_x$ is consequently fully extended. On the contrary, when pH rises and reaches neutrality, the tertiary amine units become deprotonated and hydrophobic. Macromolecules collapse to monomolecular polysoap globules due to hydrophobic interactions and water exclusion. The permanently ionized quaternary ammoniums stabilize the globular structure, through electrostatic repulsions, provided $10 < x < 20$. The monomolecular hydrodispersed globules have a uniform size of ca. 8 nm. The tertiary amino groups form a lipophilic microphase, the core of the globules, which is able to accommodate by physical entrapment, lipophilic molecules. These substances are microdispersed in water despite their inherent insolubility in aqueous media[12]. When experiencing an acidic pH, the globules undergo a cooperative globule-to-coil conformational transition of the all-or-none type and thus release the entrapped substance all of a sudden. The globules, which are stable at neutral pH, were proposed to solubilize and transport drugs after parenteral administration. However, Q-P(TDAE)$_x$ that behaved adequately after intramuscular and intraperitoneal injections[13] turned out to be dramatically toxic when given intravenously[2].

For the sake of comparison in vitro, the six selected polycations were added to native RBC suspended in natural and synthetic media. A means to show the contribution of RBC membrane charges is to compare native and desialylated RBC, desialylation removing most of the negative charge-bearing sialic acid present on cell surfaces. Therefore Q-P(TDAE)$_{11}$ was also added to suspensions of desialylated RBC. The effects of the polycations on hemagglutination and on hemolysis were studied comparatively in blood protein-containing and blood protein-free media.

This paper is aimed at reviewing the main results. Data will be discussed with regard to the polycation hemotoxicity and to the interest of polycations as DNA-condensing species in gene therapy.

2. EXPERIMENTAL

2.1 Materials

2.1.1 Buffers

pH 7.4 phosphate-buffered and Tris-buffered isoosmolar media were prepared as previously described[14], and are respectively referred to as PBS and Tris buffer hereafter.

The buffered isoosmolar albumin solutions were prepared by dissolving 0.2, 0.4, 5 or 40 g/l of bovine serum albumin (BSA, fraction V, Sigma, St Quentin-Fallavier, France) into PBS. The 5 g/l albumin solution was used as suspending medium for RBC desialylation by neuraminidase at pH 7.0. Before to be used to suspend native or desialylated RBC, the 0.2, 0.4 and 40 g/l albumin buffer had their pH adjusted to 7.4 by adding sodium hydroxide.

2.1.2   Enzyme preparation

Neuraminidase from *Clostridium perfringens* (E.C. 3.2.1.18) was provided by Sigma (type V, St Quentin-Fallavier, France).

2.1.3   Polymers

19 and 124 kDa PLL and 73 kDa dextran were purchased from Sigma (St Quentin-Fallavier, France). 500 kDa DEAE-dextran came from Pharmacia (Uppsala, Sweden), PEL from Chisso Corp. (Japan), medium MW PDDAC from Aldrich (St Quentin-Fallavier, France) and P(DMAEMA) from Rohm & Haas. Q-P(TDAE)$_{11}$ was prepared as previously described[15,16]. This copolymer had a molar mass of 130,000 kDa as determined by Dynamic Laser Light Scattering.

2 mM and 10 mM polycation solutions with concentrations expressed in terms of average residue (monomol) were used.

19 and 124 kDa PLL, DEAE-dextran and Q-P(TDAE)$_{11}$ were tested at both 2 et 10 mM concentrations. The three other polycations, namely PEL, PDDAC and P(DMAEMA) were tested at the 10 mM concentration only. All polycations were studied in suspensions of native RBC except for Q-P(TDAE)$_{11}$, which was also studied in the presence of desialylated RBC.

2.1.4   RBC preparation

RBC were conditioned as already described[14]. In most cases, 0.8 ml of packed RBC collected after the third washing were suspended in 1.2 ml of the selected experimental medium, namely autologous plasma, autologous serum, albumin buffer or Tris buffer, resulting in a 40% hematocrit. Tris buffer was used to wash the RBC to be suspended in Tris buffer, plasma or serum. RBC to be suspended in 40g/l albumin buffer were washed with PBS. The suspensions were incubated at 37°C for 10 min. Aliquots of 0.2 ml of the solutions to be tested comparatively, namely Tris buffer, 2mM or 10mM polycation, were then added to RBC suspensions. The final mixtures were further incubated at 37°C for 15 min and then centrifuged at 1910 × g for 10 min. In a few cases, 0.2 ml aliquots of the same tested solutions were added to 1.2 ml of 40g/l albumin buffer. The resulting mixtures were incubated at 37°C for 10 min prior to addition of 0.8 ml of packed RBC collected after the last washing. The suspensions were further incubated at 37°C for 15 min and then centrifuged at $1910 \times g$ for 10 min.

2.2   Methods

2.2.1   Microscopy

2.2.1.1   Light microscope observations

Light microscope observations of RBC suspensions were carried out using a Nikon Labophot microscope ($\times 400$ magnification) equipped with a Nikon F501 camera.

RBC were observed through the microscope immediately after addition of the solution to be tested (Tris buffer or polycation solutions) when added after RBC introduction, or, alternatively, after addition of the washed RBC when the tested solution was added to the suspending medium before RBC. Another observation of each suspension was carried out after 15-min incubation at 37°C.

2.2.1.2   Fluorescence microscope observations

Diphenylhexatriene (DPH), provided by Sigma, once excited at 350 nm, fluoresces with a maximum at 452 nm.

Fluorescence microscope observations of RBC suspensions were performed using an epifluorescence Ortholux II Leitz microscope ($\times 1250$ magnification) equipped with an excitation filter (340-380nm), a dichromatic mirror Leitz RKP400, an emission filter Leitz LP430 and an Orthomat Leitz camera.

About 10 mg of DPH crystals were immersed in 3 ml of a 10 mM (Q-P(TDAE)$_{11}$ solution, and this preparation was sonicated for 1 h (Branson 200, 19 W) in order to saturate the hydrophobic core of the polycation with the fluorescence marker. After a 20-min centrifugation, absorbance of the supernatant was measured at 350 nm in order to check the presence of DPH.

The observations were carried out on RBC suspensions with a final hematocrit of 2.5 % and 0.25 % and a final (Q-P(TDAE)$_{11}$ concentration of 5 mM.

2.2.2   Hemolysis

2.2.2.1   Potassium assessment

The potassium released from RBC was assessed in the supernatant of centrifuged RBC suspensions using a flame photometer (IL 943 IP, Instrumentation Laboratory)[14].

2.2.2.2   Hemoglobin assessment

The hemoglobin present in the supernatant of centrifuged RBC suspensions was assessed by spectrophotometry (spectrophotometer DU Serie 7000 Beckman) using the Cripps differential method[17] as previously described[14]. The released hemoglobin (rHb) was defined as the percentage of hemoglobin present in the supernatant as referred to the total hemoglobin present in the RBC suspension.

2.2.3   RBC mechanics

The membrane elasticity (shear elastic modulus denoted Es) was determined by viscometry using a Couette viscometer (RV 100 / CV 100 Haake). RBC were suspended (hematocrit of 9 %) in a dextran solution (175 g/l) with a viscosity $\mu = 11.5$ mPa.s at 37°C.

2.2.4   Polycation-plasma protein interactions

To 1.2 ml of plasma or of albumin buffer incubated for 10 min at 37°C, 0.2 ml of Tris buffer or Q-P(TDAE)$_{11}$ (2 mM or 10 mM) or 19 kDa PLL (10 mM) or 124 kDa PLL (10 mM) or DEAE-dextran (10 mM) was added. The proteic medium-polycation (or Tris buffer) mixture was then incubated for 10 min at 37°C. Packed RBC (0.8 ml) were added to the mixture. The resulting suspension was then observed through the microscope before being allowed to incubate for 15 min at 37°C. At this time, a second microscope observation was performed, and rHb and potassium concentration were measured in the supernatant.

2.2.5   Desialylation of RBC

Washed RBC (1.5 ml) were suspended in 5g/l albumin buffer (4.25 ml) and incubated with 0.25 ml of the neuraminidase solution for 90 min at 37°C under gentle shaking, the final concentration in enzyme being 83 mIU/ml of

RBC suspension. After incubation, neuraminidase was eliminated by washing the RBC three times with Tris buffer or PBS depending on the final suspension medium.

2.2.6   Statistical analysis

In order to evaluate the significance of the results, StatGraphics® was used to carry out an analysis of variance when the hypothesis of variance equivalence could not be rejected. Otherwise, the test of Kruskal-Wallis was used.

3.   RESULTS

Polycation solutions with two different concentrations (2 and 10 mM repeating unit concentration respectively) were prepared in Tris buffer. The in vitro physico-chemical characteristics of the final RBC suspensions, namely pH 7.4, 310 mOsm and 37°C, were selected to mimic blood physiological conditions.

3.1   Effect of Q-P(TDAE)$_{11}$ on the deformability of the RBC membrane

In order to quantify the effect of Q-P(TDAE)$_{11}$ on RBC membrane elasticity, RBC were first suspended at 40 % hematocrit in various media, namely plasma, serum and Tris buffer. After addition of the solution to be tested, namely Tris buffer or Q-P(TDAE)$_{11}$ 2 mM or 10 mM, the mixture was incubated at 37°C for 15 min and then centrifuged. In order to increase the shear stress exerted on the RBC, the resulting RBC packs were resuspended in a dextran solution whose viscosity was about twice the viscosity of RBC cytoplasm. As expected[18], dextran did not induce agglutination. The hematocrit (9%) of the suspension in the presence of dextran was sufficiently low to preclude RBC interactions.

Due to RBC deformability, the apparent viscosity of the suspension decreased when the shear stress increased. The shear-thinning behavior of the suspension was compared to the predictions of a microrheological model[18] that was shown to relate the evolution of the apparent viscosity to Es, which is characteristic of the ability of RBC to deform while keeping a constant surface area.

Regardless the suspending medium, Q-P(TDAE)$_{11}$ had no significant effect on Es, which remained in the normal range (2-3.5 µN/m). Basically membrane mechanical properties depend on the cytoskeletal network and on its interactions with the membrane integral proteins[19]. Since no deformability variation was observed, it was concluded that the membrane cytoskeletal network was not affected by Q-P(TDAE)$_{11}$.

3.2   Light microscope observations

During the study of the effect of the different polycations on agglutination of RBC suspended in autologous natural media (plasma and serum) and in synthetic media (albumin et Tris buffers), the terms small, medium and large agglutinates reflected qualitatively the size of these agglutinates.

3.2.1   Addition of polycations to suspensions of RBC

Native RBC were introduced in the selected media, and the polycation solution at 2 or 10 mM, or Tris buffer as control, was then added to the suspensions, to mimic actual intravenous injection. RBC were observed microscopically immediately after final mixing and after a 15-min incubation at 37°C. Agglutination of desialylated RBC was tested only in the presence of Q-P(TDAE)$_{11}$.

3.2.1.1   Agglutination in synthetic media

Except for the 2 mM 19 kDa experiment, PLL induced an instantaneous agglutination, which could be observed with the naked eye. 10 mM 19 kDa PLL gave medium and large agglutinates, whereas 2 mM 124 kDa PLL produced medium agglutinates. For 10 mM 124 kDa PLL, a single agglutinate was observed in the whole microscope field.

PEL induced a very slight agglutination only.

DEAE-dextran induced an immediate agglutination, which could be observed with the naked eye too. The number and size of the agglutinates increased with the increasing DEAE-dextran concentration and their size was still larger after incubation.

PDDAC and P(DMAEMA) gave very large agglutinates, which could be observed visually, and some smaller ones around.

Q-P(TDAE)$_{11}$ induced a strong agglutination, which increased with the increasing concentration. The proportion of ghosts in the agglutinates increased and the agglutinates could be observed with the naked eye.

The agglutination of desialylated RBC by Q-P(TDAE)$_{11}$ was similar to that of native RBC.

In summary, except for PEL and 2 mM 19 kDa PLL, which induced a very slight agglutination, all other polycations induced immediate formation of agglutinates, more strongly for 124 kDa PLL, PDDAC, P(DMAEMA), and still more strongly for Q-P(TDAE)[11].

3.2.1.2 RBC suspended in anticoagulated plasma

Upon addition of Tris buffer to native RBC suspended in plasma, typical rouleaux[20,21] were observed as well as reseaux, i.e. 3-dimensionally branched and piled rouleaux regardless of the incubation time.

Mainly rouleaux were observed in 2 mM 19 kDa PLL. With 10 mM 19 kDa PLL, the size and number of rouleaux decreased to give a majority of isolated cells after incubation. For both concentrations, agglutinates were rarely observed. Upon addition of 10 mM 124 kDa PLL, medium agglutinates formed, which could be observed with the naked eye. Finally, for a given concentration, the molar mass increase induces a decrease of agglutination by fibrinogen, whereas PLL-induced hemagglutination increased.

In the presence of PEL, the only difference with control was the rare occurrence of very small agglutinates made of RBC or of small rouleaux.

Observations in the presence of 2 mM DEAE-dextran were similar to those obtained for the control notwithstanding some small agglutinates made of rouleaux. On the contrary, upon addition of 10 mM DEAE-dextran, small and medium agglutinates made of rouleaux could be observed with the naked eye. Then, hemagglutination by fibrinogen was little modified since rouleaux were always present and agglutinates were only made of rouleaux.

In the presence of 10 mM PDDAC, the number and size of the rouleaux were strongly reduced, thus showing a clear diminution of agglutination by fibrinogen. Moreover, small and large agglutinates were instantaneously formed that could be observed visually. Nevertheless some isolated cells were visible.

With 10 mM P(DMAEMA), small rouleaux were seldom observed. Moreover their number decreased during incubation. RBC were mostly isolated or formed a few medium agglutinates.

When 2 mM Q-P(TDAE)[11] was added to native RBC suspended in plasma, monodispersed cells and isolated shortened rouleaux were immediately observed after mixing[14]. In the case of the 10 mM Q-P(TDAE)[11] solution, native RBC led to large agglutinates with ghosts in the center after the incubation period.

Upon addition of Tris buffer, the desialylated RBC aggregated without forming rouleaux, in contrast to the behavior of native RBC, but in agreement with observations made under similar conditions[23]. Upon addition of 2 mM Q-P(TDAE)[11] to desialylated RBC suspended in plasma, the cells aggregated right after mixing and formed networks, which differ from reseaux because the latter are usually formed from interconnected rouleaux. In the case of the 10 mM Q-P(TDAE)$_{11}$ solution, much smaller agglutinates were observed in the case of desialylated RBC than for native RBC.

In summary, it was observed that only Q-P(TDAE)$_{11}$ induced the formation of ghosts in the agglutinates. Furthermore agglutination depended on the molar mass of the polycation as well as on its concentration and type. Agglutination may be negligible (PEL), progressive in a few minutes (Q-P(TDAE)$_{11}$ and P(DMAEMA)) or instantaneous (124 kDa PLL, DEAE-dextran and PDDAC). Moreover, the formation of rouleaux by fibrinogen may be disturbed by polycations (Q-P(TDAE)$_{11}$, PLL, PDDAC and P(DMAEMA)).

3.2.1.3   RBC suspended in serum

In serum, RBC do not form any rouleau[22]. Upon addition of Tris buffer, RBC behaved similarly. Only the presence of rare agglutinates differentiated the observations made in the presence of 2 mM 19 kDa PLL from those of the control. In the case of 10 mM 19 kDa PLL, RBC gave primarily agglutinates of small and medium sizes with a few cells remaining isolated. In the case of 124 kDa PLL, some small hemagglutinates were already detected for the 2 mM concentration whereas, for the 10 mM concentration, many large size agglutinates were visible with the naked eye. 2 mM DEAE-dextran gave little difference with respect to the control, except the presence of a few small agglutinates. In the presence of 10 mM DEAE-dextran, many small and medium size hemagglutinates and a few isolated cells were observed. The addition of PDDAC in RBC suspensions produced many very large agglutinates. In contrast, only a very few agglutinates appeared in the presence of PEL. P(DMAEMA) generated small and medium size hemagglutinates together with isolated cells whose number was smaller after 15 min incubation. Upon addition of Q-P(TDAE)$_{11}$, RBC formed agglutinates, ghosts and isolated cells, depending on the polymer concentration. The greater the concentration, the greater the size and the number of the agglutinates, the greater the ghost number and the smaller the isolated cell number.

In contrast to native RBC, desialylated RBC suspended in serum formed networks upon addition of Tris buffer. When 2 and 10 mM Q-P(TDAE)$_{11}$ were added to desialylated RBC, agglutinates, networks and monodispersed cells were observed. The size of agglutinates was concentration-dependent as shown by comparing the effects of both polymer solutions.

In conclusion, as in the case of the plasma medium, only Q-P(TDAE)$_{11}$ induced the formation of ghosts in agglutinates of RBC suspended in serum, the agglutination and the ghost formation being molar mass- and concentration-dependent. From all the tested polycations, only PEL and 2 mM 19 kDa PLL and DEAE-dextran did not cause agglutination. In most cases agglutination was observed with the naked eye, except for 2 mM 124 kDa PLL and Q-P(TDAE)$_{11}$, which required the microscope.

3.2.2   Addition of RBC to preformed mixtures polycation/plasma and polycation/albumin

The formation of polyelectrolyte complexes being known as mixing condition-dependent, the order of addition of the components of polycation-medium-RBC suspensions was modified. For the experiments reported so far, the polycation was added to whole blood or to RBC already suspended in the selected suspension medium. At the present point, selected polycations, namely (Q-P(TDAE)$_{11}$, 19 and 124 kDa PLL and DEAE-dextran, were added to two of the previous suspension media, namely plasma and albumin buffer, prior to RBC, the final concentration in polycation being 10 mM.

3.2.2.1   In plasma

In the case of 19 kDa PLL, agglutinates of small and medium sizes were observed but they were formed exclusively of rouleaux in contrast to what was observed for the reverse order of addition. In the case of 124 kDa PLL, a large number of rouleaux and of reseaux of rouleaux were observed. Moreover, agglutinates were rare and mainly composed of rouleaux. With DEAE-dextran, rouleaux and reseaux of rouleaux were always observed. After the 15-min incubation, the number of reseaux decreased and the number of agglutinates increased. In the case of Q-P(TDAE)$_{11,}$ some rouleaux were detected together with isolated cells parallel. Qualitatively, agglutination was not as important as for the reverse order of addition and agglutinates were composed of normal RBC in the absence of ghosts. The contact between the polycations and the proteins led to significant protection of the RBC. The interactions between polyanionic proteins and polycation in the absence of RBC was confirmed by the collection of a white pellet upon centrifugation of mixtures plasma-Q-P(TDAE)$_{11}$ and plasma-124 kDa PLL previously incubated for 10 min at 37°C. In the case of DEAE-dextran, the protecting effect of plasma proteins was smaller than for Q-P(TDAE)$_{11}$ and 124 kDa PLL. Moreover, agglutination increased with time, and no pellet was observed after centrifugation of a plasma-DEAE-dextran mixture. For the 19 kDa PLL, an intermediate behavior was found since there were both an increase of rouleaux formation, usually attributed to fibrinogen, and an increase of agglutination, although centrifugation did not led to pellet formation.

3.2.2.2   In albumin buffer

When 19 or 124 kDa PLL and DEAE-dextran were mixed with albumin prior to RBC, no significant difference was observed as compared with the reverse order of addition, except agglutinate size increase in the case of 19 kDa PLL. In contrast, Q-P(TDAE)$_{11}$ induced a strong decrease of the number of isolated cells and a correlative agglutination increase. However, agglutinates were not visible with the naked eye.

3.3   Fluorescence microscope observations

Advantage was taken of the ability of Q-P(TDAE)$_{11}$ globules to entrap physically water insoluble molecules to show the interactions between polycations and the membrane of RBC. Accordingly a fluorescent marker, namely DPH was dissolved within the globule core up to saturation in order to have enough fluorescence. The final polymer concentration was set at 5 mM and low hematocrits were used to prevent Q-P(TDAE)$_{11}$–induced agglutination.

Under these conditions, a blue fluorescence was observed at cell surfaces, thus demonstrating the close contact between the fluorescent carrier polycation and the cells. However, given the resolution of the fluorescence microscope, it was impossible to decide whether the fluorescent carrier was at the surface of the RBC or included within the lipid bilayer.

3.4   Hemolysis

3.4.1   Addition of polycations to suspensions of RBC

3.4.1.1   RBC suspended in Tris buffer medium

PEL did not induce any significant hemolysis. DEAE-dextran and PDDAC caused a very little but significant ($p < 0.01$) release of hemoglobin as compared with the control. In the presence of 19 and 124 kDa PLL and P(DMAEMA), hemolysis was significantly greater as referred to the control ($p < 0.01$) but smaller than in the presence of Q-P(TDAE)$_{11}$. Furthermore, comparison between potassium and hemoglobin releases obtained with 19 kDa and 124 kDa PLL showed that hemolysis depended on molar mass. Last but not least hemolysis also depended on the polycation concentration. At 10 mM, the hemolysis caused by Q-P(TDAE)$_{11}$ was larger for desialylated RBC than for the native ones (p < 0.05).

3.4.1.2 RBC suspended in anticoagulated plasma

We have previously shown that Q-P(TDAE)$_{11}$ hemolysis is concentration-dependent[14]. On the other hand, the releases of potassium (rK$^+$ between 0.07 and 0.18 mM) and hemoglobin (rHb between 0.29 and 0.41 %) from RBC suspended in plasma in the presence of 19 and 124 kDa PLL, DEAE-dextran, PEL, PDDAC and P(DMAEMA) were not significanly different from the release observed for the control medium (rK$^+$ = 0.16 ± 0.04 mM and rHb = 0.28 ± 0.03%). In the case of desialylated RBC, Q-P(TDAE)$_{11}$-induced hemoglobin and potassium releases were not significantly different from those caused by the control.

3.4.1.3 Desialylated RBC suspended in serum

Both hemoglobin and potassium releases were lower for desialylated RBC than for native ones

3.4.1.4 RBC suspended in 40 g/l albumin buffer

PEL, DEAE-dextran and P(DMAEMA) did not cause any significant release of hemoglobin since rHb remained lower than 1%, i.e. the background, in all cases. The presence of 19 kDa PLL, 124 kDa PLL, PDDAC or Q-P(TDAE)$_{11}$ led to significant hemoglobin releases (p<0.001 as compared to control), the effects being maximum for Q-P(TDAE)$_{11}$ and the high molar mass PLL. Similar trends were observed from potassium release data, except in the case of P(DMAEMA), which led to a significant release of potassium as compared to the control (p<0.001). From a general viewpoint, the polycation-induced hemolysis increased with the polymer concentration, as already observed for Q-P(TDAE)$_{11}$[14]. In the presence of 10 mM Q-P(TDAE)$_{11,}$ the amount of potassium released by desialylated RBC was significantly lower (p<0.05) than in the case of the Tris buffer.

3.4.2 Addition of RBC to preformed mixtures composed of polymer solutions and plasma or albumin

In order to study the effect of the order of addition of the polymer, resuspension medium and RBC on hemolysis, polycation solutions (10 mM

Q-P(TDAE)$_{11}$, 10 mM PLL 19 kDa and 124 kDa and 10 mM DEAE-dextran) were added to plasma or albumin buffer suspending media prior to RBC. Data were used to compare the two media

3.4.2.1   In anticoagulated plasma

As previously described[14], the addition of RBC within plasma-Q-P(TDAE)$_{11}$ mixture led to a much smaller hemolysis. On the other hand, DEAE-dextran and 19 and 124 kDa PLL did not cause any hemolysisas in the case of the inverse order of addition.

3.4.2.2   In 40 g/l albumin buffer

Hemolysis was greater when Q-P(TDAE)$_{11}$ was added to albumin buffer prior the RBC than when medium and RBC were mixed prior to be in contact with the polycation. Hemolysis was significantly smaller when 19 and 124 kDa PLL were added to albumin buffer prior the RBC introduction ($p<0.05$ for rHb in the presence of the two polycations, $p<0.05$ and $p<0.01$ for rK$^+$ in the cases of 19 kDa PLL and 124 kDa PLL respectively). In the presence of DEAE-dextran, hemolysis was not significantly different from that of the control.

4.   DISCUSSION

The resuspension of washed RBC in various media, including plasma and serum proteins, and the comparison between different polycations provided further information on the factors, which can affect the agglutination behavior of RBC and the hemolysis. As already mentioned in literature and shown specifically for the Q-P(TDAE)$_{11}$ polycation, the behavior of RBC in the presence of polycations depends primarily on polyanion-polycation-type interactions. All the features underlined in this presentation, namely dependence on the medium of suspension, polycation concentration-dependence, effect of the order of addition on agglutination, ghosts formation and hemolysis-related releases of potassium and hemoglobin well agree with the occurrence of polyanion-polycation-type interactions. In polymer science, such interactions are known to depend on the relative characteristics of both polyelectrolytes (molar mass, acid and basic strength). We did find this type of features for the selected polycations, at least qualitatively. One could also find features typical of polyanions for the plasma and serum proteins and the RBC cell surface.

The behavior of the pool of plasma proteins could be deduced from the comparison between Tris buffer and plasma. The most important finding was the remarkable protecting effect of the proteins present in plasma as compared with the protein-free Tris buffer medium, the effect being larger for the desialylated RBC than for the native ones. The protecting effect of the plasma proteins was formerly assigned to the competition between RBC and the pool of proteins capable of interacting with the polycation[14]. Therefore, decreasing the surface negative charge of the cells acted in favor of protein-polymer interactions, thus leading to greater protection of the cells against the hemolysing effect of polycation. On the other hand changing from one polycation to the other did change agglutination and hemolysis.

The effect of fibrinogen can be deduced from the comparison between plasma and serum. Based on the differences, it was previously concluded for Q-P(TDAE)$_{11}$ that fibrinogen, which is the only difference between the two media, contributed at least in part to the protecting effect. Desialylation led to a slightly smaller protecting effect of the serum proteins as compared with plasma proteins. As in the case of plasma, the decrease of the surface negative charge due to desialylation acted in favor of the interaction between fibrinogen and the cationic macromolecules. The comparison between the selected polycations show that the protecting effect of fibrinogen was polycation-dependent.

Comparing the behavior of native and desialylated RBC suspended in albumin when 10 mM Q-P(TDAE)$_{11}$ was added, showed no significant difference in hemoglobin and potassium releases, thus showing that each protein has its own behavior. The polycations, which caused a weak hemolysis in Tris buffer, gave similar results in albumin buffer. The effect produced by the other polycations were similar to the data obtained with Q-P(TDAE)$_{11}$.

One of the most striking phenomenon discovered in the case of Q-P(TDAE)$_{11}$ was the effect of the order of addition of the charged species present in resuspended RBC on agglutination and hemolysis. Some of the studied polycations behaved similarly whereas others did not show any sensitivity. Hemolysis of desialylated RBC suspended in albumin was not significantly affected by the order of addition, thus showing that the sialic acids did play a significant complexing role, as borne out by the fixation of the fluorescence labeled Q-P(TDAE)$_{11}$ polycationic molecules at the cell surface.

Based on the present understanding of polyanion-polycation complexes and given the complexity of the composition of plasma in proteins with various polyelectrolyte properties, one can consider that a synthetic polycation faces a great number of competing polyanionic systems, including RBC, when introduced in whole blood. The identification of the contribution of fibrinogen was made possible by comparing data collected for plasma and serum. However, the behavior in a pool of polyanionic components is much more difficult to analyse. Albumin being the most important single component among the plasma proteins, it was studied separately. In the case of Q-P(TDAE)$_{11}$, this protein did not exhibit any protecting effect on native RBC and on desialylated ones, except in the case of desialylated RBC when hemolysis was monitored through K$^+$ release. Surprisingly enough, the introduction of the RBC into a mixture of the Q-P(TDAE)$_{11}$ polymer with albumin led to a greater hemolysis than when the polymer was added to the suspension of RBC in the albumin solution. We have no explanation for this phenomenon, which was not observed systematically for the other polycations. Anyhow, the behavior of Q-P(TDAE)$_{11}$ shows that a polyanion-polycation complex can affect RBC much more than the polycation alone insofar as hemolysis is concerned.

Another interesting feature issued from this work is the generalisation of the rapidity with which protein-polymer interaction set up since one minute of protein-polymer contact was sufficient to result in the protective effect of protein. The collection of a white pellet by centrifugation of plasma and Q-P(TDAE)$_{11}$ or PLL in the absence of RBC appeared as another support to the occurrence protein-polycation complexation.

Presently it is difficult to correlate the cell toxicity of polycations, as shown in the present work, to the structural characteristics of the synthetic macromolecules and to the systemic toxicity of polycations given intravenously (i.v.). Q-P(TDAE)$_{11}$ and high molar mass PLL are known to be very toxic in vivo. Little information is available in literature as to the in vivo behavior of the other polycations. Insofar as RBC toxicity features like agglutination and hemolysis are concerned, it should be possible to tentatively classify the polycations according to their perturbing effects. Presently, hydrophobicity and permanent quaternary ammonium charges seem to be significant factors.

It appears from the above findings that polycations interact with RBC to give rise to two more or less related phenomena, namely agglutination and hemolysis, the latter showing that the RBC cell membrane integrity is very much perturbed, in agreement with the occasional observation of ghosts.

Insofar as cell surface perturbation is concerned, agglutination could be the source of emboli and finally thrombosis, thus explaining the clotting of the lung capillary bed clotting observed previously for Q-P(TDAE)$_{11}$ and the related in vivo lethality[x].

As for the lipid bilayer permeation, several phenomena could be responsible, namely dissolution of the hydrophobic part of the polycation into lipids causing disorder and permeability, or induction of lipid flip-flop because of the electrostatic interactions, or dissolution of the lipids into the hydrophobic part of the polycation causing defects in the membrane and permeability. Reality might be a combination of such phenomena.

Anyhow, the polycation-induced RBC toxicity depends definitely on the nature and the structure of the polycations. In other words, RBC might be a useful tool to anticipate the in vivo relative toxicity. However, in vivo, the interactions between RBC and proteins and polycations also depend on other factors such as the polycation local concentration and the blood flow rate, and thus on the experimental injection conditions. Such a behavior has already been observed once since a defined solution of Q-P(TDAE)$_{11}$ was found less lethal when slowly perfused than when given i.v.[x]. Similarly to the effect of the order of mixing the interacting species, the influence of the i.v. administration conditions well agree with the involvement of a thermodynamically unbalanced formation of polyelectrolyte complexes.

5. CONCLUSIONS

In this presentation, it was shown that the behavior of RBC resuspended in different media can be used to investigate comparatively more or less toxic polycations. The reported findings show that the phenomena depend primarily on polyelectrolytic interactions involving several charged species, namely the polycation, plasma proteins and RBC. All these species are in competition and it seems that the balance of the competition depends very much on the relative strengths of these different charged species regarding each other. This could explain why a given protein such as albumin can stimulate the RBC toxicity of one polycation and inhibits that of another polycation under similar conditions. Beside the intrinsic characteristics of the polycations, namely chain structures, nature of the cationic charges, basic strength, charge density, hydrophobicity, etc., administration conditions are likely to play a significant role via the injection and blood flow rates.

Last but not least, this contribution suggests that efforts should be made to standardise the protocol of investigation of the bioactivity of synthetic polycations, especially regarding pharmacological activity, drug transport and gene transfection. For the latter, profitable information might be withdrawn from a systematic examination of the consequences of DNA complexation with cell membrane-perturbing polycations in the presence of full blood.

ACKNOWLEDGMENTS

The authors thank Dr X. Delavenne and the biochemistry laboratory of the Compiègne hospital for potassium assessment through flame photometry and Prof. G. Cassanas for her help with statistical analysis. They also gratefully acknowledge the Comité de l'Oise de la Ligue Nationale Contre le Cancer for financial support.

REFERENCES

1. Seiler, M. W., Venkatachalam, M. A., and Cotran, R. S., 1975, *Science* 189: 390-393.
2. Illum, L., Huguet, J., Vert, M., and Davis, S.S., 1985, *Intern. J. Pharm.* 26: 113-121.
3. Choksakulnimitr, S., Masuda, S., Tokuda, H., Takakura, Y., and Hashida, M., 1995, *J. Contr. Rel.* 34: 233-241.
4. Behr, J.-P., 1993, *Acc. Chem. Res.* 26: 274-278.
5. Jallet, V., 1994, *Masquage de charges cationiques d'une polybase globulaire bifonctionnelle en vue de minimiser sa toxicité vis-à-vis du sang.* Ph. D. Thesis, Rouen, France.
6. Rubini, J.R., Stahmann, M.A., and Rasmussen, A.F. Jr., 1951, *Proc. Soc. Exptl. Biol. Med.* 76: 659-662.
7. Katchalsky, A., Danon, D., Nevo, A., and De Vries, A., 1959, *Biochim. Biophys. Acta* 33: 120-138.
8. Marikovsky, Y., Danon, D., and Katchalsky, A., 1966, *Biochim. Biophys. Acta* 124: 154-159.
9. Nevo, A., De Vries, A., and Katchalsky, A., 1955, *Biochim. Biophys. Acta* 17: 536-547.
10. Liptay, S., Weidenbach, H., Adler, G., and Schmid, R.M., 1998, *Digestion* 59: 142-147.
11. Cherng, J.Y., van de Wetering, P., Talsma, H., Crommelin, D.J., and Hennink, W.E., 1996, *Pharm. Res.* 13: 1038-1042.
12. Huguet, J., and Vert, M., 1985, *J. Contr. Rel.* 1: 217-224.
13. Illum, L., Huguet, J., Vert, M., and Davis, S.S., 1986, *J. Contr. Rel.* 3: 77-85.
14. Moreau, E., Ferrari, I., Drochon, A., Chapon, P., Vert, M., and Domurado, D. 2000, *J. Contr. Rel.*, 64: 115-128
15. Vallin, D., Huguet, J., and Vert, M., 1980, *Polym. J.* 12: 113-124.
16. Chapon, P., 1998, *Recherche de stéréospécificité dans l'activité catalytique de polybases globulaires à base de polyamines tertiaires partiellement quaternisées*, Ph. D. Thesis, Montpellier, France.
17. Cripps, C.M., 1968, *J. Clin. Pathol.* 21: 110-112.
18. Drochon, A., Barthès-Biesel, D., Lacombe, C., and Lelièvre, J.C., 1990, *J. Biomech. Eng.* 112: 241-249.
19. Nash, G.B., and Gratzer, W.B., 1993, *Biorheology* 30: 397-407.
20. Fåhraeus, R., 1929, *Physiol. Rev.* 9: 241-274.
21. Skalak, R., 1984, *Biorheology* 21: 463-476.
22. Shiga, T., Imaizumi, K., Harada, N., and Sekiya, M., 1983, *Am. J. Physiol.* 245: H252-H258.
23. Böhler, T., and Linderkamp, O., 1993, *Clin. Hemorheol.* 13: 775-778.

INFLUENCE OF PLASMA PROTEIN AND PLATELET ADHESION ON COVALENTLY COATED BIOMATERIALS WITH MAJOR SEQUENCES OF TWENTY REGIOSELECTIVE MODIFIED N- OR O-DESULPHATED HEPARIN/ HEPARAN DERIVATIVES

H. Baumann and M. Linßen
Macromolecular Chemistry and Textile Chemistry, Haemocompatible and Biocompatible Biomaterials, University of Technology, RWTH-Aachen, Worringer Weg 1, D-52074 Aachen, Germany

1.   INTRODUCTION

It seems to be generally accepted that protein adsorption is the first event occurring when polymer surfaces or biomaterials are contacted with blood before platelets may attach to the surface[1-5]. It is depending of the type of surfaces, reactive, hydrophilic, hydrophobic, amphiphilic microdomain structured and with different topology[1-4,6-8], which of the plasma proteins will be adsorbed or not adsorbed to the surface from more than hundreds of plasma protein components. Among those plasma proteins can be proteins of contact activation system, like XII, high molecular weight kininogen (HMWK) and prekallikrein[9,10]. When the polymer surface is negatively charged to a certain content serin proteases of coagulation system will be activated on the surfaces and at the end of the cascade fibrin clot formation takes place[11]. Surface bound XIIa may also activate plasminogen for lysis of fibrin on surfaces. The plasminogen will be adsorbed preferentially on lysinated surfaces[12]. XIIa may also activate via neutrophils the alternative pathway of complement system. This complement activation will be

Biomedical Polymers and Polymer Therapeutics
Edited by Chiellini *et al.*, Kluwer Academic/Plenum Publishers, New York, 2001

triggered by hydrophilic OH-group containing polymer surfaces with sufficient OH-groups such as in cellulose, too[13]. Some special classes of plasma proteins such as fibrinogen, fibronectin, vitronectin, and von Willebrand factor interact strongly with its hydrophilic or hydrophobic domains to hydrophilic or hydrophobic surface and then they react with specific integrin receptors of platelets which causes platelet adhesion, aggregation, activation and spreading[14,15].

Also several serinprotease inhibitors will be adsorbed to surfaces. They may inactivate activated serinproteases via complex formation resulting in antithrombogenic surfaces. These reactions may be catalytically modulated by surface bound polysaccharides such as heparan sulphate, heparin, dermatan sulphate and covalently bound chondroitin sulphate in thrombomoduline[16]. Most of the heparan sulphates on the luminal surface of blood vessels, in the glycocalix, have been reported to be AT III inactive[17]. The endothelial cell surface heparan sulphate (ESHS) isolated by us is AT III inactive and shows inertness to platelets[18-21]. Their function could be at least the downregulating of the adhesion of platelet adhesive plasma proteins to a no level effect.

An athrombogenic surface should show no complement activation[11], no platelet adhesion, no adhesion of sufficient amounts of contact activation proteins and no adhesion of plasma proteins which may cause platelet adhesion and reaction via the integrin receptors[14,15]. Among the different concepts of surface modification of biomaterials the endothelial cell surface heparan sulphate (ESHS) originally located at the glycocalyx, the interface between blood vessels and blood, seems to show the highest haemocompatibility standard; no platelet adhesion, no complement activation, no contact activation and no adsorbtion of platelet activating proteins such as fibrinogen, fibronectin, vitronectin and von Willebrand factor. ESHS was immobilised to 10 different artificial surfaces[18-21] and represents the concept of regioselective arranged functional groups in the repeating units of ESHS as well as domain like arrangements of those groups along the polymer chain[21]. The biosynthesis of such polysaccharides with heparan backbone precursors has been studied by Lindahl et al.[22]. The regioselective enzymatic modification of heparan precursor in the golgi leads to following reaction products (see Fig.1).

Unsulphated heparan sulphate domains will be formed as a result of non enzymatic modification of heparan precursors at distinct areas of the polymer chain. Mixed GlcNAc and GlcNSO$_3$ domains and high sulphated domains will be formed depending on completeness of former reaction step[23].

It depends on the cell type which other variation of the modification reaction step in low or high sulphated domains takes place[23].

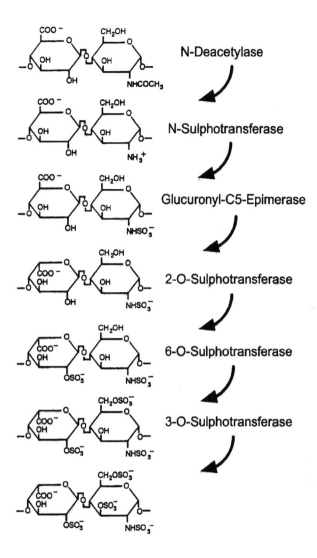

Figure 1. Biosynthesis of heparin and heparansulfates, with different possible enzymatic modification steps resulting in 17 (possibly 18) different hexosamine-glycosamine-sequences[22]

ESHS consists of two non sulphated domains, two low sulphated domains and one high sulphated domain[21]. Heparin contains 30% of completely sulphated ATIII binding sequences, the end product of biosynthesis (Fig.1, bottom). The sequence structures of heparin and domain like structures of ESHS can now be synthesised from heparin via specific regioselective and stereoselective 6-O- or 2-O- or 3-O- or N-desulphation or -sulphation and N-de- or re-acetylation reaction, solely or in combination for each group. 20 different completely and intermediate modified derivatives were synthesised and immobilised to surface which will be used for plasma protein adhesion studies with serum albumin, fibrinogen, fibronectin and IgG in order to show if special regioselective groups are responsible for downregulating plasma protein adhesion. The measurements will be done with fluorescence labelled proteins. In comparison to this the influence of platelet adhesion with citrated blood will be measured. The measurements will be done with single proteins under static and dynamic conditions and in some cases with two proteins in combination for showing competition reactions. The used combination of protein concentration are extreme low and comparable to literature conditions, mostly 100 or 1000 times lower than the concentration in plasma. The studies may help to understand structure function of ESHS domains and that of the sequences of heparin.

2.   MATERIALS AND METHODS

2.1   Chemicals

Chemicals were obtained from Merck (Darmstadt, Germany) and Sigma (Munich, Germany) in p.a. quality. Heparin grade I was from Serva (Heidelberg, Germany).

Cellulose dialysis tubings were from Serva (Heidelberg, Germany).

2.2   Preparation of Regioselective Modified Heparin Derivatives

All desulphation reactions, except the total 6-O-desulphation, have been done as earlier described[24]. Also combined desulphation of two or three groups have been done as earlier described. Total 6-O-desulphation has been done according to the method of Takano[25]. The reaction products have been analysed by ^{13}C-NMR.

2.3   Carboxylation of Cellulose Membranes and Introduction of Oligoamide Spacers

Cellulose membranes were reacted with 3-Aminopropyltriethoxysilan. The amino groups have been reacted with adipic acid as earlier described[19,20].

2.4 Immobilisation of Regioselective Modified Heparin Derivatives

Carboxyl groups terminated oligoamid spacers of cellulose membranes have been activated with N-cyclohexyl N`-2-morpholinoethyl-carbodiimidemethyl p-toluene sulphonate as earlier described[20].

2.5 Isolation of Fibronectin

Fibronectin was isolated from plasma and purified according to the method of Engvall and Ruoslathi[26].

2.6 Labelling of proteins (Albumin, Fibrinogen, Fibronectin and IgG) with Fluorescence Reagents

Fluorescein-5-isothiocyanate (FITC) labelling (fluorescence emmission 485 nm, extinction 535 nm) was done according to Maeda et al.[27] and Faraday et al.[28] Texas Red and Texas Red labelled albumin are commercially available from Molecular Probes, Leiden, Netherlands. Dansylchlorid labelling of proteins was done according to Miekka and Ingham[29].

2.7 Protein Adsorption Studies

3mg/100ml fluorescence labelled plasma proteins were dissolved in a physiological buffer (0.05 m Tris, 0.154 m NaCl, pH 7,4). The measurements were made under static conditions at 0.25, 0,5, 0,75, 1, 2, 3.5, 5, 10, 20, 30, 60 min and under dynamic condition with shear rates at 1050 sec^{-1} and at room temperature.

2.8 Platelet Adhesion Test

Platelet adhesion tests on the modified cellulose membranes were done in the modified Baumgartner chamber[30] under shear rates of 1050 sec^{-1} with citrated whole blood as described earlier[31]. Platelet adhesion tests with silicon tubes were done with citrated whole blood without any contact to unmodified surface. Luminal modified silicone tubes with regioselective modified heparin derivatives were prepared as earlier described[31]. The platelet adhesion has been measured before and after the test via percentage loss of platelets.[31]

2.9   Surface Concentration of Immobilised Regioselective Modified Heparin Derivatives

The membranes have been hydrolysed with 8 N HCl and the released glucosamine content has been detected via HPLC with pulsed amperometric detection as previously described[18]. Surface concentration is in all cases higher than a monolayer.[19]

3.   RESULTS

The regioselective modified heparin derivatives and heparin were immobilised on cellulose according to following scheme.

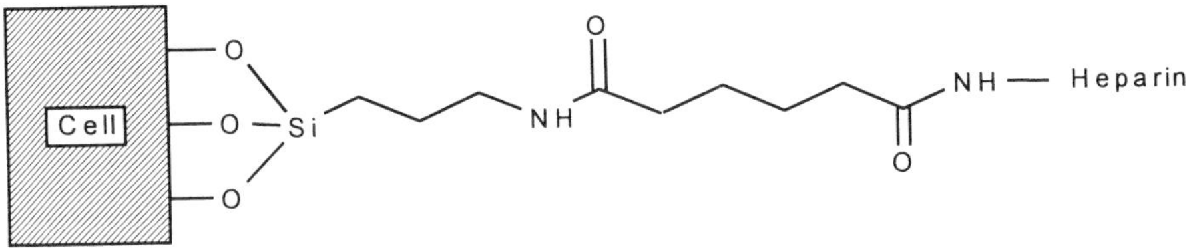

Figure 2. Heparin is immobilised via an amide bond on cellulose surface, using adipic acid as a spacer.

Surface concentrations were measured in the range of 10-30 pmol/cm^2, which is 1,5 - 5 times higher than mean side to side coordination for a monolayer coating.

Three fluorescence markers (FITC, Texas Red, dansyl chloride) have been used for labelling the four selected plasma proteins albumin, IgG, fibrinogen and fibronectin. The three fluorescence markers will be used for combined estimation of different labelled plasma proteins on the modified membranes, so that fluorescence can be measured at different for the marker specific wavelengths. In plasma, albumin has the highest concentration of 3500-5500 mg/100ml serum followed by IgG 800-1800 mg/100ml plasma, fibrinogen 200-450 mg/100ml plasma and fibronectin with 30 mg/100ml plasma. The minimum concentration for monolayer formation of fibrinogen, fibronectin and albumin are reached with these diluted solutions. Under experimental conditions only 3 mg/100ml physiological buffer were used for each plasma protein which is a concentration mostly used in literature. A new experimental analytical arrangement for measuring fluorescence labelled proteins spectroscopically on membranes has been developed. 18 spots of each modified membrane has been measured on microtiter plate at different wavelengths.

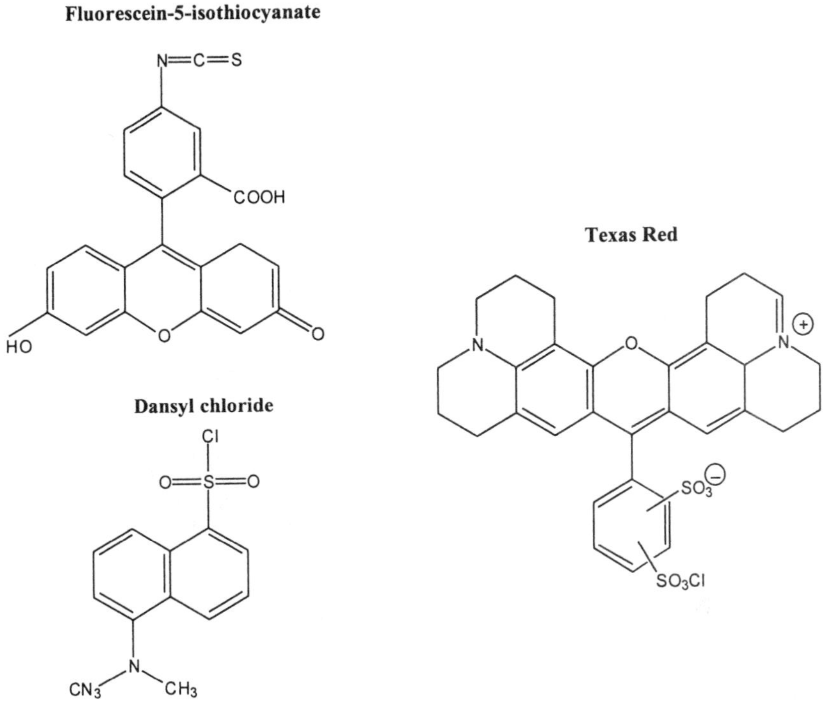
Fluorescein-5-isothiocyanate
N=C=S
COOH
HO
O
O
Texas Red
N
O
N
SO3
SO3Cl
Dansyl chloride
Cl
O=S=O
N
CN3
CH3

Figure 3. Structure of used fluorescence labelling agents for plasma proteins

Two modified membranes and one unmodified membrane has been used for getting the results. The static protein adhesion was done by dipping the membrane into physiological protein solution at different times. For dynamic protein adhesion measurement the modified Baumgartner chamber[30] was used as a perfusion system with the labelled plasma proteins in physiological Tris-buffer. To reduce the influence of the surface of the tube, an artificial surface in the chamber, the system was preadsorbed with the protein that should be measured. Then fluorescence measurements with membranes were done as described above.

3.1 Influence of Adsorbed Plasma Proteins on Heparin Coated Membranes under Static and Dynamic Conditions

In Figure 4 the time dependent protein adsorption of the four proteins on heparinised cellulose surfaces under static and dynamic conditions are shown.

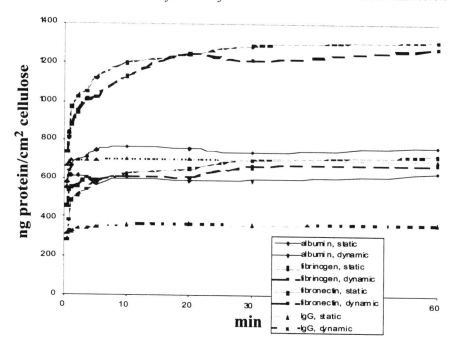

Figure 4. Static and dynamic adsorption of FITC-albumin, FITC-fibrinogen, FITC-fibronectin and FITC-IgG on cellulose, covalently immobilised with heparin.

At dynamic condition at the equilibrium the highest amount of adsorbed plasma proteins represents fibronectin followed by albumin with 45% lower adsorption respectively fibrinogen with 50% and IgG with nearly 70% lower adsorption. There are two types of adsorption curves, one type which is nearly independent of the adsorption time (IgG) and the other type which increases asymptotically in the first minutes to a nearly constant equilibrium value (albumin, fibrinogen, fibronectin). The two types of adsorption curves are nearly independent of the used static or dynamic adsorption conditions. There is no significant difference between adsorbed plasma proteins at the equilibrium observed between dynamic and static conditions for fibrinogen and fibronectin whereas albumin shows a nearly 20% reduced adsorption under static conditions and IgG shows a nearly 50% reduced adsorption at dynamic conditions compared to static conditions.

3.2   Influence of Individual Groups of Regioselective Modified Heparin on Plasma Protein Adhesion

There was no significant influence observed with fibronectin and IgG adsorption on regioselective modified heparin derivatives at dynamic conditions (figure not shown). In contrast to this albumin showed a significant influence on type of coatings with regioselective modified heparin derivatives (Fig.5). The highest albumin adsorption was observed with N- and O-oversulphated heparin. Completely selective 6-O-desulfation of high sulphated glucosamine repeating units of heparin according to methods of Takano[25] resulted in the highest reduction of albumin adsorption. Selective N-desulphation or 2-O-desulfation leads to a lower but nearly equal reduction of albumin adsorption for each group. 3-O-desulfation showed the lowest reduction of albumin compared to heparin. Similar trend could be observed with fibrinogen adsorption (not shown). This is very important because fibrinogen is known to be a cocatalyst for platelet aggregation.

6-O-desulphation was done in the case nor completely but only partially with a method according to Ayotte[in24] with 60% reduced 6-O-desulfated groups. It might be expected when we take in account the trend of partial 6-O-desulfation that the total 6-O-desulphation may lead to the highest reduction in fibrinogen adsorption.

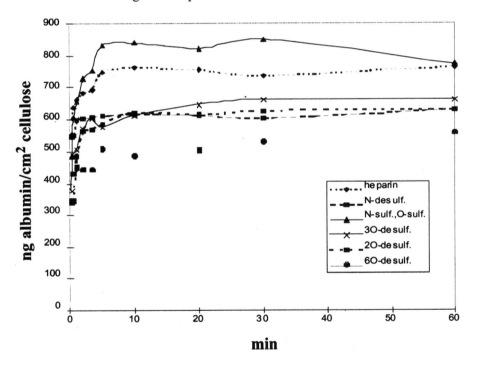

Figure 5. Comparison of the time dependent dynamic adsorption of FITC-albumin on cellulose covalently immobilised with heparin and regioselective modified heparin derivatives.

3.3   Influence of Individual Groups of Regioselective Modified Heparin in the Initial Phase of Albumin Adhesion

The initial phase of plasma protein adhesion has been reported to be very important for understanding the total adsorption process of plasma proteins. The results of the influence of each individual group in the repeating units of heparin on albumin adsorption after 1 min adsorption time under static conditions are seen in Figure 6.

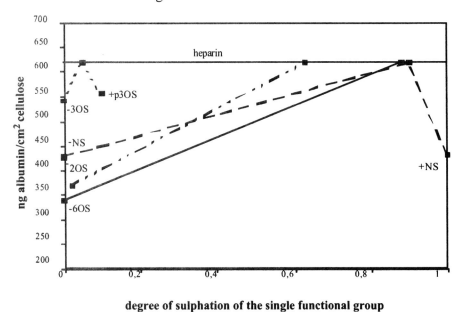

Figure 6. Dependence of static FITC-albumin-adsorption (1 min) on the degree of sulphation of a single functional group of regioselective modified heparin derivatives, covalently immobilised on cellulose.

The highest reduction of albumin adsorption has been found with 6-O-desulfation of heparin. This result is consistent with those of equilibrium adsorption at longer time. The influence of iduronic acid 2-O-desulfation shows also a similar but reduced reduction of albumin adsorption compared to 6-O-desulfation. The influence of N-desulphated heparin on reduction of albumin adsorption was also pronounced but lower than that at equilibrium and 3-O-desulfated heparin showed the lowest influence. Removal of NAc-groups and resulphation surprisingly leads to a reduced albumin concentration. The results of albumin adsorption at dynamic conditions, not shown here, show a similar trend. Also the results on fibrinogen adsorption show a similar trend at the equilibrium (not shown).

3.4 Influence of the Type of Fluorescence Labelling Agent on Plasma Protein Adhesion

The labelling of proteins, polysaccharides etc. with the fluorescein-5-iso-thiocyanate (FITC) reagent has often been used for adsorption studies in literature. However the labelling reagent has a high molecular weight and is hydrophobic in character. Therefore it may have an influence on the adsorption properties of proteins. This is important when three labelling reagents will be used at the same time to estimate the influence of three proteins on adsorption to surfaces. We could show that FITC and Texas Red showed comparable adsorbed amount of albumin whereas dansyl chloride with lower molecular weight showed a reduced albumin adsorption of 1/3 (not shown).

3.5 Adsorption of Two Different Proteins with FITC and Texas Red Labelling Reagents

Because of the fact that the labelling reagents FITC and Texas Red show comparable relation influences on adsorption of plasma proteins, 1:1 mixtures of Texas Red-albumin and FITC-IgG were used to show the simultaneous adsorption properties of two proteins.

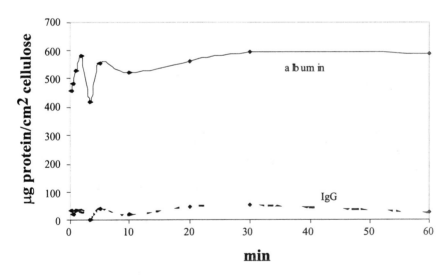

Figure 7. Adsorption (static) of Texas Red-albumin and FITC-IgG from an 1:1 mixture of the two proteins out of physiological buffer on cellulose, modified with heparin.

In the 1:1 mixture (IgG and Albumin) IgG will be practically not adsorbed on heparin coating whereas albumin shows a similar high adsorption level compared to that of albumin alone. The results are shown in Figure 7. 6-O-desulphated heparin shows the same trend (not shown).

3.6   Platelet Adhesion on Membranes Coated with Regioselective Modified Heparin Derivatives

The concentration of used plasma proteins in plasma is in three cases 100–1000 times higher than with our model reaction.

The amount of percentage platelet loss with citrated whole blood by the loss of thrombocytes before and after the test run is shown in Figure 8.

Again it can be seen a strong influence of each functional group of the various regioselective desulphated heparin derivatives. Partial 6-O-desulfation (60 %) shows a similar strong reduction in platelet loss as N-desulphated heparin, whereas 2-O-desulfation and 3-O-desulfation does not show an influence compared to unmodified heparin.

Combined totally 6-O- and partially 2-O-desulphation or complete O-desulphation and partial reacetylation showed practically no platelet loss. The platelet adhesion increases slightly with completely desulphated, deacetylated, and N-resulphated heparin.

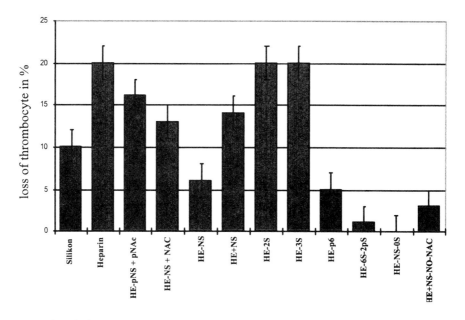

Figure 8. Relative percentage thrombocyte loss of coated silicone surfaces modified with heparin (HE) and totally or partially (p) 6O-, 2O-, 3O-, and N-desulphated (-) or N-deacetylated, N-reacetylated and O- or N-resulphated heparin derivatives (+).[31]

4. DISCUSSION

When a polymeric surface is contacted with blood first events may be protein adsorption from more than 100 plasma proteins onto the material surface within seconds and these proteins may be displaced by a series of others within a few minutes[1-4]. This so called Vroman effect plays an important role for haemocompatibility of biomaterials because it seems to be unlikely that platelets ever contact the bare surface of an artificial material[1-3,5]. The concentrations of the dissolved plasma proteins and their diffuse mobility are so far in excess to those of platelets that surface must become coated with proteins before platelets can reach it[4]. The presence of platelets aggregating or activating plasma proteins such as fibrinogen, fibronectin, vitronectin α 2 macroglobuline and van Willebrandt factor, the cofactor for platelet recruitment and aggregation, have specific significance to haemocompatibility of biomaterials in the adsorbed state[32] since platelets may react with its receptors to the adsorbate[14,15]. The thrombogenicity of a surface correlates directly with the amount of fibrinogen on the surface relative to other proteins[33]. Adsorbed contact activation proteins on polymer surface are important for triggering serin proteases of coagulation cascade and fibrin clot formation. ESHS does not show such effect whereas heparin triggers contact activation system. Complement activation will be induced by hydrophilic OH- group containing surfaces which may also have an important additional influence to haemocompatibility[13]. ESHS does not show complement activation. This molecular and cellular reaction may be modulated by polysaccharides, the mentioned glycosaminoglycans and by several plasma proteins on polymer surfaces[16, 17] or on luminal side of natural blood vessels.

In contrast to this, serum albumin seems to passivate polymer surfaces against platelet adhesion[32]. Albumin with the highest concentration of all plasma proteins (about ten times higher than fibrinogen) seems to be frequently the first adsorbed protein which may also be substituted by fibrinogen, IgG, fibronectin, high molecular weight kininogen, and factor XII.[1,2] The sequence of adsorption, desorption and displacement of plasma proteins seems also to be influenced by the type of biomaterial surface. The topology of the surface and hydrophilic / hydrophobic microdomain structure[6] does have also an important influence for separate adhesion sequence of the plasma proteins similar to either hydrophilic or hydrophobic surfaces[7]. Hydrophobic surfaces show high protein adsorption potential[34] and higher platelet adhesion. They interact predominantly with more hydrophobic plasma proteins as HDL or with hydrophobic domains of serum albumin and fibrinogen[8]. The displacement reaction seems to be slower, in some cases proteins will be adsorbed irreversible with conformational change of proteins. In contrast to this, hydrophilic surfaces with their high water uptake adsorb relative low amount of plasma proteins, the displacement reaction seems to be quicker. The conformational change of proteins seems to be neglectible and platelet adhesion is relatively low. They may be classified as biomaterials with a relatively high haemocompatibility standard[3]. ESHS belongs to the class of substances with high water binding capacity. Those surfaces with hydrogel character show the lowest effects of plasma protein adhesion and platelet adhesion with whole blood. A lot of adhesion studies have been done with single proteins at concentrations much more lower than in nature. These model adsorption studies will give information about adsorption properties to special surfaces of biomaterials but is not representative for real situation in vivo. It is regrettable that most surfaces are not analysed and defined enough to get comparable results due to the influence of defined structural components of polymer surfaces and the possible multiple events described above.

However we used the concept of regioselective modified heparin derivatives which enables to vary defined functional groups (6-OS, 3-OS, NS, NAc in glucosamine and 2-OS in uronic acid) in each position of the repeating units of the heparan backbone. These arrangements of groups represents mayor domains or sequences of natural heparan sulphates. Also partial modification of each mentioned group have been done synthetically. Up to twenty regioselective modified derivatives were synthesised and only one immobilisation procedure was used under defined mild conditions to fix these derivatives on the polymer surfaces. We made sure that mean values for monolayer covering of surfaces will be reached based on our analytical results combined with literature known calculations[19]. This system seems to be a good approach to get really good defined surfaces because the immobilised sequences of polysaccharides were characterised with NMR and other methods in solution before immobilisation. With selected immobilised plasma proteins, albumin, fibrinogen, fibronectin, and IgG, the structures show a clear influence of regioselective arranged groups in the cases of albumin and fibrinogen whereas IgG and fibronectin does not show a significant regioselective influence on surface for protein adhesion. The most important reduction of fibrinogen and albumin adsorption has been shown with 6-O-desulphated sequences followed by N- and 2-O-desulphated sequences compared to unmodified heparin. The 3-O-sulphation has a lower effect. The regioselective influence is independent of the fact weather static or dynamic measurements are used. All the adsorption studies lead to at least a monolayer adsorption of the plasma proteins. This is because of the low concentration of plasma protein in the model adsorption studies[8]. Simultaneously adsorption of two proteins out of a 1:1 mixtures lead to different results. There was found no IgG adsorption although the albumin adsorption was as high as in single adsorption studies.

Earlier results, with different experimental adhesion studies described above under physiological conditions, show a massive reduction of AT III binding, when only 3-O-positions are desulphated in heparin[35]. The influence of regioselective desulphated and pure heparin shows also an important influence to platelet adhesion with citrated whole blood. Again 6-O-desulphation reduces more drastically the platelet adhesion than N-desulphated does. 3-O-S and 2-O-S groups does not show an influence whereas totally 6-O-desulphated derivatives lead to nearly platelet inert surfaces. Also totally desulphated and N-reacetylated derivatives lead to platelet inertness. There is a good correlation between downregulating of fibrinogen adsorption and percentage platelet adhesion most important for 6O-desulphated and less important for N-desulphated heparin derivatives. The role of acetyl groups is not clear yet. However we will get some more information about properties of ESHS and artificial surfaces with the studies concerning platelet adhesion and protein adhesion. The studies will be expanded to get more information about ESHS properties and how nature regulates at least minimal or no platelet reactive portein adsorption with ESHS coating. However we should be carefully when the in vitro results will be transferred to the in vivo situation. Perhaps as stated by Brash the most notable difference between artificial and biological situations is the initial phase of the process when the adsorption of protein is known to occur within seconds[4].

However high but real individual plasma protein concentrations in nature and their real natural ratio to each other, not used during adhesion studies, may suppress adsorption at least of platelet adhesive plasma proteins additionally to the optimal regioselectively influence of modified heparanstructures on surfaces to a no level effect.

ACKNOWLEDGMENTS

The authors acknowledge major financial support from Deutsche Forschungsgemeinschaft DFG (project Ba 630/8-1 and Ba 630/8-2) and some additional financial support by the Gesellschaft der Freunde der Aachener Hochschule. The authors are indepted to Miss Nicole Schaath for skillful measurements, Dr. Philip de Groot for helpful discussions and the instructions for fibronectin preparation and Dr. Bernd Huppertz for helpful instructions for synthesis of regioselective modified heparin derivatives.

REFERENCES

1. Vroman L., 1988, *Bull NY Acad Med*, 64: 352.
2. Vroman L., 1983, *Biocompatible Polymers, Metals, and Composites*, (Szycher M, ed.), Technomic Publ., Lancaster, Pennsylvania, pp. 81.
3. Salzmann E.W., Merrill E.W., Kent K.C., 1994, *Hemostasis and Thrombosis: Basic Principles and Clinical Practice*, (Colman R.W., Hirsh J., Marder V.J., Salzmann E.W., eds.), Lipincott Comp., Philadelphia, pp.1469.
4. Brash J.L., 1983, *Biocompatible Polymers, Metals, and Composites* , (Szycher M., ed.), Technomic Publ., Lancaster Pennsylvania, pp. 35.
5. Baier R.C., Dutton R.C., 1969, *J. Biomed. Mater. Res.*, 3: 191.
6. Iwasaki Y., Fujiike A., Kurita K., Ishihara K., Nakabayashi N., 1996, *J. Biomater. Sci. Polym. Ed.*, 8: 91.
7. Okano T., Aoyagi T., Kataoka K., Abe K., Sakurai Y., Shimada M., Shinohara I., 1986, *J. Biomed. Mater. Res.*, 20: 919.
8. Turbill P., Beugeling T., Poot A.A., 1996, *Biomaterials*, 17: 1279.
9. De la Cadena R.A., Wachtfogel Y.T., Colman R.W., 1994, *Hemostasis and Thrombosis: Basic Principles and Clinical Practice*, (Colman R.W., Hirsh J., Marder V.J., Salzmann E.W., eds.), Lipincott Comp., Philadelphia, pp.219.
10. Scott C.H.F., 1991, *J. Biomater. Sci. Polym. Ed.*, 2: 173.
11. Brash J.L., Scott C.H.F, ten Hove P., Wojcieckowski P., Colman R.W., 1988, *Blood*, 71:932.
12. Woodhouse K.A., Weitz J.I., Brash J.L., 1994, *J. Biomed. Mater. Res.*, 28: 407.
13. Deppisch R., Göhl H., Ritz E., Hänsch G.M., 1998, *The complement system*, (Rother K., Till G.O., Hänsch G.M., eds.), Springer Verlag, Heidelberg, pp. 487.
14. Sixma J.J., van Zanten H., Banga J.D., Nievenhuls H.K., de Groot P., (1995), *Seminars in Hematalogy*, 32: 89.
15. Charo I.F., Kieffer N., Phillips D.R., 1994, *Hemostasis and Thrombosis: Basic Principles and Clinical Practice*, (Colman R.W., Hirsh J., Marder V.J., Salzmann E.W., eds.), Lipincott Comp., Philadelphia, pp. 489.
16. Bourin M.C., Lindahl U., 1993, *Biochem. J.*, 289: 313.
17. Rosenberg R.D., Bauer K.A., 1994, *Hemostasis and Thrombosis: Basic Principles and Clinical Practice*, (Colman R.W., Hirsh J., Marder V.J., Salzmann E.W., eds.), Lipincott Comp., Philadelphia, pp.837.
18. Baumann H., Müller-Krings U., Keller R., 1997, *Sem. Thromb. Hemost.,* 23: 203.
19. Baumann H., Keller R., 1997, *Sem. Thromb. Hemost.*, 23: 215.
20. Baumann H., Keller R., Baumann U., 1999, *Frontiers in Biomedical Polymer Applications - Vol. 2*, (Ottenbrite R.M., Chiellini E., Cohn D., Migliaresi C., Sunamoto J., eds.), Technomic Publ., Lancaster, Basel, pp. 159.
21. Otto V., Wolff C., Baumann H., Keller R., 1997, *Werkstoffe für die Medizintechnik*, (Breme J, ed.), DGM-Informationsgesellschafts-Verlag, Frankfurt/M, pp. 161.
22. Lindahl U., Kjellen L., 1987, *Biology of ExtracellularMmatrix. A Series: Biology of Proteoglycans* (Wight T.N., Mecham R.P., eds.), pp. 59.
23. Gallagher J.T., Lyon M., 1989, *Heparin, Chemical and Biological Properties, Clinical Applications*, (Lane D.A., Lindahl U., eds.), Edward Arnold, London pp.135.
24. Baumann H., Scheen H., Huppertz B., Keller R., 1998, *Carbohydr. Res.,* 308: 381.
25. Takano R., Ye Z., Van Ta T., Hayashi K., Kariya Y., Hara S., 1998, *Carbohydr. Letters*, 3: 71.
26. Engvall E., Ruoslathi E., 1977, *Int. J. Cancer*, 20: 1.
27. Maeda H., Ishida N., Kawauchi H., Tuzimura K., 1969, *J. Biol. Biochem.*, 65: 777.

28. Faraday N., Goldschmidt-Clermont P., Dise K., Bray P.F., 1994, *J. Lab. Clin. Med.,* 123, 728.

29. Miekka S.I., Ingham K.C., 1980, *Arch. Biochem. Biophys.,* 203, 630.

30. Sakariassen K.S., Aarts P.A.A.M., de Groot P.G., 1993, *J. Lab. Clin. Med.*, 102: 522.

31. Huppertz B., Baumann H., Keller R., 1999, *Frontiers in Biomedical Polymer Applications - Vol. 2*, (Ottenbrite R..M., Chiellini E., Cohn D., Migliaresi C., Sunamoto J., eds.), Technomic Publ., Lancaster, Basel, pp.115.

32. Young B.R., Lambrecht L.K., Cooper S.L., Mosher D.F., 1982, *Adv. Chem. Ser.*, 199: 317.

33. Pitt W.G., Park K., Cooper S.L., 1986, *J. Colloid Interface Sci.*, 111: 343.

34. Clark W.R., Macias W.L., Molitoris B.A., Wang N.H.L., 1995, *Kidney Intern.,* 48: 481.

35. Huppertz B., Keller R., Baumann H., 1994, *Int. J. Exp. Pathol.*, 75: A 83.

PREPARATION AND SURFACE ANALYSIS OF HAEMOCOMPATIBLE NANOCOATINGS WITH AMINO- AND CARBOXYL GROUP CONTAINING POLYSACCHARIDES

[1]H. Baumann, [1]V. Faust, [1]M. Hoffmann, [1]A. Kokott, [1]M. Erdtmann, [2]R. Keller, [1]H.-H. Frey, [1]M. Linßen, and[1]G. Muckel
[1]*Macromolecular Chemistry and Textile Chemistry, Haemocompatible and Biocompatible Biomaterials, University of Technology, RWTH-Aachen, Worringer Weg 1, D-52074 Aachen, Germany;* [2]*Clinical Centre of Cologne, Ostmerheimer Str. 200, D-51109 Cologne, Germany*

1.   INTRODUCTION

The topmost molecular layer of solid materials normally determines their properties to the environment, e.g. to body fluids and their reactivity to chemical systems such as of catalysts. Therefore, surface modification of solid polymers with ultrathin layers of reactive, hydrophilic, hydrophobic and amphiphilic components or polymers, solely or in combination, is chosen to improve surface and interface properties to other systems.

In case of biomaterials such as artificial organs, implants or extracorporal systems which are necessarily in contact with blood, they have to have the highest known standard of biocompatibility, called haemocompatibility. That means compatibility and inertness to platelets and to other cellular components of blood, no contact activation and also no adhesion of plasma proteins which are known to trigger cellular reactions when they are adsorbed on surfaces and interact with their receptors. Additionally it means no level effect of complement activation and no cytotoxicity.

Biomedical Polymers and Polymer Therapeutics
Edited by Chiellini *et al.*, Kluwer Academic/Plenum Publishers, New York, 2001   195

In nature at the interface between blood and luminal side of blood vessels is located an endothelial cell surface heparan sulphate (ES-HS) - a polysaccharide glycosaminoglycan type 2 - with a layer thickness[1] of 1 pmol/cm^2. This polysaccharide seems to be responsible for the high standard of haemocompatibility[2, 3]. However it is not known whether ES-HS will be attacked and degraded with time by leucocytes, radicals and other blood components and how often ES-HS will be regenerated by biosynthesis. Therefore, artificial surfaces or organs modified with monolayers of ES-HS or ES-HS analogous sequences (regioselective modified heparin derivatives) may need, for long term implantation probably, more than a monolayer of covalent bound ES-HS which allows safety for a longer implantation time if drastically degradation processes may occur.

In the present contribution different covalent coating procedures will be reviewed and described (Fig.1). For ES-HS and heparin, this results in different thickness of nanocoatings and different attachment to polymer surfaces.

Various immobilisation methods lead to side to side, single point (II, III), and multi point (IV, V) attachment resulting in mono- (II to V) and multilayers (VI) of the bonded polysaccharides. For these concepts we used carboxylated or aminated surfaces to which glycosaminoglycans type 2 (ES-HS, heparin HE) or regioselective modified heparin derivatives are covalently bonded via residual amino groups or via carboxyl groups. The resulting amide bonds are stable against hydrolysis under physiological conditions.

Also a new end point attachment of nondegraded glycosaminoglycan type 2 (HE) or regioselective modified heparin derivatives will be described: the reducing end is oxidised by electrolysis and the aldonic acid lactone formed reacts with amino groups of the polymer surface. Also coatings by ionic immobilisation (HE, ES-HS) will be described.

The mean amount of immobilised glycosaminoglycans (GAGs) was analysed via hydrolytic released glucosamine and quantitatively determined by AE-HPLC by pulsed amperometric detection (PAD). A new quantitative method for the determination of surface bonded GAGs by ESCA will be described using ESCA sensitive markers: ESCA sensitive heparin derivatives were obtained by labelling heparin with fluorine containing marker molecules (pentafluorobenzyl bromide (PFB) and heptafluorobutyl amine (HFB)). Literature known calculations based on charge repulsion or space requirement of HE[4, 5] are compared with our experimental results for calculating the mean thickness of the coatings.

Furthermore the hydrolytic stability of the covalent coatings and spacers and the stability against sterilisation will be reported.

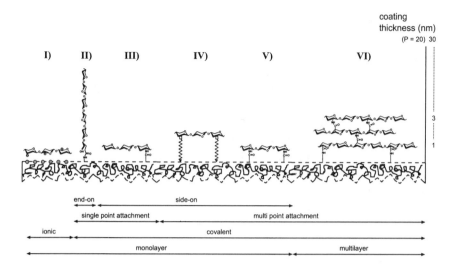

Figure. 1. 6 measure made immobilisation concepts for the attachment of GAG typ 2 (HE, ES-HS) and regioselective modified heparin derivatives on polymer surfaces to get different coating thickness (in nm).

2.   MATERIALS AND METHODS

2.1   Chemicals

Chemicals were obtained from Merck (Darmstadt, Germany) and Sigma (Munich, Germany) in p.a. quality. HE grade I was from Serva, Heidelberg, Germany.

ES-HS has been isolated from medium of endothelial cell cultures via several chromatographic methods as described earlier[2,6]. Regioselective modified heparin derivatives were prepared as described earlier[7,8].

PVC, silicone, polyurethane (PU), polyether sulphone (PESU) and polycarbonate (PC) were obtained from Fresenius AG, St. Wendel, Germany. Polypropylene (PP, Vestolen[TM]) was from Hüls AG, Marl, Germany. Polyethylene (PE, Lupolen[TM]) was from BASF, Ludwigshafen, Germany. Polystyrene (PS) was from Amelung GmbH, Lemgo, Germany. Cellulose dialysis tubings were from Serva, Heidelberg, Germany.

2.2   Carboxylation of Polymers

a) Carboxylation with heterobifunctional photoactivatible reagents:

Polyethylene and polypropylene samples were defatted with acetone and hexane and stirred with a solution of N-(4-azido-2-nitrophenyl)-11-aminoundecanoic acid (ANPAU) or p-azidobenzoic acid in ethanol. Polymer samples were removed and air dried. After irradiation with a mercury high pressure lamp at 260 nm the polymer samples were washed with ethanol.

b) Carboxylation with 3-iodopropionic acid:

Polyurethane sheets were stirred with a solution of 0.1 M sodium isopropylate in iso-propanol at 4°C. Two hundred mg of 3-iodopropionic acid were added and stirred for 2.5 h. Samples were rinsed with deionised water, 0.1 M HCl and again with deionised water.

c) Carboxylation with dicarbonic acid dichlorides:

PA 6.6 membranes were stirred with a solution of 1.8 mmol N-ethyldiisopropyl amine and 1.8 mmol dicarbonic acid dichloride in 3 ml ether for 30 min. Residual acid chloride groups were hydrolysed with 0.5 M $NaHCO_3$ solution. The membranes were rinsed with deionised water.

d) Carboxylation with phenylsulphonyl isocyanate:

PESU sheets were stirred with a solution of 0.3 g phenylsulphonyl isocyanate in 100 ml anhydrous ether at RT. Samples were rinsed with ether, ethanol and deionised water. The polymer sheets were hydrolysed with 8 M HCl at 100°C, rinsed with water, 5% NaOH and deionised water.

e) Carboxylation with chlorosulphonic acid and 6-aminocapronic acid:

18 ml chlorosulphonic acid were dissolved in 600 ml ethanol and cooled to -5°C. Polymer samples (PESU, PC) were added to the solution, stirred for 4 h at RT and stirred with water.

The samples were stirred overnight in a solution of 1.2 g 6-aminocapronic acid in 600 ml 0.5 M $NaHCO_3$, rinsed with 4 M NaCl and deionised water.

2.3   Amination of Polymers

a) Amination with 3-aminopropyltriethoxy silane:

Silicone tubes were aminated with 3-aminopropyltriethoxy silane as described earlier[9].

b) Amination with nitratic acid and reduction:

Polystyrene was stirred for 2 - 6 h at room temperature or 70°C with 600 ml nitratic acid. It was rinsed with tap water. 10 ml hydrazine hydrate in 600 ml 0.5 M $NaHCO_3$ and a spatula tip of Raney Nickel were added and stirred for 24 h at room temperature. Finally the polystyrene probes were rinsed with water, a solution of 0.02 M imidazole and 2 M NaCl in water and again with water.

c) Amination with chlorosulphonic acid and 1,12-diaminododecane:
Chlorosulphonation was carried out as described in carboxylation method e) with PS samples. The premodified polystyrene was added to a solution of 18 g 1,12-diaminododecane in 600 ml ethanol for 18 h and stirred for 18 h at RT. The polystyrene was rinsed with water, ethanol and imidazole/NaCl solution.

2.4   Preparation of Spacers

a) Preparation of carboxyl group terminated spacers with methacrylic acid: PVC, PU, PESU or PC samples were stirred with an aqueous solution of Fenton's Reagent (56 mg $FeSO_4$ x 7 H_2O, 50 mg $Na_2S_2O_5$, 2 ml H_2O_2 30 %) and methacrylic acid (2 ml) for 2 h at room temperature and rinsed with tap water for 30 minutes and another 2 h with deionised water.

b) Preparation of carboxyl group terminated spacers with dicarbonic acids: Aminated polymers (silicone, PU, and PA 6.6) were reacted with dicarbonic acids as described earlier[9].

c) Preparation of amino group terminated spacers with oligomethacrylic acid and 1,6-diaminohexane: the oligomethacrylic acid premodified sheet (PVC) is stirred with 100 ml 0.05 M solution of 1,6-diaminohexane in 0.1 M MES-buffer (pH 4.75) and 300 mg CME-CDI for 16 h at 4°C. The polymer sheets are rinsed with 4 M NaCl solution and deionised water.

2.5   Cationisation of Cellulose

Cellulose was treated with N,N,N-3-chloro-2-hydroxypropyl trimethyl-ammonium chloride as described earlier[2, 10, 11].

2.6   Immobilisation of GAG Type 2 (HE, ES-HS) and Regioselective Modified Heparin Derivatives

Method I:
Heparin res. ES-HS were immobilised ionically to cationic cellulose as described earlier [2, 10, 11].

Method II:
Heparin (300 mg) was dissolved in 100 ml of 20 mM acetic acid/acetate-buffer after the addition of 50 mg $CaBr_2$. The solution was electrolysed for 1 h at a current of 0.1 A. The solution was perfused through a cation exchange column (Cellex CMC), dialysed and lyophilised. The residual product was dissolved in deionised water and stirred for 48 h at 40°C with aminated polymer samples (PVC, silicone). The polymers were rinsed with deionised water, 4 M NaCl and deionised water.

Method III - V:
Carboxylated polymers were activated with a water soluble carbodiimide (CME-CDI) and reacted with GAG type 2 (HE, ES-HS) and regioselective modified heparin as described previously[3, 9].

Method VI:
Aminated polystyrene was stirred with a solution of 600 mg HE and 3 g CME-CDI in 600 ml 0.1 M MES buffer (pH 4.75) at RT overnight. Probes were rinsed with tap water, 0.02 M imidazole/2 M NaCl and deionised water.

2.7   Analytical Methods

2.7.1   Determination of Polymer Bonded Carboxyl Groups

Polymer samples were incubated with 10% HCl for 20 min and rinsed with bidistilled water until the conductivity drops below 2 µS/cm. A sample is placed in 20 ml bidistilled water and titrated conductometrically with 0.01 M NaOH.

2.7.2   Determination of Polymer Bonded Primary Amino Groups

The primary amino groups were determined by the ninhydrin method as described by Moore and Stein[12].

2.7.3   Determination of Polymer Bonded GAG Type 2 (HE, ES-HS) and Regioselective Modified Heparin with AE-HPLC-PAD

Surface modified polymers were hydrolysed with HCl and the liberated glucosamine was determined by quantitative anion exchange HPLC with pulsed amperometric detection (PAD) as described elsewhere[10,13].

2.7.4   Surface Analysis of Heparin Coated Polymers by ESCA

ESCA investigations were carried out on a M - Probe Surface Spectrometer SSX 100, Surface Science. HE ionically immobilised on cationised cellulose membranes was analysed quantitatively by means of the sulphur signal in the ESCA spectrum. Surface HE content was calculated from the peak integration of the sulphur peak. For increased ESCA sensitivity HE was labelled with pentafluorobenzyl bromide (PFB) or heptafluorobutylamine (HFB). Labelled HE was immobilised covalently to silicone and quantified by ESCA by means of the fluorine signal.

2.8 Stability Test of Covalent HE Coating on Silicone

HE coated silicone tubes were stored for 4, 7, and 14 days at 50, 70, and 90°C in physiological phosphate buffered saline (PBS). The HE content before and after storage was measured by the HPLC method described above.

2.9 Stability of HE against γ-Sterilisation

Heparin was exposed to various doses of γ-rays of a ^{60}Co-source. The sulphation degree and sulphation pattern were analysed by ^{13}C-NMR-spectroscopy (Pulse NMR-Spectrometer AC 300, Bruker, Karlsruhe, Germany).

3.   RESULTS

3.1   Immobilisation Concepts for Attachment of GAG Typ 2 (HE, ES-HS) and Regiosective Modified HE Derivatives on Polymer Surfaces

The results of surface coating 10 different polymers with HE res. ES-HS can be seen in Table 1. GAG type 2 and regioselective modified heparin derivative immobilisation was achieved by amide bond formation between carboxyl groups on the polymer surface and amino groups in the GAG chain or vice versa carboxyl groups were activated by the water soluble carbodiimide N-cyclohexyl-N′-(2-morpholinoethyl)carbodiimide-methyl-p-toluene sulphonate (CME-CDI).

The immobilisation results may be divided into 6 groups according to the immobilisation procedure:

I) Ionic immobilisation

Ionic immobilisation of heparin is a relative simple and rapid method for the preparation of short duration antithrombotic surfaces. Maximum anticoagulant activity of the immobilised molecules is obtained because of high flexibility of the chains. Because of relatively weak surface bonding, ionic immobilised HE, ES-HS and regioselective modified HE derivatives are not intended for long time applications.

II) End point attachment

End point attachment of heparin has been described by Larm et al.[14]. They degraded HE by HNO_2 treatment forming reactive anhydromannose residues followed by reaction with free amino groups of the polymer surface. The cleavage by HNO_2 takes place specifically at glucosamine -N-sulphate moieties or glucosamine. The degradation process does not only reduce the molecular weight of HE but also leads to loss of glucosamine -N-sulphate containing domains in HE. Both features may change the biological activity of the HE molecules.

Table 1. Immobilisation of heparin and ES-HS onto 10 different polymers. Polymers are functionalised either with carboxyl or amino groups before GAG type 2 immobilisation. Roman numbers refer to the immobilisation concepts in Figure 1. Abbreviations: PVC: polyvinyl chloride, PU: polyurethane, PE: polyethylene, PP: polypropylene, PA 6.6: polyamide 6.6, PESU: polyethersulphone, PS: polystyrene, PC: polycarbonate, ANPAU: N-(4-azido-2-nitrophenyl)-11-aminoundecanoic acid, CM-PEG: carboxymethyl-polyethylene glycol, QUAB: N,N,N-3-chloro-2-hydroxypropyl trimethylammonium chloride.

	Functionalization			Surface coverage	
				Heparin 13 kD	ES-HS 36kD
Polymer	Introduced functional group (reagent)	Immobilis-action concept	mol/cm^2	$pmol/cm^2$	
PVC	-COOH (ANPAU)	IV	-	7,2-12,2	-
	-COOH (Oligo-methacrylic Acid)	IV	12,2	4,8	0,4-0,9
	-NH$_2$ (Oligo-methacrylic Acid / 1,6-Diaminohexane)	II	11,9	21,2-38,9	-
Silicone	-COOH (Adipic Acid)	IV	4,6	10,7-12,0	1,5-3,4
	-COOH (CM-PEG)	IV	3,3-3,9	3,9-4,4	1,1-1,5
	-NH$_2$ (3-Aminopro-pyltriethoxysilane)	II	5,6	9,1-23,3	-
PU	-COOH (Adipic Acid)	III/V	1,6	8,0	-
	-COOH (Oligo-methacrylic Acid)	IV	2,4	6,3	1,2
	-COOH (Iodopropionic Acid)	III/V	1,7	8,0	-
PE	-COOH (ANPAU)	IV	2,5	49,4	5,7
	-COOH (p-Azidobenzoic Acid)	III/V	1,7	23,0	-
PP	-COOH (ANPAU)	IV	3,0	67,6	8,0
	-COOH (p-Azidobenzoic Acid)	III/V	2,3	25,0	-
PA 66	-COOH (Suberic Acid)	III/V	9,3	45,0	-
	-COOH (Succinoyl Choride)	III/V	9,7	62,5	-
	-COOH (Sebacinoyl Chloride)	IV	10,7	68,0	38,0
PESU	-COOH (ANPAU)	IV	2,1	10,4	-
	-COOH (p-Azidobenzoic Acid)	III/V	2,0	8,3	-

	-COOH (Oligomethacrylic Acid)	IV	2,5-2,7	2,6-11,9	2,3
	-COOH (Phenyl-sulphonylisocyanat)	III/V	2,4-2,8	6,6-8,3	-
	-COOH (Chlorosulphonic Acid/6-Amino-capronic Acid)	III/V	2,9-15,6	2,8-17,1	-
PS	-NH$_2$ (Chlorosulphonic Acid/1,12-Diaminododecane)	VI	81,6	130	-
	-NH$_2$ (nitrating Acid/ Raney Ni)	VI	120,4	130	-
PC	-COOH (Oligomethacrylic Acid)	IV	1,8-2,3	3,5-13,8	-
	-COOH (Chloro-sulphonic Acid / 6-Aminocapronic Acid)	III/V	2,9-15,6	2,8-17,1	-
Cellulose	-NR$_3^+$ (QUAB)	I	247-2830	112-5080	6-55
Endothelial Cell Surface 5,4 x 10^{11} saccharide chains/cm^2			-	-	0,89[1]
		Side-on		3,6[4]	1,1
Theoretically calculated		Side-on		7,7[5]	2,4
		Side-on		157,0	157,0

We have developed a new method for end point attachment. The reducing end of ES-HS and HE was oxidised by electrolysis to an aldonic acid followed by lactonisation according to a method by Emmerling and Pfannemüller[15] for oligomaltose. The aldonic acid lactone reacted immediately with primary amino groups on the polymer surfaces.

Under optimised reaction conditions neither chain degradation nor loss of sulphate or acetyl groups takes place. This method enables for the first time to immobilise intact HE res. ES-HS or regioselective modified HE derivatives end on to polymer surfaces.

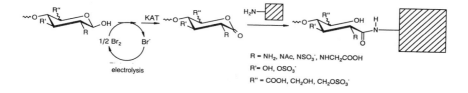

Figure 2. Electrolytic oxidation of reducing terminus of HE and ES-HS. The aldonic acid was formed converted into a lactone and reacted by ring opening with primary amino groups on polymer surfaces.

III) - V): Side on coordinated covalent immobilised monolayers of HE, ES-HS and regioselective modified HE derivatives.

One point side on attachment may be achieved using only one residual free amino group and carboxyl group terminated spacers as oligomethacrylic acid (type IV), carboxymethyl PEG (type IV) or heterobifunctional azides (type III/V). Amino group determination by the ninhydrin method showed, on the average, only one free amino group per three HE molecules. The polymer surface were activated with CME-CDI and then HE, ES-HS and regioselctive modified HE derivatives were added. Additional amino groups may be liberated by N-desulphation in acidic media[16] which lead to two, three or multipoint attachments during immobilisation depending on the number of released amino groups per GAG chain.

Different types of spacers of the carboxylation and amination reagent have been used. The spacer length varied between directly carboxylated polymers without any spacer (phenylsulphonylisocyanate followed by hydrolysis), short spacers with 3 atoms in the spacer chain (iodopropionic acid), and 413 (CM-PEG-spacer on silicone) atoms in the spacer chain. The length of the oligomethacrylic acid spacer is not known.

The spacer length is known to be important for getting a sufficient tertiary structure of HE with full anticoagulant activity.

VI) Multilayer formation.

During the immobilisation procedure, reaction between free amino groups of the glucosamine and carboxyl groups of the uronic acids may occur and lead to crosslinking between the GAG chains. In the previous case, a monolayer with side on mono- and multipoint attachment, this crosslinking reaction was avoided by a two step process with separate activation of the carboxyl groups on the polymer surface and subsequent addition of HE, ES-HS and regioselective modified HE derivatives.

In the present case, a multilayer of HE is formed on polystyrene equipped with amino group terminated spacers with multi point attachment, because of the high content of carboxyl groups in the GAG molecules. This immobilisation result is obtained by simultaneous addition of the activating reagent CME-CDI and HE. The carbodiimide activated carboxyl groups of GAGs were then reacted with both types of amino groups, the aminated surfaces and the free amino groups of other GAG chains.

3.2   Calculated Mean Thickness of Coating

Theoretical calculations based on space requirement of heparin molecules[4, 5] have been used to decide weather monolayers or multilayers will be formed. All immobilisation procedures despite of concept I) and concept IV) yield heparin amounts, which are in the range of the theoretically calculated amount for a monolayer.

The coating thickness of end point attached GAG depends on the length of the repeating unit of 30 nm for 20 disaccharides, considering that the polymer chains are in an extended conformation. The mean coating thickness of side to side oriented mono- and multilayers of GAGs may be calculated on the assumption of Bokros[2] that the thickness of a flat and uncoiled extended GAG chain is about 1 nm for each layer. Additionally the spacer length in extended conformation can be calculated.

3.3   Quantitative Determination of the Surface Specificity of the Immobilisation of HE by ESCA

A new method was developed for calculating the quantitative surface distribution of at least four different components as the polymer, plasticiser, spacer, and heparin from ESCA data. The amounts of the components were calculated from the proportions of the elements in the ESCA survey spectrum, using a model of the surface concerning the repeating units of the carrier polymer and the coatings. For improving the sensitivity of the method, ESCA sensitivity of heparin was increased by labelling with markers [pentafluorobenzyl bromide (PFB), heptafluorobutyl amide (HFB)].

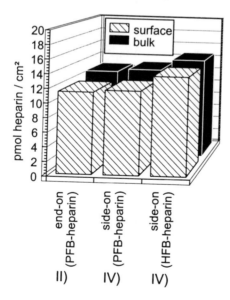

Figure 3. ESCA data (hatched) of HFB- res. PFB labelled heparin covalently immobilised on silicone tubes versus analytical data obtained by HPLC after hydrolysis (black). Agreement of the data show that the immobilised heparin is located completely in the ESCA sensitive layer of the polymer.

The information depth of the ESCA method is about 10 nm. In the present case results for heparin surface and bulk concentration of different immobilisation procedures are compared. Investigations of three different heparin immobilisation procedures (ionic, covalent end-on, covalent side-on) show, that heparin is located completely on the surface only in the case of covalent immobilisation to silicone (Fig.3).

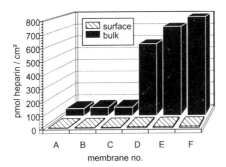 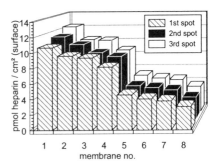

Figure 4. ESCA data of heparin ionically immobilised on cationised cellulose membranes. Left: ESCA data (hatched) versus HPLC data (black) of 6 different membranes show that heparin penetrates into the polymer. Right: Uniformity of heparin coverage was investigated at 6 different heparin surface concentrations. Three spots were taken of each membrane which showed the uniform heparin coverage.

More than 90% of the ionically immobilised heparin was not located at the ESCA sensitive layer of the cationic cellulose (Fig.4, left). The surface sensitivity of the HE immobilisation was calculated by creating the quotient of the surface amount, measured by ESCA and the bulk amount measured by HPLC after hydrolysis. Uniformity of heparin coverage was investigated on cationic cellulose membranes with different heparin surface concentrations. Three spots were measured on each membrane and show the uniformity of heparin surface coverage within the accuracy of measurement (Fig.4, right).

3.4   Hydrolytic Stability of HE Coatings

For stability tests under physiological conditions heparinised silicone probes have been stored in phosphate buffered saline (PBS) at different temperatures up to 14 days. Independent of temperature the surface concentration of heparin drops to a value half of the initial value within 7 days. After this period the value remains constant (Fig.5 top). The stability of heparin molecules has been investigated under steam- and gamma-sterilisation conditions. Under gamma-sterilisation conditions the molecular weight and the sulphation pattern did not change up to a dose of 50 kGy (normal sterilisation dose in europe is 25 kGy) (Fig.5 bottom). Three cycles of steam sterilisation also did not affect the molecular weight, sulphation pattern of the heparin molecule itself, and the surface coverage of a heparinised polymer probe[4].

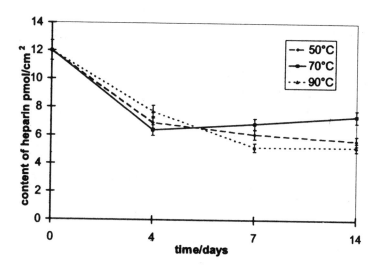

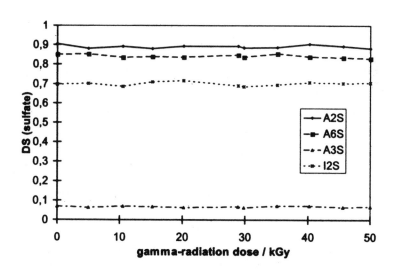

Figure 5. Stability test of heparin under different hydrolytic and sterilisation conditions. Top: Storage of heparin covalently immobilised on silicone at different temperatures in PBS buffer. Bottom: γ-sterilisation of heparin up to a dose of 50 kGy. The sulphation pattern was not affected by the sterilisation process.

4.   DISCUSSION

In the present paper some immobilisation results have been reviewed and new results are added. The advantage of uronic acid and amino sugar containing repeating units such as glucosamine or galactosamine containing repeating units is that at least three different functional groups, -COOH, -NH₂, -OH, may be used for selective covalent immobilisation of polysaccharides to polymer surfaces. This combination of functional groups with different reactivity enables building up defined layers such as monolayers with monoattachment, diattachment, triattachment and so on. Furthermore, the reducing end may be used to get an end point attachment. A multilayer attachment will be obtained if aminated surfaces will be coupled with carbodiimide and HE or ES-HS in a one step reaction or if HE or ES-HS will be bonded electrostatically. The covalently immobilised monolayers of ES-HS may be used for long term implants with the highest haemocompatibility standard similar to nature. The amide bond was stable against hydrolysis. Also ES-HS may be heat sterilised (130°C) or γ-sterilised (50 kGy) without any degradation of the polymer chain and without loosing the biological properties. The election of the ultrathin nanocoating depends on the duration of implantation of biomaterials.

For long term implantation we have developed multilayers or nanocoatings with higher thickness. This type of coating probably gives additional safety when radicals, leukocytes or macrophages attack the outermost layer of ES-HS. In this case, a reservoir of many other layers of ES-HS may protect the implant probably for a very long time before it will be completely denuded from ES-HS.

Surface analysis of cellulose ionically coated with HE and silicone with covalently immobilised HE were carried out by means of ESCA. Comparison between ESCA surface analysis and HPLC bulk analysis data show that ionically cellulose bonded HE penetrates into the partially quaternised cellulose membrane, because cellulose membranes will be degraded by introducing quaternised amino groups via the reagent QUAB (N,N,N,-3-chloro-2-hydroxypropyltrimethylammonium chloride). The reaction with less degraded cellulose membranes leads to more specific surface reactions.

Contrary to these results more than 90 % of the HE covalently attached to silicone were located at the surface as shown by comparising the ESCA[17, 18] and HPLC data. The amounts of HE bonded to the surface are in the same range as theoretically calculated for HE monolayers[4, 5]. The ESCA investigations of surface modified silicone tubes were done with fluorine labelled HE derivatives to increase the ESCA sensitivity of the HE molecules. In the past fluorine containing labelling reagents have been used for qualitative detection of functional groups e.g. OH, CO, COOH, NH_2[19, 20]. We developed a new method for quantitative determination of small amounts of surface bonded HE by means of fluorine labelled HE derivatives for the immobilisation reactions. The relatively high amounts of HE ionically immobilised on cationised cellulose allows the quantitative determination of HE by the S 2p-peak which is too insensitive for the smaller amounts obtained by covalent attachment of HE.

Regioselective modified heparin derivatives may also be used in a similar manner for different nanocoatings (Fig.6).

These regioselective modified heparin derivatives may represent sequences of ES-HS and thus will not only be used for structure - function studies but also in the future as a semisynthetic ES-HS.

Also other polysaccharides containing carboxyl and amino groups such as chondroitin sulphate, dermatan sulphate[21], N-carboxymethyl chitosan[22], and some bacteria polysaccharides [23, 24] may be used in a similar manner to build up ultrathin coatings on surfaces with defined attachments and mean thickness.

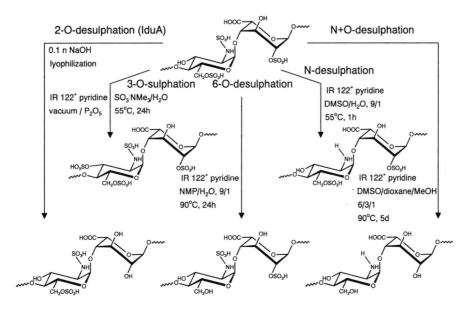

Figure 6. Regioselective modification of heparin. 2-O-desulphation, 6-O-desulphation, N-desulphation, N+O-desulphation and 3-O-sulphation were carried out solely or in combination to obtain a wide spectrum of regioselective modified heparin derivatives[7, 8].

If polysaccharides will be used carrying only one functional group in the repeating unit, either an amino group or a carboxyl group, the immobilisation variations will be reduced to one type of functionalised polymer surfaces. But also defined surface layers may be prepared with single point or multipoint attachment. For all polysaccharides with an intact reducing terminus, end point attachment should be possible.

The calculated thickness of the ultrathin GAG type 2 layers are mean values, based on models from the literature. The AFM may show the topology of the GAG layers as well as of the functionalised polymer surfaces. The measurements are now in preparation and they should give more information about the ultrathin coatings.

5.   CONCLUSION

In the present paper ultrathin layers of glycosaminoglycans type 2, in the range of about 1 to 30 nm thickness, with different attachments on different functionalised polymer surfaces have been reviewed and synthesised. Among these are covalently bonded side to side coordinated monolayers of - 1, 2, 3, and multipoint attachments and multilayers. Different types and length of spacers were used. Also a new method for end point attachment of GAGs type 2 without degradation of the chain length or the functional groups was described. These reactions were via electrolytic oxidation of the reducing terminus to aldonic acid, followed by lactonisation and reaction with primary amino groups of the polymer. A new quantitative ESCA method using ESCA sensitive markers was developed. Additionally the GAG type 2 content of the bulk has been measured by means of HPLC determining the GlcN content, which is a part of the repeating unit of the GAGs type 2. In the case of covalently immobilised HE, the ratio of surface bonded GAG, determined by ESCA and GAG in the bulk of the polymer correlated well and was used to determine information about the surface specificity of the immobilisation reaction. The immobilised GAG type 2 coating was shown to be stable against hydrolysis under physiological conditions and γ-sterilisation. The described coating procedures may also be used for preparing defined nanocoatings of regioselective modified heparin derivatives and also other desoxyamino polysaccharides.

ACKNOWLEDGMENTS

We thank the Deutsche Forschungsgemeinschaft, the Gesellschaft der Freunde der Aachener Hochschule, the Fresenius St. Wendel GmbH, and the Bundesministerium für Bildung und Forschung for financial support of our project No. 03M1506 3.

REFERENCES

1. Williams M.P., Streiter H.B., Wusteman F.S., Cryer A., 1983, *Biochem. Biophys. Acta* 756: 83.
2. Baumann H., Müller-Krings U., Keller R., 1997, *Sem. Thromb. Hemost.* 23: 203.
3. Baumann H., Keller R., 1997, *Sem. Thromb. Hemost.* 23: 215.
4. Eloy R., Belleville J., Rissoan N.C., Baguet J., 1988, *J. Biomater. Appl.* 2: 475.
5. Bokros J.C., Gott V.L., La Grange L.D., Fadall A.M., Vos K.D., Ramos M.D., 1969, *J. Biomed. Mater. Res.* 3:497.
6. Otto V., Wolff C., Baumann H., Keller R., 1997, *Werkstoff für die Medizintechnik*, (Breme J, ed.), DGM-Informationsgesellschafts-Verlag, Frankfurt/M, pp. 161.
7. Huppertz B., Baumann H., Keller R., 1999, *Frontiers in Biomedical Polymer Applications - Vol. 2*, (Ottenbrite R..M., Chiellini E., Cohn D., Migliaresi C., Sunamoto J., eds.), Tecnomic Publ., Lancaster, Basel, pp.115.
8. Baumann H., Scheen H., Huppertz B., Keller R., 1998, *Carbohydr. Res* 308: 381.
9. Baumann H., Keller R., Baumann U., 1999, *Frontiers in Biomedical Polymer Applications - Vol. 2*, (Ottenbrite R.M., Chiellini E., Cohn D., Migliaresi C., Sunamoto J., eds.), Technomic Publ., Lancaster, Basel, pp. 159.
10. Emonds M., Ruzicka E., Muckel G.H., Keller R., Müller U., Baumann H., 1991, *Clin. Mater.* 8: 47.
11. Baumann H., Keller R., Ruzicka E., 1991, *J. Membrane Sci.* 61: 253.
12. Moore S., Stein W.H., 1954, *J. Biol. Chem.* 211: 907.
13. Erdtmann M., Keller R., Baumann H., 1994, *Biomaterials* 15:1043.
14. Larm O., Larsson R., Olsson P., 1983, *Biomat., Med. Dev., Art. Org.* 11: 161.
15. Emmerling W.N., Pfannemüller B., 1978,. *Makromol. Chem.* 179: 1627.
16. Verstraete M., 1983, *Heparin- New biochemical and chemical aspects.* (Witt I., ed.) de Gruyter & Co., Berlin, New York, pp. 275.
17. Briggs D., Seah M.P., 1983, *Practical surface analysis by Auger and X-ray photoelectron spectroscopy.* John Wiley & Sons, Chichester, New York, Brisbane, Toronto, Singapore.
18. Siegbahn K., (1982), *Science* 217: 111.
19. Ertel S.I., Chilkoti A., Horbett T.A., Ratner B.D., 1991, *J. Biomater. Sci. Polym. Ed.* 3: 163.
20. Chatelier R.C., Gengenbach T.R., Vasic Z.R., Griesser H.J., 1995, *J. Biomater. Sci Polym. Ed.* 7: 601.
21. Scott J.E., Heatley F., Wood B., 1995, *Biochem.* 34: 15467.
22. Muzzarelli R.A.A., Tanfani F., Emanuelli M., Mariotti S., 1982, *Carbohydr. Res.* 107: 199.
23. Casu B., Grazioli G., Hannesson H, Jann B., Jann K., Lindahl U., Naggi A., Oreste P., Razi N., Torri G., Tursi F., Zoppetti G., 1994, *Carbohydr. Lett.* 1: 107.
24. Casu B., Grazioli G., Razi N., Guerrini M., Naggi A., Torri G., Oreste P., Tursi F., Zoppetti G., Lindahl U., 1994, *Carbohydr. Res.* 263: 271.

ISOLATION AND CHARACTERISATION OF ENDOTHELIAL CELL SURFACE HEPARAN SULPHATE FROM WHOLE BOVINE LUNG FOR COATING OF BIOMATERIALS TO IMPROVE HAEMOCOMPATIBILITY

[1]M. Hoffmann, [1]R. Horres, [2]R. Keller, and [1]H. Baumann
[1]*Macromolecular Chemistry and Textile Chemistry, Haemocompatible and Biocompatible Biomaterials, University of Technology, RWTH-Aachen, Worringer Weg 1, D-52074 Aachen, Germany;* [2]*Clinical Centre of Cologne, Ostmerheimer Str. 200, D-51109 Cologne, Germany*

1. INTRODUCTION

Heparan-sulphate was isolated for the first time as a byproduct of commercial heparin isolation in 1948 and described as a low sulphated heparin by Jorpes and Gardell[1]. During its` early stage some researchers thought heparan-sulphate to represent a new class of glycosaminoglycans[2]. Others supposed it to be an intermediate of the heparin-biosynthesis[3].

In contrast to heparin, which is found in the granules of mastocytes[4], heparan-sulphate occurs ubiquitous on cell surfaces[5, 6]. This fact clarifies former distinguishing problems among the two glycosaminoglycans who nowadays represent the so called type 2 class of GAGs[7]. Heparin as well as heparan sulphate obtain the same backbone structure which is composed from an alternating copolymer of uronic acid and glucosamine[8]. Apart from the different occurrence in mammalians the type 2 GAGs also differ significantly e.g. according to their degree of sulphation, acetylation and especially according their structural heterogeneity which is exposed in the distinct domain structure of the heparan sulphates[9]. The first researcher who described heparane sulphate localized on the luminal surface of endothelial

Biomedical Polymers and Polymer Therapeutics
Edited by Chiellini *et al.*, Kluwer Academic/Plenum Publishers, New York, 2001

cells was Bounassisi who also considered that this substance might participate in some of the mechanisms that regulate haemostasis[10]. Further investigations according to the distribution of the different proteogylcans synthesized by vascular endothelial cells in culture (heparan- and chondroitin- sulphate) were reported by Oohira and Wight[11]. During the last two decades the knowledge concerning heparan sulphates synthesised by endothelial cells increased drastically: Different types of the so called core-proteins to which the heparan sulphate chains are attached were identyfied[12, 13]. Structure- function relationships as well as the localization of heparan sulphates on the luminal or albumin surface of endothelial cells were investigated[14,15].

For the last 15 years our group focused on the development of haemocompatible artificial surfaces such as implants or extracorporeal blood contact devices using the heparan sulphate isolated from bovine aortic endothelial cell culture as coating substance[16, 17, 18]. Covalently immobilized on 10 different polymers used as biomaterials the haemocompatibility was enhanced to the highest level which is given by standard of natural blood vessels[19]. Up to now endothelial cell surface heparan sulfate (ES-HS) is the only known substance for generating a thrombogenic biomaterials by covalent immobilisation on polymer -, glass -, ceramic - or metal - surfaces which show the same haemocompatibility standard as the luminal side of natural blood vessels[20]

For technical application of this partially substituted alternating copolymer of hexuronic acid and glucosamine some investigations were performed to develop more effective alternatives to the complicated and expensive isolation procedure of ES-HS from primary cell cultures of bovine aortic endothelial cells[21]. Since the lung is the organ with the highest content of endothelial cells and the largest blood contact surface it was chosen for the isolation of ES-HS.

2. MATERIALS AND METHODS

2.1 Chemicals

Chemicals were obtained from Merck (Darmstadt, Germany) and Sigma (Munich, Germany) in p.a. quality.

Cellulose dialysis tubings were from Serva, (Heidelberg, Germany). Papain crystal suspension, chondroitinase ABC and heparin lyases were achieved from Sigma (Munich, Germany).

Antithrombin III immobilized onto agarose was also obtained from Sigma.

Fractogel EMD-DMAE 650-S was purchased from Merck (Darmstadt, Germany).

All other chromatography media were purchased from Pharmacia Biotech (Uppsala, Sweden)

2.2 Preparation of Lung Tissue

Bovine lung was achieved from the local slaughterhouse and immediately frozen. It then was cut into little pieces and freeze-dried. The lung was powdered by passing through meat mincer. The resulting bovine lung powder was defatted by stirring overnight in the equivalent volume of acetone. In the following the suspension was passed through a buechner funnel and washed well with the same solvent. After suction to dryness residues of acetone were removed by evaporating overnight. For the extraction with 4M guanidine hydrochloride solution the dry lung powder was mixed with the dry guanidine hydrochloride and then diluted with the corresponding amount of water and stirred overnight. These procedure is essential to prevent from clumping of the organic material.

2.3 Alkaline β-Elimination

The resulting suspension from guanidine hydrochloride extraction was spun down for 30 minutes at 2000g. After decantation from the pellet the resulting dark green highly viscous liquid was cooled to 4°C adjusted to 0,5M NaOH with 1M NaOH solution and again stirred for 15h at this temperature. The extract was dialysed against deionized water until the pH was neutral and in the following again separated from the precipitate occurring during dialysis by centrifugation and decantation.

2.4 Expanded Bed Adsorption on DEAE-Streamline

5 liters of the β-eliminated and dialysed lung tissue extract were adjusted to 0,3M NaCl, 0,01M Tris, 1g/l NaN$_3$, pH 7,3 in a total volume of 10 liters and applied to DEAE-Streamline using the Expanded Bed Adsorption technique that will be described later. The elution was carried out with a linear NaCl-gradient from 0,3M NaCl up to 0,8M NaCl in a total volume of 2,7 liters. Fractions with a salt content between 0,4M and 0,5M NaCl were pooled, again dialysed and prepared for enzymatic treatment.

2.5 Enzymatic Digestions of the Crude Glycocaminoglycan Pool

The first degradation reaction was carried out with 1 U chondroitinase ABC in 0,3M NaOAc and 0,73M tris-hydroxymethylaminomethane at pH 8 and 37°C. The reaction was controlled by observing the UV-absorption at 232nm. After a digestion time of 15h, when constant 232nm absorbance values are reached another U of the enzyme is given to the reaction mixture. The reaction is finished when the second U of chondroitinase does not result in any significant change of the 232nm values.

For the second degradation reaction the pH was adjusted to 6,2 and 1ml of papain crystal suspension and 100mg of cysteiniumhydrochloride was added. Incubation was carried out for 15h at 56°C.

The solution was then cooled to 4°C, mixed with the equivalent volume of cold 1M NaOH solution, kept at this temperature for 15h and dialysed.

2.6 Further Chromatographic purification of the Crude Lung-Heparan-Sulphate.

The enzymatic digestion solution was further purified by an additional ion-exchange-chromatography on Fractogel EMD-DMAE-650-S using the same linear salt gradient just mentioned for the DEAE-Streamline chromatography. Fractions were detected by the DMMB-method, pooled, concentrated and applied to a Superdex-200 gel filtration. Final desalting and lyophilisation results the highly purified desired product.

2.7 Analytical Methods

2.7.1 Determination of Molecular Weight of Lung-Heparan-Sulphate

A HEMA-Bio Linear column (8mm X 300mm) was calibrated with keratansulphate- and dermatansulphate- standards of the molecular weights 52,1 kD, 25,6 kD and 20,7 kD. The sample volume was 20μl at a concentration of ca. 0,1 g/l. The column was eluted at a flow rate of 1 ml/min with a tris buffer (0,3M tris, 1g/l NaN_3, pH7,3). Detection of the peaks was carried out via UV-absorption at 214nm.

2.7.2 Determination of Contaminating Amino-Acids and Amino-Sugars

A 40 μg sample was dissolved in 100μl of water, 42μl of 32% HCl was added and heated under N_2 for 15 h at 105°C. The sample was dried at 60°C in a stream of nitrogen. The resulting solid was dissolved in 500μl loading buffer and 50μl of this solution was chromatographed on an Amino Acid Analyzer Alpha Plus. Automatised comparison of the integration of the peaks of a chromatogram of a standard amino acid mixture with the integrations of the peaks of the chromatogram of the sample resulted in the concentration of each amino acid and amino sugar.

2.7.3 Determination of Disaccharide Composition of Bovine Lung-Heparan-Sulphate.

300μg of bovine ES-HS in 1ml of Acetate buffer, 0,1M sodium acetate 0,1mM calcium acetate pH 7 were dissolved in a quartz cuvette. One unit of heparinase III (EC 4.2.2.8.) dissolved in 100μl of water was added and the increase of the UV adsorption at 232nm was measured during a period of 100 minutes. In the following the cuvette was incubated for 15h at 37°C to complete the reaction. The same procedure was repeated with one unit of heparinase I (EC 4.2.2.7) and one unit heparinase II (without EC number). After finishing the enzymatic digestion with the three heparin-lyases the disaccharides were chromatographed on a Sephadex G 25SF column (100cm, 1,0cm) with water as eluent. At a flow rate of 0,14 ml/min the disaccharides could be detected via UV adsorption at 232nm.The product-containing fractions were collected freeze dried and dissolved in 50μl of water.

The disaccharide composition of bovine lung ES-HS was determined by anion exchange HPLC using a Spherisorb 5 SAX column. At a flow rate of 1ml/min the 50μl sample was injected and the system equilibrated with aqueous HCl at pH 3,5. The elution was carried out by applying a linear NaCl gradient for 45min reaching a final salt concentration of 0,5M NaCl. In the following the column was washed with 1m NaCl solution and equilibrated again. The column was calibrated with the following disaccharide standards: α-ΔUA-(1-4)-GlcNAc

α-ΔUA-2S-(1-4)-GlcNAc
α-ΔUA-(1-4)-GlcNAc-6S
α-ΔUA-(1-4)-GlcNS-6S
α-ΔUA-2S-(1-4)-GlcNAc-6S
α-ΔUA-2S-(1-4)-GlcNS-6S

Identification of bovine lung ES-HS derived disaccharides was done by comparison of the retention times with the retention times of the standard disaccharides to the lung derived product. The amount of the different disaccharides was calculated from the corresponding peak areas.

2.7.4 Determination of Antithrombin-III-Affinity of Bovine Lung-Heparan-Sulphate.

Affinity chromatography was carried out on a column of 5ml antithrombin III-agarose with an inner diameter of 1cm. 300µg of heparan-sulphate were dissolved in 1ml of 0,05M NaCl solution and applied to the column. The elution was carried out stepwise first with 20ml of the loading solution followed by 20ml of 0,154M NaCl solution and finally by 50ml of a linear NaCl gradient up to an end concentration of 2,3M NaCl. 18 fractions of 5ml were collected and detected with DMMB for heparan-sulphate[22].

2.7.5 Determination of the Degree of Acetylation by ^{1}H-NMR-Spectra.

The degree of acetylation was determined from integration of the N-acetyl signal of ^{1}H-NMR spectra at 2 ppm in relation to the sum of integrations of all other signals. ^{1}H-NMR spectra were measured on a Bruker AC 300.

2.8 Covalent Immobilisation of Bovine Lung-Heparan-Sulphate on Cellulose-Membranes

Cellulose dialysis tubing was kept in stirred 100 ml of ethanol/water (1/1) containing 2ml of 3-amino-propyl-triethoxysilane for 15h at 45°C. The membrane was washed well with plenty of water and immersed in a solution of 0,9mg bovine lung ES-HS in 80ml of 0,1M 2-(N-morpholino)-ethanesulfonic acid buffer pH 4,75. 200mg of N-cyclohexyl-N`-2-(morpholinoethyl)-carbodiimide-methyl-p-toluolsulfonate were added in portions of 10mg during 6 hours. The coupling solution was stirred for 15h at 4°C. The membrane was washed with 4M NaCl solution for 2h and with deionised water.

2.9 Platelet Adhesion Test

Platelet rich plasma was prepared from citrated blood of a healthy young man by centrifugation for 20min at 100g. The amount of platelets was determined in a Coulter counter to 325G/l. These cellulose membranes were perfused in a Baumgartner chamber modified according to Sakariassen[23] at 37°C and a shear rate of 1050s^{-1} for 30min. After perfusion the membranes were rinsed with 0,01M HEPES buffer pH 7,3 and fixated with 0,5% glutaraldehyde solution for 20min. In the following the membranes were rinsed with methanol for 5min and stained with May-Grünwald solution for 5min. After rinsing with methanol for 5min the membranes were stained with Giemsa solution for 20min, rinsed twice with methanol and viewed under a light microscope at 400 fold magnification.

3.   RESULTS

3.1   Isolation of Bovine Lung Endothelial-Cell Surface Heparan-Sulphate

As the yield of ES-HS derived from cell culture (0,02mg per liter of conditioned medium) is very low we have tried to isolate it from more easily available sources. Since the lung is the organ with the largest blood contact surface in comparison to its weight it was chosen as starting material. The beyond shown isolation procedure had best results:

Lyophilisation

Fat Removal with Acetone

Guanidine Hydrochloride Extraction

Alkaline β Elimination

Dialysis

Expanded Bed Adsorption on DEAE – Streamiline
And Ion Exchange Chromatography

Enzymatic Digestion

Ion Exchange Chromatography

Gel Filtration

Desalting

Lyophilisation

Scheme 1. Main isolation steps of bovine lung endothelial cell surface heparan sulphate.

As a detection of heparan-sulphate with UV-light or with DMMB was impossible because of disturbances caused by nucleic acids, proteins and other impurities, the first two ion exchange chromatographies were pooled according to the ionic strength at which the product is eluted. This range of ionic strength is known from the isolation of cell culture ES-HS. The first chromatographic step that could be detected with DMMB is shown below:

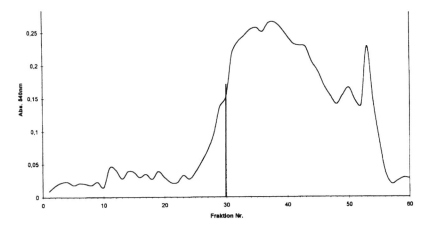

Figure 1. Preparative gel filtration on Superdex-200. DMMB-positive fractions on the left of the indicated bar were pooled.

A yield of 3mg ES-HS per bovine lung could be obtained. Referring to an estimated blood contact surface of $300m^2$ per bovine lung this corresponds to 3% of the theoretical yield. The reasons for this limitation are not fully understood yet. Chelating reactions of the heparan sulphate with calcium-ions was excluded by addition of EDTA to the extraction solution. Enzymatic degradation during the first purification steps could not occur as the lung tissue was frozen, dry or under denaturing conditions since all enzymes are destroyed. Oxidative degradation was excluded when the extraction was carried out under nitrogen atmosphere.

3.2   Expanded Bed Adsorption

Expanded bed adsorption is a single pass operation in which macromolecules are recovered from e.g. crude tissue extract and transferred to process for chromatographic purification without the need for the separate functions of clarification and concentration. Streamline gel and columns are specially designed for this purpose. The gel matrix consists of 6% crosslinked agarose beaded around quartz particles with a specific density of 1,2g/ml. A further advantage of this method generated during the adsorption to the gel can plug the column. Using a Streamline-50 column filled with 500ml DEAE-Streamline 5 liters of still turbid alkaline degraded extract could be purified in a single pass.

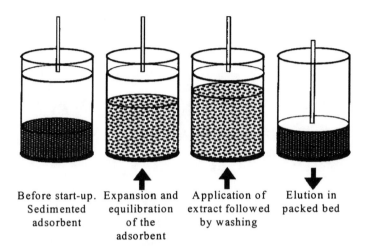

Before start-up. Expansion and Application of   Elution in
Sedimented   equilibration extract followed packed bed
adsorbent   of the   by washing
           adsorbent

Figure 2. The principle of expanded bed adsorption

3.3   Analytical Characterisation of Bovine Lung Endothelial-Cell Heparan-Sulphate

Molecular weight was determined by gel permeation chromatography on a HEMA Bio Linear column which was calibrated with keratan sulphate standards of known molecular weight. ES-HS derived from bovine aortic endothelial cell culture shows a molecular weight of 35kD while the bovine lung derived preparation exhibits a slightly different molecular weight of 30kD.

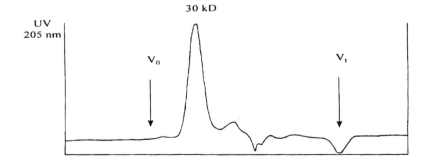

Figure 3. Analytical gel permeation chromatography on HEMA Bio Linear column for molecular weight determination.

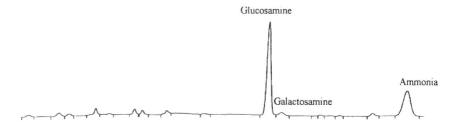

Figure 4. Amino acid and amino sugar analysis of bovine lung ES-HS.

To prove the absence of galactosamine (derived from contaminating chondroitin-sulphate) and amino acids the heparan-sulphate was hydrolysed and subjected to amino acid analysis that resulted in the following elution profile.

Amino acids and amino sugars were detected by a UV/VIS spectrometer at 570nm using the ninhydrin method which forms the intensively coloured Ruhemanns-violet. The above chromatogram shows one major peak which belongs to the glucosamine moiety of the purified heparan-sulphate. Solely neglectable impurifications of galactosamine and amino acids could be detected demonstrating the preparations` high degree of purity.

Apart from the amino acid analysis the absence of characteristic absorbance maxima at 260nm for nucleic acid contamination) and 280nm (for protein contamination) in UV spectra testified the purity of the preparation too.

According to the accuracy of measurement the disaccharide composition was identically for cell culture derived ES-HS and ES-HS from bovine lung.

Table 1. Disaccharide composition of bovine lung endothelial surface heparan sulphate.

Disaccharide	Percentage Content
UA-GlcNAc	54,3%
UA-GlcNH2	
UA-GlcNS	12,5%
UA-GlcNAc(6S)	
UA(2S)-GlcNAc	21,6%
UA-GlcNS(6S)	9,5%
UA(2S)-GlcNS	
UA(2S)-GlcNAc(6S)	
UA(2S)-GlcNS(6S)	2,1%

The elution profile of the AT III affinity chromatography again shows the same result for both preparations. The major amount (more than 90%) was eluted during application of the 0,05M NaCl solution; the rest of the sample was eluted under physiological salt concentration. These results demonstrate in contrast to former result of another working group[24], i.e. that ES-HS does not have any binding sites for AT III[17].

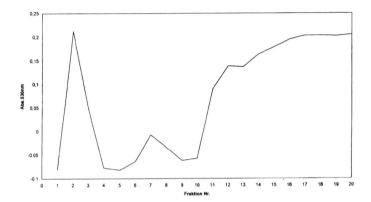

Figure 5. AT III affinity chromatogram of bovine lung ES-HS. The absorption of fraction 11-20 is caused by the increasing salt concentration.

The degree of acetylation was determined from the relation of the integration of the N-acetyl signal of ^{1}H-NMR spectra at 2ppm and sum of the integration of the other signals. Residing signals confirm the identity of this heparan sulphate to ES-HS from bovine aortic endothelial cell culture as a fingerprint comparison.

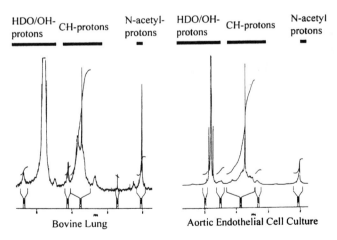

Figure 6. The left of the above shown spectra belongs to the described preparation while the right one is derived from aortic endothelial cell surface heparan sulphate from cell culture. The conformity between the two spectra again indicates the equality of the two substances. The degree of acetylation was determined to 52,6% in comparison to the cell culture derived material for which a degree of acetylation of 56,1% was measured.

3.4   Covalent Immobilisation of Bovine Lung Endothelial-Cell Heparan-Sulphate

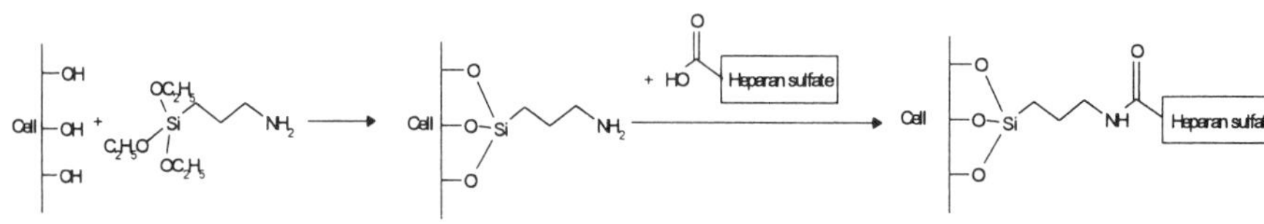

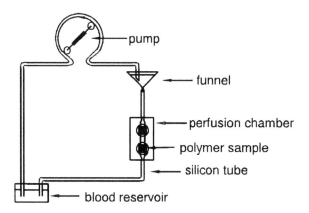

Figure 7. Activation of cellulose membrane and immobilisation of bovine lung ES-HS

Immobilisation of bovine lung endothelial surface heparan sulphate was carried out by forming an amide bond between the glycosaminoglycan uronic acid moiety and the amino groups of partially substituted cellulose membranes. For this purpose the membranes were rinsed with 3-aminopropyltriethoxysilane in ethanol/water resolution in an amino group concentration of $10,2 nmol/cm^2$. Carboxylic groups of the glycosaminoglycan were activated by a water soluble carbodiimide (CME-CDI) and stirred with the partially substituted cellulose membranes leading to a GAG surface concentration of $4,5 pmol/cm^2$.

3.5   Platelet Adhesion Test

Figure 8. Schematic drawing of the blood testing system

For in-vitro testing of haemocompatibility, bovine lung ES-HS was covalently immobilized according to the above described procedure on cellulose membranes. These membranes were perfused with platelet rich plasma in a Baumgartner chamber modified by Sakariassen at high shear stress for 30 minutes, stained and examined for adhering platelets. Exactly like ES-HS from bovine aortic cell culture lung ES-HS showed no platelet adhesion.

4. CONCLUSION

In the present paper a new method for the isolation of bovine lung endothelial cell surface heparan sulphate is described. The aim of this work was to investigate for an alternative source leading to ES-HS as described from bovine aortic cell culture. Because of the excellent properties cell culture ES-HS coating of biomaterials has shown in "in vitro" as well as in "in vivo" experiments, a commercially available source should be found. The solution presented in this work leads to sufficient amounts for technical application of the coating system, especially the isolation procedure bears the potential of a high degree of automation for industrial purposes.

The identity between aortic cell culture and bovine lung heparan sulphate is not fully proven yet. Apart from the determination of the molecular weight all analytical methods presented in this study, especially the in vitro blood testing indicates the high degree of similarity of the two substances.

The slight difference in molecular weight between ES-HS isolated from aortic endothelial cell culture (35kD) and whole lung ES-HS (30kD) is obviously not caused by the isolation procedure but by the difference between aortic endothelial cells and microvascular endothelial cells that mainly occur in lung which is proven by the above mentioned analytical methods. Referring to the in vitro testing of lung ES-HS this difference in molecular weight does not have any influence on the biological potency of the substance. So the isolation procedure described in this study represents a new powerful tool offering a new and economic access to haemocompatible biomaterials.

With the procedure described in the present paper there is an economical and technical approach available to produce sufficient amounts of ES-HS for large scale coating for biomaterials (e.g. implants directly exposed to blood, haemodialyzers, oxygenators, plasma separators, or catheters) to improve haemocompatibility.

In the near future further investigations will be carried out such as testing of cytotoxicity, different methods of sterilisation (steam, γ, EO) and animal experiments to become a better insight to the potential of bovine lung ES-HS. All these investigations already have been carried out successfully for the cell culture derived material and according to the similarity between the two substances we are hopeful to gain the same results as for bovine aortic endothelial surface heparan sulphate.

ACKNOWLEDGMENTS

We thank the Fresenius St. Wendel AG, and the Bundesministerium für Bildung und Forschung for financial support of our project No. 03M1506 3. For the generous gift of the keratan sulphate standards we also thank Prof. Dr. H.W. Stuhlsatz.

REFERENCES

1. Jorpes J. E., Gardell S., 1948, *J. Biol. Chem.* 176: 267.
2. Cifonelli J. A., King J. A., 1977, *Biochemistry* 16: 2137.
3. Taylor R. L., Shiveley J. E., Conrad H. E., Cifonelli J. A., 1973, *Biochemistry* 12: 3633.
4. Lindahl U., Kjellén L., 1987, *Biology of proteoglycans*, Wight T. N., Mecham R. P. (eds.), Academic Press inc., Orlando, 59.
5. Höök M., Kjellén L., Johansson S. and Robinson J., 1984, *Ann. Rev. Biochem.* 53 847.
6. Fransson L.-Å., 1987, *TIBS* 12: 406.
7. Kjellén L. and Lindahl U., 1991, *Annu. Rev. Biochem.* 60: 443.
8. Howell W. A., Holt E., 1918, *Am. J. Physiol.* 47: 328.
9. Hiscock D. R. R., Canfield A. and Gallagher J. T., 1995, *Biochimica et Biophysica Acta* 1244 104.
10. Buonassisi V., 1973, *Experimental Cell Research* 76: 363.
11. Oohira A., Wight T. N. and Bornstein P., 1983, *The Journal of Biological Chemistry* 258: 2014.
12. Saku T. and Furthmayr H., 1989, *The Journal of Biological Chemistry* 264: 3514.
13. David G., 1993, *The FASEB Journal* 7 1023.
14. Keller R., Silbert J. E., Furthmayr H. and Madri J. A., 1987, *American Journal of Pathology* 128: 268.
15. Keller R., Pratt B. M., Furthmayr H. and Madri J. A., 1987, *American Journal of Pathology* 128: 299.
16. Erdtmann M., Keller R. and Baumann H., 1994, *Biomaterials* 15: 1043.
17. Jerg K. R., Emonds M., Müller U., Keller R. and Baumann H., 1991, *Polymer Preprints* 32: 243.
18. Emonds M., Ruzicka E., Muckel G. H., Keller R., Müller U. and Baumann H., 1991, *Clinical Materials* 8: 47.
19. Otto V., Wolff C., Baumann H., Keller R., 1997, in *Werkstoffe für die Medizintechnik*, (Breme J., Ed.), DGM-Informationsgesellschafts-Verlag, Franfurt/M 161.
20. Baumann H., Müller U., Keller R., 1997, *Sem. Thromb. Hemost.* 23: 203.
21. Baumann H., Keller R., 1997, *Sem. Thromb. Hemost.* 23: 215.
22. Chandrasekhar S., Estrman M. A. and Hoffmann H. A., 1987, *Analytical Biochemistry* 161: 103.
23. Sakariassen K. S., Aarts P. A. M. M., De Groot P. G., Houdijk W. P. M. and Sixma J. J., 1983, *J. Lab. Clin. Med.* 102: 522.
24. Marcum J. A., Reilly C. F. and Rosenberg R. D., 1987, in: *Biology of Proteoglycans* Wight T. N. and Mecham R. P.(eds.) Academic Press, Orlan

COMPOSITE MATERIALS AS SCAFFOLDS FOR TISSUE ENGINEERING

Luigi Ambrosio, Paolo Netti, Biagio Santaniello, and Luigi Nicolais
Institute of Composite Materials Technology, CNR, and Interdisciplinary Research Centre in Biomaterials, University of Naples Federico II, Piazzale Tecchio 80, 80125 Naples, Italy

1. INTRODUCTION

Tissue engineering represents a novel frontier of modern biotechnology arising from the integration of knowledge from the fields of cell and molecular biology, materials science and surgical reconstruction to engineer new tissue[1].

The biomaterials used to interface with cells can be both of natural or synthetic origin, or hybrid materials. The design of the optimal scaffold for tissue engineering is one of the major challenges of this field. Indeed, the ideal scaffold must combine several structural and functional properties. From the structural point of view, scaffolds must possess many key characteristics such as high porosity and surface area, structural strength and specific three-dimensional shape. From the functional point of view, the scaffold has to be able to direct the growth of cells migrating from surrounding tissue or of cells seeded within the porous structure of the scaffold. The scaffolds must therefore provide a suitable substrate for cell attachment, migration, proliferation and differentiation[2].

Biodegradable polymeric scaffolds have been extensively used as matrix for the regeneration of skin[3], cartilage, bone[4] and other tissues[5]. Many natural and synthetic biodegradable polymers such as collagen, poly(a-hydroxyesters) and poly(anhydrides) have been used[6]. Porous polymer scaffolds have been prepared by numerous techniques, including solvent casting-salt leaching, thermally induced phase separation, solvent evaporation, fibre forming unwoven structure[7]. These porous scaffolds are designed to provide a suitable environment for cellular growth and remodelling, however there are still several aspects of cellular-material interactions that should be optimized to provide the desirable tissue regeneration.

An interesting feature of the environmental influence on development, maintenance and tissue remodelling is represented by the ability of cells to recognize the stiffness of their environment. Indeed, recent studies demonstrate that cellular forces acting on the adhesive contacts made with the extracellular matrix (ECM) contribute significantly to cell shape, viability, signal transduction and motility[8]. The cytoskeleton generates a level of tension against the ECM that is proportional to ECM stiffness. The strength of this tension exerted against the ECM affects the migratory speed of the cell. Furthermore, not only cells can generate stress but, in turn, stress can affect cell behaviour. The importance of mechanical forces on cell growth and remodelling is particularly evident in tissues such bone, cartilage, blood vessels, tendons, muscles, where the physical stimuli, such as mechanical loading, deformation, pressure, fluid flow, are transduced into electrical or biochemical messages. This process has been defined as mechanotransduction and represents a new interesting area of multidisciplinary research applied to tissue engineering[9].

The optimisation and the modulation of the mechanical properties of the scaffolds and cells interaction may be obtained by designing a 3D scaffolds following the structural organisation of the natural tissue. This would permit to apply mechanical stimuli to the cell trough the scaffold in a similar physiological condition.

All soft tissue is constituted of highly complex structures. The mechanical response of these natural systems are non-linear and it has been shown that this behaviour is related to the elaborate structural hierarchy[10]. The stress-strain curve generally presents a toe region, characterised by a J-shaped dependence, followed by a linear region and then by yielding and finally by a failure region. The normal physiological responses of these natural systems entail the non-linear "toe" and early linear regions of tensile response[10,11]. Previous papers have shown that it is possible to emulate the structure and mechanical properties by using composite materials[11,12].

The fundamental elements used in our approach consist of applying the composite theory in order to design composite biodegradable scaffolds based on the structural organisation and performance of the living tissue.

By changing the components of the composite structures it is possible to obtain a scaffold which is able to control factors such as cell-interaction, controls of the biomechanical characteristics change over time, control of the material degradation, possibility to apply mechanical stimuli in a physiological range.

A composite structure, obtained with helical winding of fibers, is proposed to design scaffolds for tubular tissue. The mechanical response of such a system is correlated with the deformation mechanism that is related to geometrical variations of the composite structure during a loading process.

2.   EXPERIMENTAL

2.1   Preparation of the Composite Scaffolds

Poly(L-lactic acid) (PLLA) and poly(glycolic acid) (PGA) continuous fibers were wound helically by a filament winding machine on polyethylene hoses with a desired outer diameter.

Fibers coated with the matrix solution were carried out simultaneously along with fiber winding. The matrix solution is composed of sieved salt particles dispersed in a Hydrothane®/DMAC (Dimethylacetamide) solution. After winding, the impregnate fibers were removed from the machine and inserted in a glass cylinder containing ethyl alcohol to remove the DMAC. Once the DMAC was removed the composite structure is immersed in water to leach out the salt.

Finally, the polyethylene hose was removed to obtain a hollow 3-D composite scaffold.

Degradation study was performed on both composite scaffolds in Ringer solution at T=37°C and mechanical properties were analysed after different immersion time.

2.2   Tensile Properties

Uniaxial load-strain experiments were made by using a standard Instron testing machine Mod. 4204 at room temperature in distilled water. A strain rate of 0.1 min^{-1} was used for all tests. The tests were performed on swollen sample to measure the degradation rate and the effect of porosity of the matrix on the mechanical properties.

2.3 Morphology

Morphological analysis was performed on fractured surface of composite scaffolds by using a scanning electron microscope (SEM) to evaluate the pore-structure and fiber matrix interface.

3. RESULTS AND DISCUSSION

The composite samples were designed by using the approach reported in previous papers[12,13] to prepare 3D biodegradable scaffolds for soft tissue replacement. Hydrothane® matrix was reinforced with PLLA and PGA fibers to control the mechanical properties and the degradation rate of the composite. To verify the effect of porosity on the mechanical properties of the scaffolds, the tensile properties of the composite materials with porous and no-porous structure were measured on systems with similar winding angle. The results are reported in Figure 1.

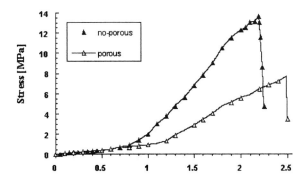

Figure 1. Stress-strain curves of porous and no-porous composites.

By comparing the elastic modulus and stress at break for porous and non porous material, the results showed a decrease of about 50% for the porous structures. The observed decrease is due to both the presence of pores and the effect of fiber-matrix interaction. The mechanical results are presented in Table 1.

Table 1. Mechanical properties of the composite structures

Structure	E(Mpa)	Ob(Mpa)	Eb(mm/mm)
	S.D	S.D	S.D
No-porous	11±0.48	13.10±0.8	2.10±0.15
Porous	5.6±0.6	7.8±2.5	2.55±0.22

Both systems showed the peculiar stress-strain curve of soft tissue[10,11] which include the initial toe region due the properties of the matrix and its interaction with fibers; a linear region and increasing of the stress is due to the alignment of the fibers until fracture.

The morphology analysis performed by SEM on porous structure are presented in Figure 2. It is evident a good distribution of interconnected pores.

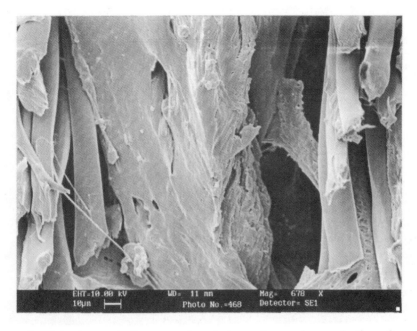

Figure 2. Photomicrograph from scanning electron microscopic examination of porous composite

To modulate the change in mechanical properties over time and to control of the material degradation, the composite scaffolds were prepared by using PLLA and PGA fibers as reinforcement. The degradation of the material in Ringer solution was followed measuring the tensile properties at given time point. The values of the elastic modulus and stress at break of two structures at different immersion time are reported in Figures 3 and 4.

As expected the degradation rate of the composite system containing PGA was faster than the one with PLLA fibers. In fact, after 20 days the PGA presented a 90% reduction of the mechanical properties, while the PLLA reinforced composite were holding 70% their properties after 80 days of bathing. It is worth nothing that the PGA reinforced composites fibers shows a higher initial stress at break, lower elastic modulus and higher degradation rate when compared to PLLA reinforced composite. This suggests that by combining both fibers it would be possible to design a scaffold with controlled degradation and mechanical properties.

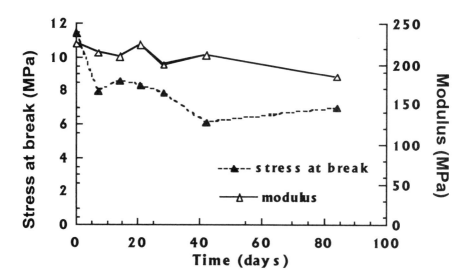

Figure 3. Elastic modulus and stress at break of Hydrothane® /PLLA system as function of immersion time in Ringer solution.

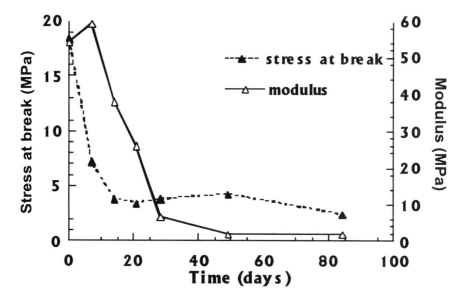

Figure 4. Elastic modulus and stress at break of Hydrothane®/PGA system as function of immersion time in Ringer solution.

4. CONCLUSION

Tissue engineering scaffolds can be suitable prepared with a wide range of structural and mechanical properties by using filament winding technology.

By changing the winding angle, porosity of the matrix and composition of reinforcing fiber it is possible to obtain materials with properties similar to that of natural tissues.

REFERENCES

1. Langer R., Vacanti J.P., 1993, *Science*, 260:920-926.
2. Vacanti J.P., Vacanti C.A., 1997, in *Principles of Tissue Engineering* (Lanza R.P, Langer R., Chick W.L. , eds.), R.G. Landes Company and Academic Press, Austin, pp. 1-5.
3. Andreassi, L., Pianigiani, E., Andreassi, A., Taddeucci, P., and Biagioli, M., 1998, *Int. J. Dermatol.*, 37:595-598.
4. Vacanti C.A., Vacanti JP, 1997, in *Principles of Tissue Engineering* (Lanza R.P, Langer R., Chick W.L., eds.), R.G. Landes Company and Academic Press, Austin, pp. 619-631.
5. Mooney D.J., Sano K., Kaufmann P. M., Majahod, K., Schloo, B., Vacanti J.P., Langer R., 1997, *J. Biomed Mater Res*, 37, 413-420.
6. Pachence J.M., Kohn J., 1997, in *Principles of Tissue Engineering* (Lanza R.P, Langer R., Chick W.L., eds.), R.G. Landes Company and Academic Press, Austin, pp. 273- 286.
7. Thomson R.C., Yaszemsky M.J., Mikos G., 1997, in *Principles of Tissue Engineering* (Lanza R.P, Langer R., Chick W.L. , eds.), R.G. Landes Company and Academic Press, Austin, pp. 263-272.
8. Galbraith CG. Sheetz MP., 1998 *Current Opinion in Cell Biology. 10(5):566-71.*
9. Gooch, K. J., Blunk, T., Vunjak-Novakovic, G., Langer, R., Freed, L., and Tennant, C. J., 1998, in: *Frontiers in Tissue Engineering*, (C. W. Patrick, jr., A. G. Mikos, and L. V. McIntire, eds.) Pergamon, New York, pp. 61-82.
10. Hiltner A., Cassidy J.J., Baer E., 1985, *Ann.Rev.Mater.Sci.* 15, 455- 459.
11. Netti P.A, D'Amore A., Ronca D., Ambrosio L., Nicolais L., 1996, *Journal of Materials Science: Materials in Medicine,* 7, 525-530.
12. Iannace S., Sabatini G., Ambrosio L., Nicolais L, 1996, *Biomaterials*, Vol.16, N.9.
13. Ambrosio L., De Santis R., L. Nicolais, 1997, *Proc Instn Mech Engrs*, 212, H, 93-99

PROTEIN IMMOBILIZATION ONTO NEWLY DEVELOPED POLYURETHANE-MALEAMIDES FOR ENDOTHELIAL CELL GROWTH

[1]Paola Petrini, [2]Livia Visai, [1]Silvia Farè, [1]Chiara Liffredo, [2]Pietro Speziale, and [1]Maria Cristina Tanzi
[1]*Bioengineering Dept., Politecnico di Milano, Piazza L. da Vinci, 32, 20133 Milano, Italy,*
[2]*Biochemistry Dept., Università di Pavia, Via Taramelli 3/B, 27100 Pavia, Italy*

1.   INTRODUCTION

A stable endothelization is a desirable characteristic for blood-contacting implants as a viable layer of endothelial cells would inhibit both bacterial colonisation and thrombus formation.

Materials and design currently used for vascular repair fail in adequately support endothelial cells growth. Therefore, the development of innovative materials able to permanently support cell attachment and growth is of crucial priority.

Endothelization of materials for cardiovascular application, and in particular of polyurethanes, has been recently studied.[1-3]

Surface modification is one of the strategies promoting cell adhesion. In particular, surface immobilisation of protein containing RGD sequences (fibronectin and other protein present in the extracellular matrix) which can interact with integrins of the cellular membrane represents an interesting way to achieve cell adhesion and activation.[4,5]

Grafting of proteins should be carried out in physiological conditions, as the conformational change in their structure may result in lack of the adhesive property of the proteins.

Biomedical Polymers and Polymer Therapeutics
Edited by Chiellini *et al.*, Kluwer Academic/Plenum Publishers, New York, 2001

For these reasons, among the family of segmented polyurethanes currently used in cardiovascular applications for their desirable properties, poly-urethane-maleamides (PUMAs) appear to be of particular interest.[6]

PUMAs are obtained with dicarboxylic acids as chain extenders, instead of traditional diols and diamines. The use of maleic acid as chain extender allows the insertion in the polymer chain of activated double bonds, which can perform as grafting sites for further derivatisation in aqueous solutions and mild conditions.[7]

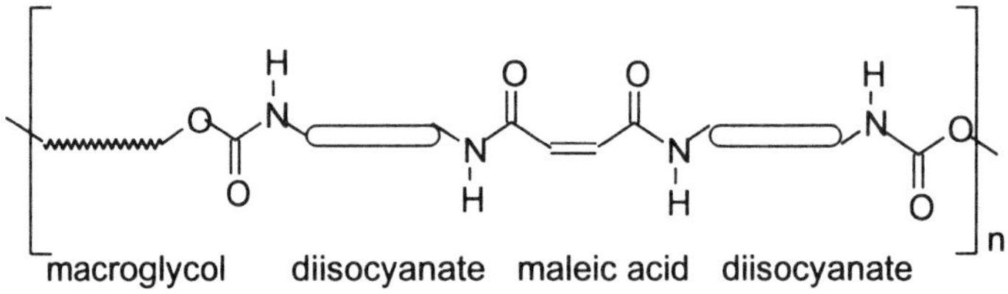

macroglycol   diisocyanate   maleic acid   diisocyanate

Figure 1. Schematic structure of poly-urethane-maleamides (PUMAs)

In this work, the interaction with fibronectin of different poly-urethane-maleamides (PUMAs) was investigated. PUMAs, carrying poly-ether or poly-carbonate soft segments, were synthesised by bulk polymerisation using aliphatic diisocyanates, and maleic acid (MA) as chain extender.[6] Figure 1 gives a schematic view of PUMAs structure, showing that the use of maleic acid as chain extender introduces double bonds, activated by the presence of amide groups in the polyurethane.

2.   MATERIALS AND METHODS

2.1   Materials

PUMAs were obtained by two-step bulk polymerization, using a polycarbonate diol (PHCD) as macroglycol, an aliphatic diisocyanate (HDI), and maleic acid ± butane diol as chain extenders. The reagents were used in the stoichiometric ratio PHCD:diisocyanate:chain extender = 1:2:1. Dibutyl-tin-dilaurate or stannous octanoate were used as catalysts. All of them were purified from residual catalyst, and solvent-cast from DMA solutions.

The obtained PUMAs were characterised by usual analytical techniques (gel permeation chromatography, GPC, differential scanning calorimetry, DSC, intrinsic viscosity analyses, [η], infrared spectroscopy, FT-IR, tensile mechanical tests).[6]

Bionate 80A, lot number 092696 (Polymer Technology Group, Berkeley, CA, USA) was used as reference material.

The chemical composition of the tested polyurethanes is listed in Table 1.

Table 1. Chemical formulation of tested polyurethanes

	soft segment	diisocyanate	chain extender	catalyst, (% w/w$_{soft\ segment}$)
PCU D2HM	PHCD	HDI	MA	DBTDL (0.05)
PCU O2HM	PHCD	HDI	MA	SnOct (0.05)
PCU D2HX	PHCD	HDI	MA+BD	DBTDL (0.05)
Bionate	PHCD	MDI	BD	---

soft segment	PHCD = poly(*1,6*-hexyl-carbonate)diol
diisocyanate	HDI = *1,6*-hexamethylendiisocyanate
	MDI = *4,4'*-methylene-bis-phenyldiisocyanate
chain extenders	MA = maleic acid, BD = *1,4*-butanediol
catalysts	DBTDL = dibutyl-tin-dilaurate
	SnOct = stannous octanoate

Disks (∅ 1.5 cm for cell growth tests or ∅ 0.5 cm for ELISA and ^{125}I-HFn tests) were cut from the sheets, extracted in ethanol (6h, r.t.) and then incubated overnight at 37°C in Phosphate Buffered Solution (PBS, pH=7.4).

2.2 Cytocompatibility tests

Cytocompatibility was checked with human skin fibroblasts (HSFB). HSFB were grown in Dulbecco's modified Eagle's medium with 10% foetal calf serum at 37°C in 5% CO_2 atmosphere.

For cytotoxicity tests, materials were extracted in PBS during 16 hours. The PBS extracts were then added to the culture medium (1:3). After 3÷72 hours, adherent cells were trypsin-digested, stained and counted by a Burker chamber.

For direct contact tests, HSFB were seeded onto polyurethane disks. After 3÷72 hours incubation, cell survival was evaluated as described for cytotoxicity tests.

2.3 HFn binding tests

For protein adhesion experiments, human fibronectin (HFn) was purified from human plasma as previously reported[8] and stored at –20°C until use. The purity of the obtained protein was checked by electrophoresis and immunoblotting.

2.3.1   Determination of adsorbed HFn by ELISA

Films of the materials were incubated 1 hour at 37°C with HFn (10 μg) in 500 μl of buffered sodium bicarbonate buffer ($Na_2CO_3/NaHCO_3$ 50 mM, pH 9.5), and accurately washed to remove the not stably bonded HFn. Additional protein binding sites were blocked by incubation with PBS containing 2% bovine serum albumin. After extensive washing the attachment of HFn to PUMAs was detected by incubation with goat antirabbit-HRP conjugated antibody. Absorbance at 490 nm (A_{490}) was evaluated. The test was made in triplicate.

2.3.2   Determination of adsorbed HFn by ^{125}I labelling

HFn was ^{125}I labelled with IODO-BEADS iodination reagent as recommended by the manufacturer (Pierce, Rockford, III). The activity of the radiolabelled fibronectin (^{125}I-HFn) resulted 3×10^4 cpm/μg.

Films of the materials were incubated 24 hours at 37° ± 1°C with ^{125}I-HFn in buffered sodium bicarbonate solution (3.3 μg of ^{125}I-HFn in 500 μl for each sample) and accurately washed to remove the not stably bonded ^{125}I-HFn.

The radioactivity associated with the incubated materials was determined in a γ-counter (CobraTM II Auto-Gamma). The test was made in triplicate.

2.3.3   Cell adhesion and proliferation onto HFn coated PUMAs

PU samples were incubated with 150 μg of HFn in 500 μl of carbonate buffer each, or in 500 μl of carbonate buffer only.

HSFB were seeded onto HFn coated- and not HFn coated PU disks. Cells adhesion was carried out in Dulbecco's modified Eagle's medium, without the addition of serum.

After 3 hours incubation, adherent cells were trypsin-digested, stained by trypan blue and counted by Burker chamber.

2.4   Scanning Electron Microscopy (SEM)

For SEM analyses, the samples were fixed with a 2.5% glutaraldehyde solution in 0.1M cacodilate buffer (pH 7.2), dehydrated through a graded series of ethanol (up to 100%), and critical point-dried using CO_2. Specimens, mounted on aluminium stubs, were sputter-coated with gold, and examined with a Leica Cambridge Stereoscan 440 microscope at 3-7 kV acceleration voltage.

3.   RESULTS AND DISCUSSION

PUMAs were extracted in ethanol, in order to remove lower molecular weight substances. After this treatment, PUMAs appear to promote cell adhesion and proliferation, in particular PCU D2HM (Table 2). The only exception is represented by PCU O2HM which inhibit cell proliferation. This result may be explained with a possible slow release of low molecular weigh substances, not extracted by ethanol, from the bulk of the material according to the high polidispersity value (d = 4.5)[6] shown by this polyurethane-maleamide.

Table 2. Cytocompatibility of PUs: direct contact tests. Data are reported as % vs control, tissue culture polystyrene.

	Adhesion	Proliferation	
	3 hours	*24 hours*	*72 hours*
PCU D2HM	77.5±5.4	88.7±3.3	95.7±1.4
PCU O2HM	73.9±0.5	19.3±1.0	16.5±1.2
PCU D2HX	58.3±3.4	93.9±9.3	65.3±6.0
Bionate	50.1±3.0	106.0±4.0	91.0±1.4

The results of fibronectin binding onto polyurethanes, as evaluated by ^{125}I-labelled fibronectin and immunoenzymatic assay (ELISA), are reported in Figure 2 and 3 respectively, and in Table 3.

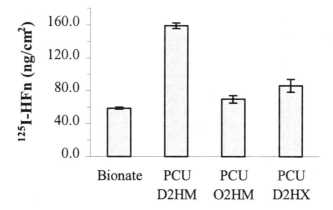

Figure 2 ^{125}I-HFn immobilisation onto polyurethanes (16 hours incubation in carbonate buffer at 37°C).

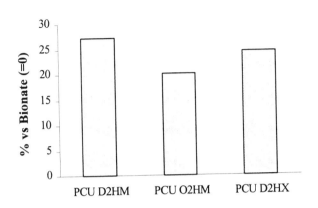

Figure 3. HFn immobilisation onto polyurethanes (16 hours in carbonate buffer at 37°C). Data reported are calculated from ELISA assay as: $(A_{490\text{-}PUMA} - A_{490\text{-}Bionate})/A_{490\text{-}Bionate} \times 100$

ELISA and ^{125}I-labelling of proteins have been seen to give results in case of fibrinogen adsorption from plasma onto glass and different copolymers.[9]

In the present work, both ELISA and ^{125}I-fibronectin adsorption tests are in agreement with the finding that PUMAs are able to absorb higher quantities of fibronectin than Bionate (PCU D2HM > PCU D2HX > PCU O2HM > Bionate). However, the binding of ^{125}I-HFn onto PCU D2HM appears to bind HFn in a higher extent than other materials, while this appears higher than that detected by ELISA. This might be explained with conformational changes of adsorbed protein, and the consequent change of epitopes facing the antibody, in ELISA assay.

Table 3. HFn binding determination by ELISA assay: effect of the incubation time. Absorbance data at 490 nm are reported.

	HFn incubation time	
	2 hours	*16 hours*
PCU D2HM	0.506 ± 0.103	1.320 ± 0.250
PCU O2HM	0.581 ± 0.281	1.229 ± 0.218
PCU D2HX	0.713 ± 0.121	1.292 ± 0.023
Bionate	0.610 ± 0.004	1.038 ± 0.024

Fibronectin binding onto PUMAs appears to be time dependent. After 2 hours of incubation, PUMAs absorb amounts of HFn comparable with Bionate (Table 3), while a 16 hours incubation promoted higher HFn-adsorption for PUMAs. This suggests that the functional groups present in the polyurethane-maleamide backbone efficiently act to promote HFn adsorption onto the materials, with a slow kinetic profile.

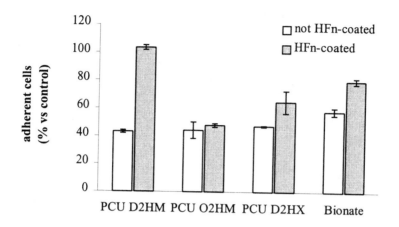

Figure 4. HSFB adhesion (3 hours) in serum free culture medium onto materials, before and after HFn-coating.

Figure 4 reports cell adhesion onto HFn-coated and uncoated PUs. In all the considered materials, cell adhesion increases after HFn coating. For PCU D2HM this effect is more evident than for the other materials indicating that the supposed conformational changes, affecting the ELISA assay, do not modify cell adhesive properties of immobilised HFn.

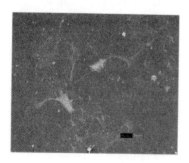
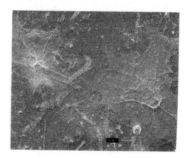

Bionate, HFn-coated   PCU D2HM, HFn-coated

Figure 5. SEM images of cells adherent onto HFn-coated Bionate and HFn-coated PCU D2HM

SEM analyses, performed to investigate morphology of adherent cells showed substantial differences between HFn-coated and not HFn-coated materials. HFn-coated samples presented well spread, interconnecting cells (Fig.5), while cells appeared mainly in the round phenotype onto not-coated materials.

From these results, it is possible to conclude that HFn immobilised onto polyurethanes retains its biological activity, promoting cell attachment and activation.

4. CONCLUSION

On most PUMAs HFn adsorption is significantly higher than onto the biomedical grade polyurethane Bionate, indicating that functional groups present in the polyurethane maleamide backbone play a role in the protein binding.

Moreover, HFn immobilisation onto PUMAs do not interfere with the cell adhesive properties of the protein.

In particular the polyurethane maleamide named PCU D2HM showed interesting properties: a high HFn binding associated to good cytocompatibility results.

To better understand the interaction of proteins with PUMAs other experiments are in progress. In particular, the kinetics of the protein binding onto PUMAs and its stability in physiological conditions, the nature of fibronectin-PUMAs bonds, and competition with other proteins present in the culture medium in now under extensive investigation.

ACKNOWLEDGMENTS

Supported by Istituto Superiore di Sanità (Italy) under the research project "Functional replacements, artificial organs and organ transplantation" (1998-1999)

REFERENCES

1. Kawamoto Y., Nakao A., Ito Y., Wada N., Kaibara M., 1997, *J. Mat. Sci.: Mat. in Med.*, 8(7); 551-557.
2. Kim J.H., Kim J., Ryu G.H., Lee K.B., Chang J.K., Min B.G., 1995, *Trans. The 21st Annual Meeting of the Society for Biomaterials*, March 18-22, San Francisco, California, USA, 409
3. Lee Y-S, Park D.K., Kim Y.B., Seo J.W., Lee K.B., Min B.-G., 1993, *ASAIO J.*, 39(3), M740-M745
4. Holland J., Hersh L., Bryhan M., Onyiriuka E., Ziegler L., 1996, *Biomaterials*, 17(22), 2147-56.
5. Lin H.B., Sun W., Mosher D.F., Garcia-Echeverria C., Shaufelberger K., Lelkes P.L., Cooper S.L., 1994, *J. Biomed. Mat. Res.*, 28, 329-342.
6. P. Petrini, S. De Ponti, S. Farè, M.C. Tanzi, 1999, *J. Mat. Sci.: Mat. in Med.*, 10(12), 711-714
7. M.C. Tanzi, B. Barzaghi, R. Anouchinsky, S. Bilenkis, A. Penhasi, D. Cohn, 1992, *Biomaterials*, 13(7).
8. M. Vuento, A. Vaheri, 1979, *Biochem. J*, 183, 331-337
9. S.M. Slack, S.E. Posso, T.A. Horbett, 1991, *J. Biomater. Sci. Polym. Edn.*, 3(1), 49-67.

FUNCTIONALIZED POLYMERS OF MALIC ACID STIMULATE TISSUE REPAIR PRESUMABLY BY REGULATING HEPARIN GROWTH FACTORS BIOAVAILABILITY

[1]Viviane Jeanbat-Mimaud, [2]Christel Barbaud, [1]Jean-Pierre Caruelle, [1]Denis Barritault, [2]Valérie Langlois, [2]Sandrine Cammas-Marion, and [2]Philippe Guérin

[1]Laboratoire de Recherche sur la Croissance Cellulaire, la Réparation et la Régénération Tissulaire, UPRESA 1813 CNRS-Université Paris XII Val de Marne, Avenue du Général de Gaulle, F-94010 Créteil, France; [2]Laboratoire de Recherche sur les Polymères, UMR C7581 CNRS - Université Paris XII Val de Marne, 2-8 rue Henri Dunant, F-94320 Thiais, France

1.   INTRODUCTION

Heparin sulfates (HS) play a key role in regulating Heparin Binding Growth Factors (HBGF) bioavailability[1]. Fibroblast Growth Factors (FGFs) represent the paradigm of HBGF and numerous studies have described HS as the natural site for the storage and the protection of FGFs in the cellular environment. *In vivo*, HS can indeed protect FGFs against proteolytic degradation, potentiate FGFs binding to its high affinity receptor and its ability to stimulate cell proliferation[2]. Similar functions are attributed to HS which are believed to play a key role in the control of HBGF bioavailability. We have recently developed functionalized dextran based biopolymers which could mimic HS functions towards HBGF, and we have shown, these compounds named RGTAs (for ReGeneraTing Agents), could stimulate tissue repair in various *in vivo* models[3]. The development of artificial biopolymers with wound healing properties presents therefore therapeutic interests[4,5].

Biomedical Polymers and Polymer Therapeutics
Edited by Chiellini *et al.*, Kluwer Academic/Plenum Publishers, New York, 2001

We now describe the synthesis of degradable polyesters with functional pendent groups selected in the goal to modulate the nature and the proportion of potential interacting sites. The possibility for building such tailor made macromolecules with a strictly controlled architecture has been already demonstrated with poly(β-malic acid). In the present case, three different malolactonic acid esters have been prepared and copolymerized in different proportions[6,7,8,9]. Three interacting sites present in natural HS related molecules have been therefore introduced directly or by chemical modification in the synthetic macromolecules (PMLA) (hydrophobic, sulfonate and carboxylate groups). More interesting, these polymeric esters, which *in vitro* protected and stabilized FGFs, could *in vivo*, stimulate bone repair or muscle regeneration. The repeatability of the PMLA synthesis and of the biological effects have been studied.

2.   SYNTHESIS OF TERPOLYMER

A water soluble and hydrolyzable polyester derived from malic acid has been synthesized by copolymerization of three different malolactonic acid esters. Functional pendent group have been selected to interact with and protect Heparin Binding Growth Factors (HBGF). In order to mimic pendant groups of CMDBS (carboxymethyl benzylamide sulfonate dextran) which interact with HBGF, the selected degradable polymer was composed of malic acid repeating units bearing three types of groups: carboxylate (65 %), sulfonate (25 %) and butyl (10 %).

Three β-substituted β-lactones have been synthesized according to the now well established synthesis route using aspartic acid as chiral precursor[6,7,8]. Three alcohols (benzyl, allyl and butyl) were used for the formation of the ester pendent group. Each malolactonic acid ester has been characterized by ^{1}H and ^{13}C NMR spectra and IR. The synthesis of terpolymer P1 has proceeded through anionic ring-opening copolymerization with benzoate tetraethylammonium as initiator[9] (Scheme 1).

The presence of three repeating units in terpolymer P1, in the same proportion as in the monomer feed, has been demonstrated by ^{1}H NMR spectrum. DSC analysis has given only one glass transition temperature (Tg) at 17 °C. Molecular weight (Mw) and polydispersity index (Ip) were determined by size exclusion chromatography (Mw = 6000, Ip = 1.4; SEC in THF, standards : polystyrene).

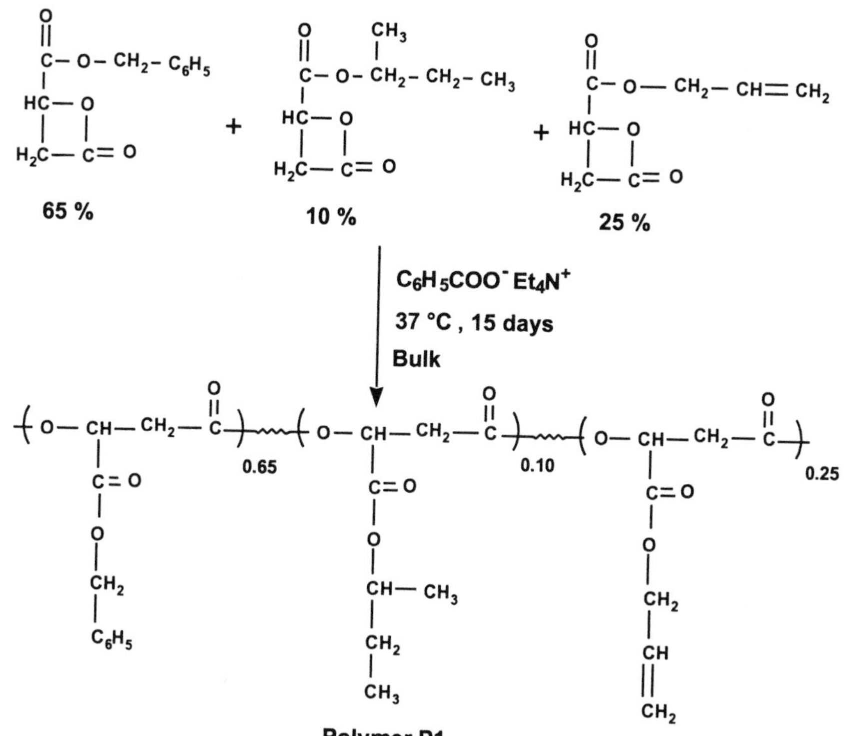

Scheme *1*. Synthesis of the terpolymer P1 via anionic ring-opening copolymerization of the three malolactonic acid esters.

Terpolymer P1 has been then subjected to three consecutive different chemical modifications (Scheme 2). The first modification has consisted in the epoxidation of the allyl function by metachloroperbenzoic acid (mCPBA). The resulting terpolymer P2 has been analyzed by ^{13}C NMR. The corresponding spectrum has shown the complete modification by the disappearance of double bonds carbon atom and appearance of carbon atoms corresponding to epoxy group. No molecular weight modification was observed (Mw = 6300, SEC in THF with polystyrene standards). The second chemical modification has consisted in the deprotection of the carboxy groups. Protecting benzyl groups were removed by catalytic hydrogenolysis according to the usual technique in the presence of Pd/C and hydrogen[10]. Quantitative reaction has led to terpolymer P3 and ^{1}H NMR spectrum has shown the total disappearance of the peaks corresponding to benzyl groups. The last reaction has consisted in polymer sulfonation by opening epoxy groups. Final terpolymer P4 was characterized by ^{1}H and ^{13}C NMR spectra. Peaks corresponding to epoxy groups have completely disappeared and the presence of sulfonate groups was displayed by elementary chemical analysis ($S = 8.7\%$).

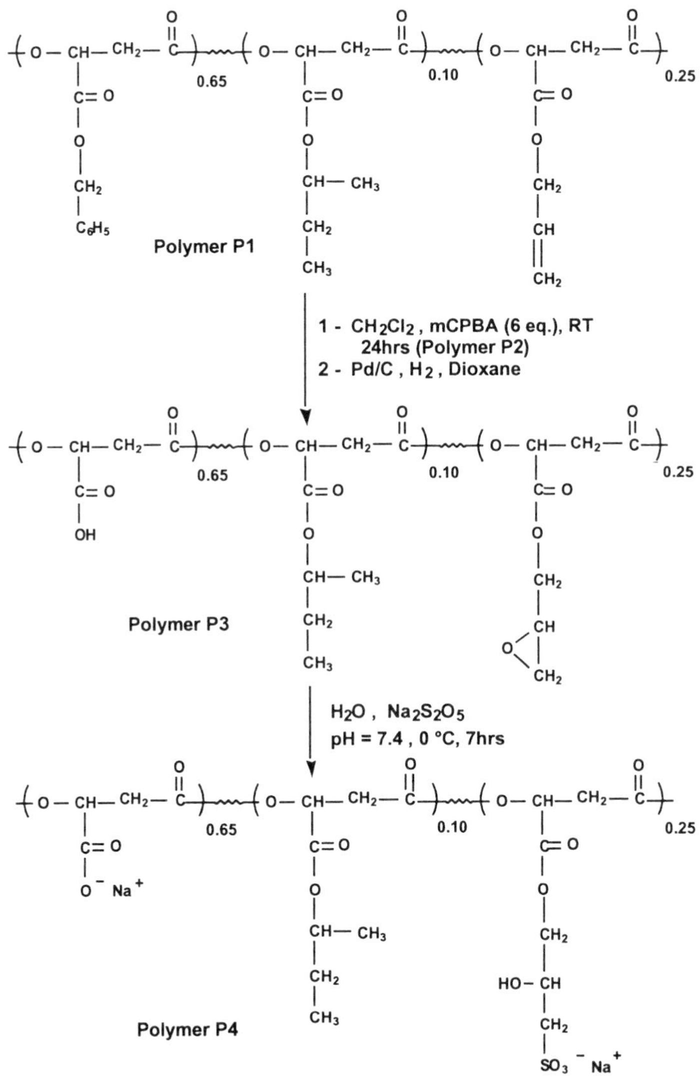

Scheme 2. Chemical modifications of terpolymer P1 leading to final terpolymer P4.

The different preparations have been leading to polymer samples with analogous characteristics. A biological evaluation of terpolymer P4 has been carried out. This polymer has been tested on bone repair and muscle regeneration.

3. BIOLOGICAL EVALUATION ON BONE REPAIR

The power of wound healing of the terpolymer P4 of malic acid derivatives was tested on a bone repair in a skull defect model in rats. Skull defects, when large enough, do not spontaneously heal in adult rats. This model has been extensively used to evaluate the bone healing potential of different products including CMDBS (Carboxymethyl benzylamide sulfonate dextrans) heparin-like polymers and HBGFs (Heparin Binding Growth Factors) such as FGFs, TGF-β[11,12,13]. As in the case of CMDBS, we have postulated that heparin-like mimicking polymers of malic acid derivatives could stimulate bone repair by interacting with the endogeneous HBGFs released in the wound site and subsequently enhance healing processes.

After a standardized trephin defect, each defect was filled with a collagen plaster soaked at 40 °C overnight in a saline buffer solution of 100 µg/ml of terpolymer P4 (A) and in a saline buffer alone (B) (Fig.1).

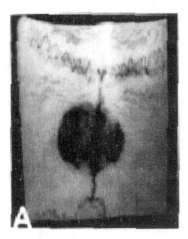

Figure 1. Effects of terpolymer P4 on new bone formation in a skull defect model. A: specimen treated with a 100 µg/ml of terpolymer P4 solution in saline buffer; B: untreated specimen

After 35 days, the radiographs show a newly formed bone (Fig.1A) which was deposited at full length of original mid sagittal sinus and some calcified new bone modules were connected to the margins of the defects[13].

As represented in Figure 2, the mean percentage of bone defect closure in collagen control group (21.7 %) was significantly inferior to values established in treated groups. Either in CMDBS-treated defects (59.78 %, p<0.001) or P4 malic acid derivatives treated (49.87 %, p<0.019) a significant bone closure was always observed. The bone closure was at least two times higher than in the collagen group. Therefore, the polymer of malic acid derivatives increases wound healing in magnitude comparable to the effects established with CMDBS.

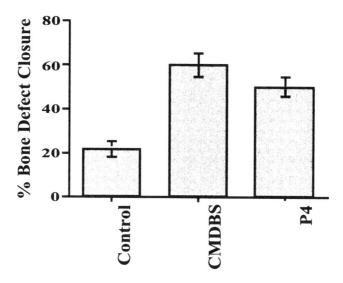

Figure 2. Radiomorphometric results of mean of bone defect closure percentages at 35 days for the three different experimental groups. Statistical analysis was performed using Anova test.

4.   BIOLOGICAL EVALUATION ON MUSCLE REGENERATION

By analogy with the results obtained with CMDBS (Carboxymethyl benzylamide sulfonate dextrans), it is interesting to know the effects of P4 terpolymer of malic acid derivatives in the regeneration model of EDL (*Extensor digitorum longus*) rat muscle[14,15].

We now present the results obtained after a single injection of terpolymer P4 (100 μg/ml) in a regenerating crushed EDL muscle. After 8 days, treated muscle and control muscle were taken and a histological analysis was realised on each muscle. Results presented in Figure 3c and 3e display a higher density of regenerated fibers than results presented in Figure 3d and 3f, demonstrating that terpolymer P4 allowed a more rapid maturation. In non-treated muscle (Fig.3d and f), fibrosis (in black & white) is in development. Histological analysis indicated that regeneration was comparable to that observed after 3 weeks without terpolymer P4.

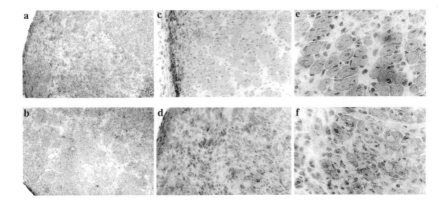

Figure 3. Effects of terpolymer P4 on regeneration of EDL rat muscles after crushing and nerve cutting. a, c, e : treated with terpolymer P4 (× 10, 20, 40 respectively); b, d, f : non-treated (× 10, 20, 40 respectively).

Hence *in vivo* , terpolymer P4 could act by protecting the heparin binding growth factors involved in the natural process of muscle regeneration and therefore favour their actions. This family of polymers should offer a new pharmaceutical potential for treating muscle atrophy and destruction and is worth studying further.

5.   STUDY OF TOXICITY OF TERPOLYMER

In vitro toxicity of final polymer P4 was investigated using Hep G2 cells. Results of the different experiments are collected in Figure 4. We can conclude that, in our conditions, this polymer does not present any effect on cell viability of Hep G2 cells within this range of concentrations (0.01-0.25 g/L).

6.   CONCLUSION

Heparan sulfates are known to increase Heparin Binding Growth Factors bioavailability and in particular growth factors such as FGFs. In order to mimic heparan sulfates, the selected degradable macromolecular chain was composed of malic acid repeating units bearing three types of group able to express the required contacts: carboxylate, sulfonate and butyl. The choice for using polymers of malic acid derivatives lies in i) non toxicity of these polymer series[16], ii) well described hydrolysis[17,18] and iii) possibility of chemical modification for accessing to complex architectures with well defined structures[19]. The use of this family of polymers is therefore very important in the therapeutic domain due to its wide versatility. With the terpolymer P4, we have observed a very good biological activity in two tissue regeneration models (muscles and bone).

Terpolymer P4

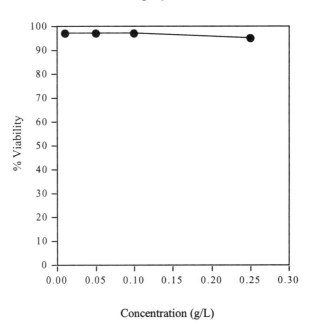

Figure 4. HepG$_2$ cells viability versus terpolymer concentration

Further *in vivo* and *in vitro* researches are under investigation and analysis of different interactions between PMLA derivatives and FGF2 is under development in order to optimize the biological activity.

ACKNOWLEDGMENTS

The authors wish to thank Pr J.L. Saffar, Pr. J. Gautron and Dr A. Meddahi for their hepfull assistance. This work was supported by the Ministère de l'Education Nationale et de la Recherche and the CNRS.

REFERENCES

1. Gallagher T., Turnbull J.E., 1992, *Glycobiology*, 2 (6): 523-528.
2. Clark R. A. F., 1996, *The Molecular and Cellular Biology of Wound Repair*, 2nd Ed., Plenum Press, New York.
3. Blanquaert F., Saffar J.L., Colombier M.L., Carpentier G., Barritault D., Caruelle J.P., 1995, *Bone*, 17 (6): 499-506.
4. Boutault K., Cammas S., Huet F., Guérin Ph., 1995, *Macromolecules*, 28: 3516-3520.
5. Cammas S., Bear M.M., Moine L., Escalup R., Ponchel G., Kataoka K.,Guérin Ph., *Int. J. Biol. Macrom.*, in press.
6. Leboucher-Durand M.A., Langlois V., Guérin Ph.,1996, *Reactive and Functional Polymers*, 31: 57-65.
7. Ramiandrasoa P., Guérin Ph., Bascou P., Hammouda A., Cammas S., Vert M., 1993, *Polym. Bull.*, 30: 501-508.
8. Guérin Ph., Vert M., 1987, *Polym. Commun.*, 28: 11-14.
9. Guérin Ph., Francillette J., Braud C., Vert M., 1986, *Makromol. Chem., Macromol. Symp*, 6: 305-314.
10. Cammas S., Renard I., Girault J.P., Guérin Ph., 1994, *Polym. Bull.*, 33: 149-158.
11. Lafont J., Baroukh B., Berdal A., Colombier M.L., Barritault D., Caruelle J.P., Saffar J.L., 1998, *Growth Factors*, 16: 23-38.
12. Blanquaert F., Barritault D., Caruelle J.P., 1999, *J. Biomed. Mater. Res.*, 00: 63-72.
13. Joyce M.E., Jingushi S., Scully S.P., Bolander M.E., 1991, *Preg. Clin. Biol. Res.*, 365: 391-416.
14. Gautron J., Kedzia C., Husmann I., Barritault D., 1995, *C.R. Acad. Sci. Paris, Sciences de la vie / Life Sciences*, 318: 671-676.
15. Aamiri A., Mobarek A., Carpentier G., Barritault D., Gautron J., 1995, *C.R. Acad. Sci. Paris, Sciences de la vie / Life Sciences*, 318:1037-1043.
16. Renard I., Cammas S., Langlois V., Bourbouze R., Guérin Ph., 1996, *A.C.S. Polymer Preprints* , 37 (2): 137-138.
17. Guérin Ph., Braud C., Vert M., 1985, *Polym. Bull.*, 14: 187-192.
18. Braud C., Caron A., Francillette J., Guérin Ph., Vert M., 1988, *A.C.S. Polymers Preprints*, 29 (1): 600-601.
19. Leboucher-Durand M.A., Langlois V., Guérin Ph, 1996, *Polym. Bull.*, 36: 35-41.

TEMPERATURE-CONTROL OF INTERACTIONS OF THERMOSENSITIVE POLYMER MODIFIED LIPOSOMES WITH MODEL MEMBRANES AND CELLS

Kenji Kono, Akiko Henmi, Ryoichi Nakai, and Toru Takagishi
Department of Applied Materials Science, Osaka Prefecture University, Sakai, Osaka, Japan

1.   INTRODUCTION

Recently, temperature-sensitization of liposomes has been attempted using thermosensitive polymers.[1-4] For example, poly(N-isopropyl-acrylamide) [poly(NIPAM)], which is a well known thermosensitive polymer, exhibits a lower critical solution temperature (LCST) at 32-35°C in aqueous solutions.[5,6] The polymer is highly hydrophilic and soluble in water below its LCST. However, it becomes hydrophobic and water insoluble above this temperature.[7] Therefore, if this polymer is conjugated to a liposome surface, the liposome is stabilized by the hydrated polymer chains below its LCST, whereas destabilization of the liposome is induced by the interaction between the liposome membrane and the hydrophobic polymer chains above the LCST. We have shown that release of the contents from liposomes coated with copolymers of N-isopropylacrylamide (NIPAM) is enhanced above the LCST of the polymers.[1,8-11]

In addition, the modification with thermosensitive polymers can provide liposomes with a temperature-sensitive surface property: the liposome surface is covered with a layer of the hydrated polymer chains below the LCST. However, the dehydrated polymer chains shrink efficiently and are adsorbed on the liposome surface above this temperature.[10,12]

Biomedical Polymers and Polymer Therapeutics
Edited by Chiellini *et al.*, Kluwer Academic/Plenum Publishers, New York, 2001

Liposome-cell interaction is known to change, depending on nature of liposome surface. For example, liposome binding to cells and liposome endocytosis by cells are influenced by the structure of head groups of lipids and surface charge density of liposomes.[13,14] Also, attachment of hydrophilic polymer, such as polyethylene glycol, to liposome surface reduces liposome uptake by cells.[15,16] Since the surface property of thermosensitive polymer-modified liposomes changes, depending on temperature, their interaction with cells is expected to be controlled by ambient temperature (Fig.1).

The object of this study is to examine temperature-induced control of affinity of liposomes to cell surfaces using thermosensitive polymers. In this study, we prepared various liposomes consisting of egg yolk phosphatidylcholine (EYPC), dioleoylphosphatidylethanolamine (DOPE), and 3β-[N-(N',N'-dimethylaminoethane)carbamoyl]cholesterol (DC-Chol), which is a cationic lipid, and modified these liposomes with thermosensitive polymers, homopolymer and copolymers of N-acryloylpyrrolidine (APr), having anchors to liposome membranes. Their interactions with model membranes and cells were investigated. Suppression and enhancement of these interactions, depending on temperature, are described.

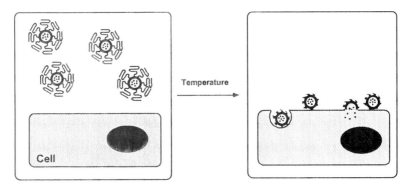

Figure 1. Schematic illustration of temperature-controlled interaction of thermosensitive polymer-modified liposomes with a cell. The interaction is suppressed by highly hydrated polymer chains below the LCST (left), but enhanced by dehydrated polymer chains and/or exposed bare liposome surface (right).

2.   POLYMER SYNTHESIS

Three kinds of polymers having anchors, copoly(APr-NIPAM)-2C$_{12}$, poly(APr)-2C$_{12}$, and a copolymer of APr and N,N-didodecylacrylamide [copoly(APr-NDDAM)] were prepared (Fig.2). Copoly(APr-NIPAM)-2C$_{12}$ was synthesized as described elswhere[9]: APr (27.2 mmol), NIPAM (6.8 mmol), 2-aminoethanethiol (1.87 mmol), and AIBN (0.36 mmol) were dissolved in N,N-dimethylformamide (15 ml) and heated at 75°C for 15 h in N_2 atmosphere. The copolymer was purified using a Sephadex LH-20 column. The amino group at the terminal of the copolymer (0.1 g) was reacted with N,N-didodecylsuccinamic acid (6×10^{-4} mol), which was prepared according to the method of Okahata et al.,[17] by using 1-ethyl-3-(3-dimethylaminopropyl)carbodiimide (6×10^{-4} mol). The copolymer was purified using a LH-20 column. The copolymer was shown to have the APr/NIPAM molar ratio of 81.6/18.4 by ^{1}H-NMR. The weight and the number average molecular weights of the copolymer were 9200 and 4100, respectively. The LCST of the copolymer was 40°C.

For the synthesis of poly(APr)-2C$_{12}$, APr (34 mmol), 2-aminoethanethiol (1.87 mmol) and AIBN (0.36 mmol) were dissolved in N,N-dimethylformamide (15 ml) and heated at 75°C for 15 h in N_2 atmosphere. The polymer was reacted with N,N-didodecylsuccinamic acid and purified as described above. The weight and the number average molecular weights of the polymer were 11500 and 5500, respectively. The LCST of the polymer was 51°C.

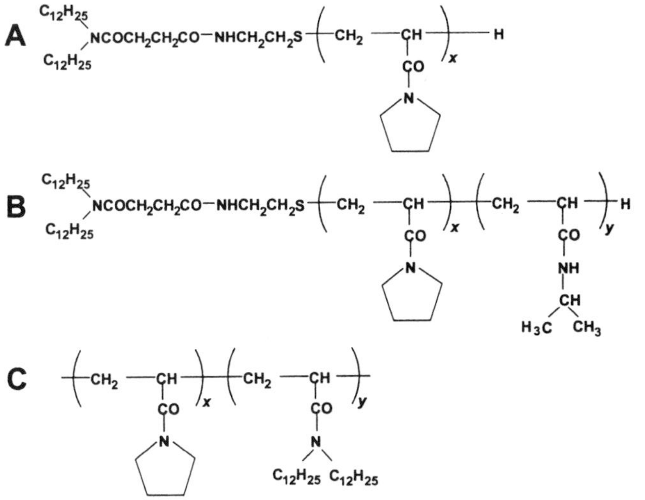

Figure 2. Structures of poly(APr)-2C$_{12}$ (A), copoly(APr-NIPAM)-2C$_{12}$ (B), and copoly (APr-NDDAM) (C).

Copoly(APr-NDDAM) was obtained according to the method previously reported.[10] APr (44 mmol), NDDAM (0.66 mmol), and AIBN (0.22 mmol) were dissolved in dioxane (88 ml) and the solution was heated at 60°C for 15 h in N_2 atmosphere. The Copolymer was recovered by precipitation with diethylether. The copolymer had the APr/NIPAM molar ratio of 97.9/2.1.

The weight and the number average molecular weights of the copolymer were 13800 and 4800, respectively. The LCST of the copolymer was 35°C.

3.   PREPARATION OF POLYMER-MODIFIED LIPOSOMES

Fixation of thermosensitive polymer to liposome membrane can be achieved either by incubation of liposomes with the polymer in an aqueous solution or by preparing liposomes from a mixture of lipid and the polymer.[11] In this study, polymer-modified liposomes were obtained by the latter method as described elswhere.[9] A thin membrane consisting of lipid (10 mg) and polymer (10-15 mg) was prepared by evaporating their solution in chloroform. The membrane was further dried under vacuum overnight and dispersed in 1.5 ml of 10 mM Tris-HCl and 140 mM NaCl solution (pH7.4) using a bath-type sonicator. For the preparation of liposomes encapsulating methotrexate (MTX), 1.5 ml of 10 mM Tris-HCl and 140 mM NaCl solution containing 1.8 mM MTX (pH7.4) was added to the lipid-polymer membrane. The liposome dispersion was extruded through a polycarbonate membrane with a pore size of 100 nm at 0°C. Free polymer and free MTX were removed by gel permeation chromatography on a Sepharose 4B column at 4°C. Liposomes were also prepared by reverse phase evaporation.[18]

The amount of polymer bound to liposome was evaluated via the method reported previously.[12] Generally, 46-65% of the amount of polymer added was fixed to the liposome membrane.

4.   INTERACTION OF THERMOSENSITIVE POLYMER-MODIFIED CATIONIC LIPOSOMES WITH MODEL MEMBRANES

Cationic liposomes can interact strongly with cell surface and deliver DNA into cells. Thus they are widely used as vehicles for gene transfer into a variety of cells. Their ability is based on nonspecific electrostatic interactions with negatively charged cell membranes. Therefore, if cationic liposomes are coated with thermosensitive polymers, the liposomes will be covered with a highly hydrated layer of the polymer chains, which may suppress their interaction with cell membranes below the LCST. However, the positively charged surface of the liposomes will be exposed by shrinkage of the polymer chains above this temperature, resulting in a strong interaction with cells via electrostatic interactions. Therefore, their affinity to cell membranes is expected to be controlled by ambient temperature. Here, the effect of temperature on interaction between thermosensitive polymer-modified DC-Chol/DOPE (1:1, mol/mol) liposomes and negatively charged liposomes consisting of EYPC and phosphatidic acid (PA) (3:1, mol/mol) is described.

4.1　Zeta Potential

Figure 3 represents zeta potentials of various cationic liposomes at varying temperatures. The unmodified cationic liposome showed roughly a constant value, ca. 12 mV, in the range of 20-60°C. The poly(APr)-2C$_{12}$-modified cationic liposome exhibited much lower zeta potential below 50°C. However, a sudden increase in zeta potential is seen between 50 and 55°C. Because the LCST of poly(APr) is ca. 51°C, it is likely that highly hydrated polymer chains taking on a extended conformation cover the liposome surface and shield positive charges on the liposome surface below the LCST. However, the polymer chains take on a dehydrated globule above the LCST. Thus the hydrated layer formed by the polymer chains disappears and then the positively charged surface of the liposome is exposed, resulting in increase in the zeta potential.

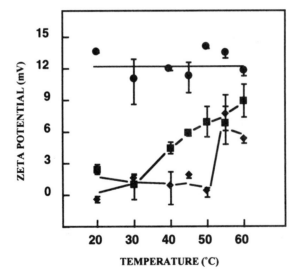

Figure 3. Zeta potentials of unmodified (●), poly(APr)-2C$_{12}$-modified (◆), and copoly(APr-NDDAM)-modified (■) DC-Chol/DOPE (1:1, mol/mol) liposomes as a function of temperature. Polymer/lipid weight ratios of poly(APr)-2C$_{12}$-modified and copoly(APr-NDDAM)-modified liposome were 0.78 and 0.65, respectively.

Similarly, the copoly(APr-NDDAM)-modified cationic liposome exhibited a low zeta potential below 30°C, but its zeta potential increased gradually above 40°C with raising temperature. It has been suggested that partial dehydration of the polymer chain occurs even below its LCST.[9] The whole chain of copoly(APr-NDDAM) might be tethered in vicinity of the membrane surface by plural NDDAM units at random points in the chain. Thus, dehydrated segments of this polymer chain might be adsorbed to the membrane through hydrophobic interactions more easily than poly(APr)-$2C_{12}$ which is bound to the membrane at its terminal point. Therefore, zeta potential of the cationic liposome modified with poly(APr-NDDAM) might be more readily influenced by temperature than that modified with poly(APr)-$2C_{12}$.

4.2   Association of Cationic Liposomes with Anionic Liposomes

Association of the cationic liposomes and the anionic liposomes was investigated by following their particle size using dynamic light scattering. Figure 4 depicts the weight average diameter of particles existing in mixed suspensions of the polymer-modified cationic liposomes and the EYPC/PA liposome at various temperatures. When the unmodified cationic liposomes were mixed with the anionic liposomes in the range of 20-60°C, large aggregates precipitated. In contrast, mixed suspensions of the polymer-modified cationic liposomes and the anionic liposomes exhibited small diameters at low temperatures. In the case of poly(APr)-$2C_{12}$-modified cationic liposome, the particle size did not change from 20 to 45°C (Fig.4A).

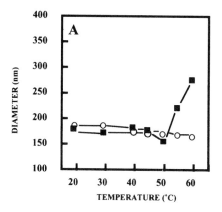 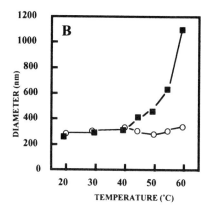

Figure 4. Diameters of poly(APr)-$2C_{12}$-modified (A) and copoly(APr-NDDAM)-modified (B) DC-Chol/DOPE (1:1, mol/mol) liposomes in the absence (O) or presence (■) of EYPC/PA (3:1, mol/mol) as a function of temperature. Polymer/lipid weight ratios of poly(APr)-$2C_{12}$-modified and copoly(APr-NDDAM)-modified liposome were 0.78 and 0.65, respectively.

However, its size increased significantly above 50°C, which is near the LCST of poly(APr). Because the diameter of the same cationic liposome did not change in the absence of the anionic liposomes in this temperature region, this result shows that the polymer-modified cationic liposomes associated with the anionic liposomes above the LCST of the polymer. Similarly, an intensive increase in diameter is seen in the case of copoly(APr-NDDAM)-modified cationic liposome above 50°C (Fig.4B).

4.3   Fusion of Cationic Liposomes with Anionic Liposomes

It has been thought that membrane fusion plays an important role in cationic liposome-mediated transfection: DNA is transferred into cytoplasm via direct fusion between cationic liposomes and plasma membrane and/or fusion between cationic liposomes and endosomal membrane after being taken up by the cell through endocytosis.[20-22] Thus, the effect of temperature on fusion between the polymer-modified cationic liposomes with the anionic liposomes was examined. Liposome fusion was detected by the change in resonance energy transfer efficiency from N-(7-nitrobenz-2-oxa-1,3-diazol-4-yl)phosphatidylethanolamine (NBD-PE) to lissamine rhodamine B-sulfonyl phosphatidylethanolamine (Rh-PE) due to dilution of these fluorescent lipids in the liposomal membrane.[23] We have already shown that the fluorescence intensity ratio of NBD to Rh (**R**) is useful to follow membrane fusion.[24]

Figure 5A shows time courses of **R** for the copoly(APr-NDDAM)-modified cationic liposomes containing NBD-PE and Rh-PE after addition of the fluorescent lipid-free EYPC/PA liposomes. The **R** value increased immediately after the addition of the anionic liposomes. While **R** increased slightly at 30°C, the increase became more significant at 60°C, indicating that the fusion occurred more intensively at this temperature. Generally, the fusion proceeded fast in the initial 2 min and then became much slower.

Figure 5 Bdepicts the increase in **R** (**ΔR**) for various cationic liposomes in the initial 3 min after addition of the anionic liposomes at varying temperatures. For the unmodified cationic liposome, **ΔR** increases monotonously from 0.27 to 0.52 with raising temperature in the region of 15-60°C, indicating that the cationic liposome fuses with the anionic liposome more intensively with temperature. For the copoly(APr-NDDAM)-modified cationic liposome, **ΔR** is kept at a much lower level below 50°C, compared to the case of the unmodified liposome. However, **ΔR** increases drastically above 50°C. This result suggests that the hydrated polymer chains covering the cationic liposome surface suppress fusion between the cationic liposome and the anionic liposome effectively. However, collapse of the tethered polymer chain enables the cationic liposome to fuse with the anionic liposome.

In contrast, fusion between the poly(APr)-2C$_{12}$-modified cationic liposome and EYPC/PA liposome was enhanced with temperature, while a very intensive enhancement of the fusion was seen around 45°C which is close to the LCST of poly(APr). Polymer chains in the poly(APr)-2C$_{12}$-modified liposome might have higher conformational freedom than those in the copoly(APr-NDDAM)-modified liposome, because of the difference of attachment to the liposome. Also, density of the polymer chain in the vicinity of surface of the former liposome might be lower than that of the latter liposome due to the same reason. The differences in mobility and in density of polymer chains might result in the different temperature-dependence of liposome fusion.

The temperature-dependent aggregation and fusion of the polymer-modified cationic liposomes with the anionic liposomes were further confirmed using an electron microscopy. The images of the polymer-modified cationic liposomes in the presence of the anionic liposomes revealed that the liposomes maintained their original size below the LCST but forms aggregates and large vesicles above the LCST, in contrast to the images of the unmodified cationic liposomes under the same conditions, in which intensive aggregation and fusion of the liposomes were observed.

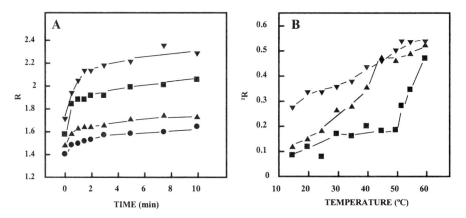

Figure 5. A: Time-courses of fusion of copoly(APr-NDDAM)-modified DC-Chol/DOPE (1:1, mol/mol) liposomes with EYPC/PA (3:1, mol/mol) liposomes at 30 (●), 50 (▲), 55 (■), and 60°C (▼). B: Fusion of copoly(APr-NDDAM)-modified (■), poly(APr)-2C$_{12}$-modified (▲), and unmodified (▼) DC-Chol/DOPE (1:1, mol/mol) liposomes with EYPC/PA (3:1, mol/mol) liposomes as a function of temperature. Ordinate (ΔR) represents the increase in R for the initial 3 min after addition of EYPC/PA liposomes. Polymer/lipid weight ratios of poly(APr)-2C$_{12}$-modified and copoly(APr-NDDAM)-modified liposomes were 0.78 and 0.65, respectively.

We examined two types of polymer-modified cationic liposomes, namely poly(APr)-2C$_{12}$-modified and copoly(APr-NDDAM)-modified cationic liposomes. While both types of liposomes exhibited similar temperature-dependence in their interactions with the anionic liposomes, the latter seems to achieve more efficient control of the interaction. As described above, in the case of the copoly(APr-NDDAM)-modified cationic liposome, the whole polymer chain might exist near surface of the liposome and, hence, this liposome has the polymer coat with higher chain density than the poly(APr)-2C$_{12}$-modified liposome. In addition, conformational freedom of the copoly(APr-NDDAM) chain is restricted more severely than the poly(APr)-2C$_{12}$ chain, because the former is fixed by the anchors at plural points. Such differences in density and mobility of the tethered polymer chain might result in the difference in their interactions with the anionic liposomes.

5.   INTERACTION OF THERMOSENSITIVE POLYMER-MODIFIED EYPC LIPOSOMES WITH CV1 CELLS

Since the tethered polymer chain changes its character between hydrophilic and hydrophobic near the LCST of the polymer, the polymer-modified liposomes are expected to reveal its affinity to cell surfaces differently, depending on temperature. We prepared EYPC liposomes modified with copoly(APr-NIPAM)-2C$_{12}$, which exhibits the LCST near the physiological temperature. Temperature-dependent interaction of the thermosensitive polymer-modified liposomes with CV1 cells, an African green monkey kidney cell line, is described.

5.1   Liposome Uptake by CV1 Cells

The influence of incubation temperature on liposome uptake by CV1 cells was examined. The cells were incubated with either the polymer-modified or the unmodified EYPC liposomes containing a fluorescent lipid, NBD-PE, at 37 or 42°C for 3 h. The amount of liposomal lipid associated with the cells after the incubation was evaluated from the fluorescence intensity of NBD-PE in the cells (Fig.6). For the unmodified liposome, approximately the same amount of liposome was taken up by the cell at both temperatures, indicating that temperature does not affect the liposome uptake in this temperature region. In the case of the polymer-modified liposome, the amount of liposome taken up by the cell at 37°C was slightly lower than that of the unmodified liposome at the same temperature. However, at 42°C the amount became two times higher than that of the unmodified liposome. In comparison with the unmodified liposome, the polymer-modified liposome was taken up by the cell less readily at 37°C and more readily at 42°C. Because the LCST of the polymer chain was ca. 40°C, the polymer chain is highly hydrated and takes on an extended conformation at 37°C. It is likely that the hydrated polymer chains cover the liposome surface and suppress the liposome-cell interaction by their steric repulsion.[15,16,25] By contrast, the tethered polymer chains become hydrophobic and are adsorbed on the liposome surface above the LCST.[10] The hydrophobic polymer chains bound on the liposome might increase hydrophobicity of the liposome surface and facilitate liposome binding to the cell and liposome recognition by the cell.

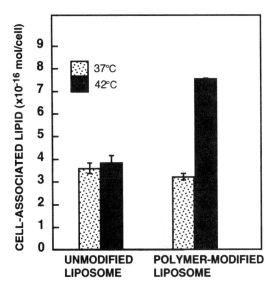

Figure 6. Uptake of copoly(APr-NIPAM)-2C$_{12}$-modified and unmodified EYPC liposomes by CV1 cells after 3 h incubation in Dulbecco's modified Eagle's medium (DMEM) supplemented with 10 % fetal bovine serum (FBS) at 37 and 42°C. Polymer/lipid weight ratio of the polymer-modified liposome was 0.71.

5.2   Liposome-Mediated Delivery of MTX to CV1 Cells

We have reported that an efficient content release occurs from the thermosensitive polymer-modified DOPE liposomes above the polymer's LCST.[8-10] However, when phosphatidylcholines are used for the liposomal lipid, only a limited portion of the contents is released from the thermosensitive polymer-modified liposomes under the same condition.[1,4,10] W examined retention of calcein in the copoly(APr-NIPAM)-2C$_{12}$-modified EYPC liposomes and found that 82 and 81 % of the loaded calcein molecules were retained after 3 h incubation in DMEM with 10 % FBS at 37 and 42°C, respectively. Also, we did not observe remarkable difference in their release behaviors at 37 and 42°C, indicating that the temperature-dependent hydrophobicity change of the polymer hardly influences the contents release from EYPC liposomes.

Since the polymer-modified EYPC liposomes were shown to retain the hydrophilic molecule at 37 and 42°C, we investigated temperature-control of delivery of MTX to CV1 cells mediated by the polymer-modified liposomes. MTX has been used for the treatment of various malignant diseases and is known to prohibit proliferation of cells by binding tightly to cytoplasmic dihydrofolate reductase.

CV1 cells were incubated with either the MTX-loaded polymer-modified or the unmodified EYPC liposomes for various periods at 37 or 42°C. After washing with PBS, the cells were further incubated in DMEM with 10 % FBS for 24 h at 37°C. Figure 7 shows the cell number after the 24 h incubation following the liposome treatment.

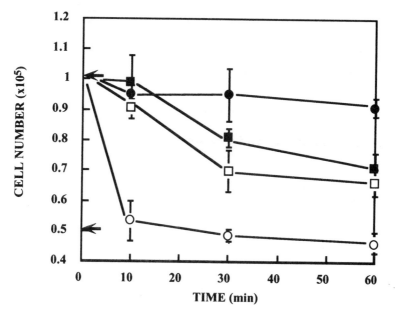

Figure 7. The effect of MTX-loaded liposomes on growth of CV1 cells. The cells were incubated with copoly(APr-NIPAM)-2C$_{12}$-modified (●, ■) or unmodified (■, □) EYPC liposomes encapsulating MTX in DMEM with 10 % FBS for 10, 30, or 60 min at 37°C (closed symbols) or 42°C (open symbols). The cells were washed with PBS and incubated in DMEM with 10 % FBS for 24 h at 37°C. The cell number after the 24 h incubation was shown as a function of the initial incubation time. The lower and upper arrows represent the initial number of cells and the number of cells after the 24 h incubation without the liposome treatment. Concentration of MTX and EYPC were 6.2×10^{-7} and 5.4×10^{-4} M, respectively. Polymer/lipid weight ratio of the polymer-modified liposome was 0.71.

For the cells treated with the unmodified liposomes, the cell growth was partly prohibited. The prohibition became more significant as the period of the liposome treatment increased. While it seems that the liposome treatment at 42°C was slightly more efficient for prohibition of the cell growth than at 37°C, this difference was not remarkable. However, for the cells treated with the polymer-modified liposomes, a marked difference in the cell growth was seen between 37 and 42°C. When the cells were incubated with the liposomes at 37°C, prohibition of the cell proliferation hardly occurred even for the cells having the 60 min incubation. In contrast, when the treatment was done at 42°C, the cell growth was completely prohibited even by the 10 min treatment. It was confirmed that free MTX of the same concentration did not inhibit cell proliferation. Thus it is unlikely that MTX leaked out of the liposomes at 42°C affected the cell growth. Therefore, it can be considered that the tethered polymer chains with hydrophobic nature promoted the liposome-cell interaction and, as a result, more MTX molecules were delivered into the cells at 42°C than at 37°C. In fact, 2.3 times larger amount of liposomal lipid was found in the cells treated with the polymer-modified liposomes at 42°C, compared to the cells treated at 37°C (Fig.6).

As is seen in Figure 6, the cells took up the polymer-modified liposomes and the unmodified liposomes approximately to the same extent at 37°C. Nevertheless, MTX encapsulated in the polymer-modified liposomes was much less effective than that entrapped in the unmodified liposomes. Since MTX must be taken up in cytoplasm to be active, it might be necessary for the anionic MTX to be delivered into low-pH compartments, such as lysosome, where MTX should be protonated and thus diffuse into cytoplasm.[26] O'Brien and collaborators showed that attachment of polyethylene glycol chains to liposome surface reduces uptake of the liposome in low-pH compartments of HeLa cell.[16] Thus, even though the same amount of liposome was taken up by CV1 cells during the incubation, a smaller fraction of the liposomes might be taken in low-pH compartments for the polymer-modified liposomes than for the unmodified liposomes. At present, the reason why the activities of these MTX-loaded liposomes are different is unclear. However, it seems that location of the liposomes in the cell affects the appearance of the MTX activity.

6.   CONCLUSION

The present study demonstrated that interactions of the thermosensitive polymer-modified liposomes with model membranes and cells were suppressed by the hydrated polymer chains attached to the liposome surface below the LCST. However, the hydrophilic-to-hydrophobic change of the tethered polymer chains above the LCST enhanced their interactions. As a result, efficient control of these interactions was achieved in response to ambient temperature. The information obtained in this study might be useful for the design of site-specific delivery systems for biologically active molecules, such as anti-cancer drugs, DNA, and oligonucleotides.

REFERENCES

1. Kono, K., Hayashi, H. , and Takagishi, T., 1994, *J. Contr. Rel.,* 30: 69-75.
2. Kitano, H., Maeda, Y., Takeuchi, S., Ieda, K. and Aizu, Y., 1994, *Langmuir*, 10: 403-406.
3. Winnik, F.M., Adronov, A., and Kitano, H., 1995, *Can. J. Chem.,* 73: 2030-2040.
4. Kim, J.-C., Bae, S.K., and Kim, J.-D., 1997, *J. Biochem.*, 121: 15-19.
5. Heskins, M. and Guillet, J.E., 1968, *J. Macromol. Sci. Chem.* A2, 1441-1455.
6. H.G. Schild, D.A. Tirrell, 1990, *J. Phys. Chem.*, 94: 4352- 4356.
7. H.G. Schild, 1992, *Prog. Polym. Sci.*, 17: 163-249.
8. Hayashi, H., Kono, K., and Takagishi, T., 1996, *Biochim. Biophys. Acta*, 1280: 127-134.
9. Kono, K., Nakai, R., Morimoto, K., and Takagishi, T., 1999, *Biochim. Biophys. Acta*, 1416: 239-250.
10. Kono, K., Henmi, A., Yamashita, H., Hayashi, H., and Takagishi, T., 1999, *J. Contr. Rel.*, 59: 63-75.
11. Hayashi, K. Kono, T. and Takagishi, 1999, *Bioconjugate Chem.* 10: 412-418.
12. Hayashi, H., Kono, K., and Takagishi, T., 1998, *Bioconjugate Chem.* 9: 382-389.
13. Lee, K.-D., Hong, K., and Papahadjopoulos, D., 1992, *Biochim. Biophys. Acta*, 1103: 185-197.
14. Allen, T.M., Austin, G.A., Chonn, A., Lin, L., and Lee, K.C., 1991, *Biochim. Biophys. Acta*, 1061: 56-64.
15. Vertut-Doi, A., Ishiwata, H., and Miyajima, K., 1996, *Biochim. Biophys. Acta*, 1278: 19-28.
16. Miller, C.R., Bondurant, B., McLean, S.D., McGovern, K.A., and O'Brien, D.F., 1998, *Biochemistry*, 37: 12875-12883.
17. Okahata, Y., and Seki, T., 1984, *J. Am. Chem. Soc.*, 106: 8065-8070.
18. Szoka, Jr., F., and Papahadjopoulos, D., 1978, *Proc. Natl. Acad. Sci.* U.S.A., 75: 4194-4198.
19. Kono, K., Nishii, H., and Takagishi, T., 1993, *Biochim. Biophys. Acta*, 1164: 81-90.
20. Singhal, A., and Huang, L., in *Gene Therapeutics*, J.A. Wolff, Ed., 1994, Birkhäuser, Boston, pp. 118-142.
21. Felgner, P.L., and Ringold, G.M., 1989, *Nature*, 337: 387-388.
22. Wrobel, I., and Collins, D., 1995, *Biochim. Biophys. Acta*, 1235: 296-304.
23. Struck, D.K., Hoekstra, D., and Pagano, R.E., 1981, *Biochemistry*, 20: 4093-4099.
24. Kono, K., Kimura, S., and Imanishi, Y., 1990, *Biochemistry*, 29: 3631-3637.
25. Du, H., Chandaroy, P., and Hui, S.W., 1997, *Biochim. Biophys. Acta*, 1326: 236-248.
26. Leserman, L.D., Machy, P., and Barbet, J., 1981, *Nature*, 293: 226-228

PART 2

POLYMERS IN DIAGNOSIS AND VACCINATION

THE USE OF POLYCHELATING AND AMPHIPHILIC POLYMERS IN GAMMA, MR AND CT IMAGING

Vladimir P. Torchilin
Department of Pharmaceutical Science, School of Pharmacy, Bouve College of Health Sciences, Northeastern University, Mugar Building 312, 360 Huntington Avenue, Boston, MA 02129, USA

1. INTRODUCTION

1.1 Diagnostic Imaging and Imaging Modalities

Whatever imaging modality is used, medical diagnostic imaging requires that the sufficient intensity of a corresponding signal from an area of interest be achieved in order to differentiate this area from surrounding tissues. Currently used medical imaging modalities include: (a) Gamma-scintigraphy (based on the application of gamma-emitting radioactive materials); (b) Magnetic resonance (MR, phenomenon involving the transition between different energy levels of atomic nuclei under the action of radio frequency signal); (c) Computed tomography (CT, the modality utilizing ionizing radiation - X-rays - with the aid of computers to acquire cross-images of the body and three-dimensional images of areas of interest); (d) Ultra-sonography (US, the modality utilizes the irradiation with ultrasound and is based on the different passage rate of ultrasound through different tissues). See also Table 1 for diagnostic moieties (reporter groups) used in different imaging modalities. Though the ultrasound imaging is widely used in clinical practice, contrast agents for ultra-sonography will not be discussed in this Chapter.

Biomedical Polymers and Polymer Therapeutics
Edited by Chiellini *et al.*, Kluwer Academic/Plenum Publishers, New York, 2001

269

Table 1. Diagnostic Imaging Modalities and Corresponding Reporter Groups

Diagnostic Imaging Modality	Reporter Group
1. Gamma-scintigraphy	Radionuclides (such as ^{111}In, ^{99m}Tc, ^{67}Ga)
2. Magnetic resonance imaging	Paramagnetic ions (such as Gd and Mn, and iron oxide)
3. Computed tomography	Iodine, Bromine, Barium
4. Ultrasonography	Gas (air, nitrogen, argon)

It is important to mention, that non-enhanced imaging techniques are useful only when large tissue areas are involved in the pathological process. In addition, though attenuations (i.e., the ability of a tissue to absorb a certain signal, such as X-rays, sound waves, radiation, or radio frequencies) of different tissues differ; however, in the majority of cases this difference is not sufficient for clear discrimination between normal and pathological areas. To solve a problem and to achieve a sufficient attenuation, contrast agents are used able to absorb certain types of signal (irradiation) much stronger than surrounding tissues. The main task in this case is to accumulate a sufficient quantity of a contrast agent in the area of interest and to keep its presence in normal tissues and organs on a minimum level. Different chemical nature of reporter moieties used in different modalities and different signal intensity (sensitivity and resolution) result in the fact that the tissue concentration that must be achieved for successful imaging varies between diagnostic moieties in broad limits (Table 2). As one can see, being rather low in case of gamma imaging it is pretty high for MRI and CT.

Table 2. Concentration of Reporter Moiety Required for Appropriate Tissue Attenuation in Different Imaging Modalities

Imaging Modality	Reporter Diagnostic Moiety	Contrast Concentration in Tissue
Gamma-scintigraphy	radionuclide	10^{-10} M
MR imaging	paramagnetic metal	10^{-4} M
CT imaging	iodine	10^{-2} M

2.　POLYCHELATING POLYMER-BASED IMMUNOCONJUGATES FOR TARGETED DIAGNOSTIC IMAGING

2.1   Synthesis of Polychelating Polymers and Their Immunoconjugates

Immunoimaging including antibody-based radioimmunoscintigraphy and MRI is now widely applied for diagnostic purposes.[1] The approach is based on the parenteral administration of a specifically labeled monoclonal antibody against a characteristic antigen of the area of interest, with subsequent registration (visualization) of the label accumulated there due to the antibody-antigen interaction. Frequently, radioisotopes and paramagnetic moieties for antibody labeling are metals. To achieve firm metal binding to an antibody, chelating moieties are introduced into a protein molecule.[2]

To improve the labeling efficacy, an individual antibody molecule has to be modified with multiple chelator residues. However, the excessive antibody modification can inactivate it.[3] To solve the problem and to couple many reporter groups to a single antibody molecule without affecting the latter, the modification of an antibody with polymers carrying multiple chelating moieties as side-groups was proposed.[4-6] The application of such chelating polymers provides the major increase in the amount of heavy metal binding sites (and metal ions) per antibody molecule with minimal influence on the antibody specific properties. Both natural and synthetic polymers can be used for this purpose, poly-L-lysine (with multiple free amino-groups, PLL) being the most often choice. To couple the chelator (using diethylene triamine pentaacetic acid, DTPA, as an example) to a polymer, the reactive intermediate of the chelator is usually used, containing mixed anhydride, cyclic anhydride, N-hydroxysuccinimide ester, etc.

To couple a chelating polymer with a protein (antibody) and to escape possible cross-linking, we have suggested and experimentally designed a new scheme of chelating polymer-antibody conjugation involving the preparation of a chelating polymer, containing a single terminal reactive group capable of interaction with an activated antibody[7,8]. The approach is based on the use of carbobenzoxy (CBZ)-protected PLL with free terminal amino-group, which is derivatized into reactive form with subsequent deprotection and incorporation of DTPA residues.

Figure 1 shows a typical process for the synthesis of a single-point modified DTPA-PLL involving the following steps: activation of terminal amino-group in CBZ-PLL with N-succinimidyl(2'-pyridyldithio)propionate (SPDP); deprotection of PDP-terminal CBZ-PLL; and interaction of free amino-groups in the main chain of PLL with DTPA mixed anhydride.

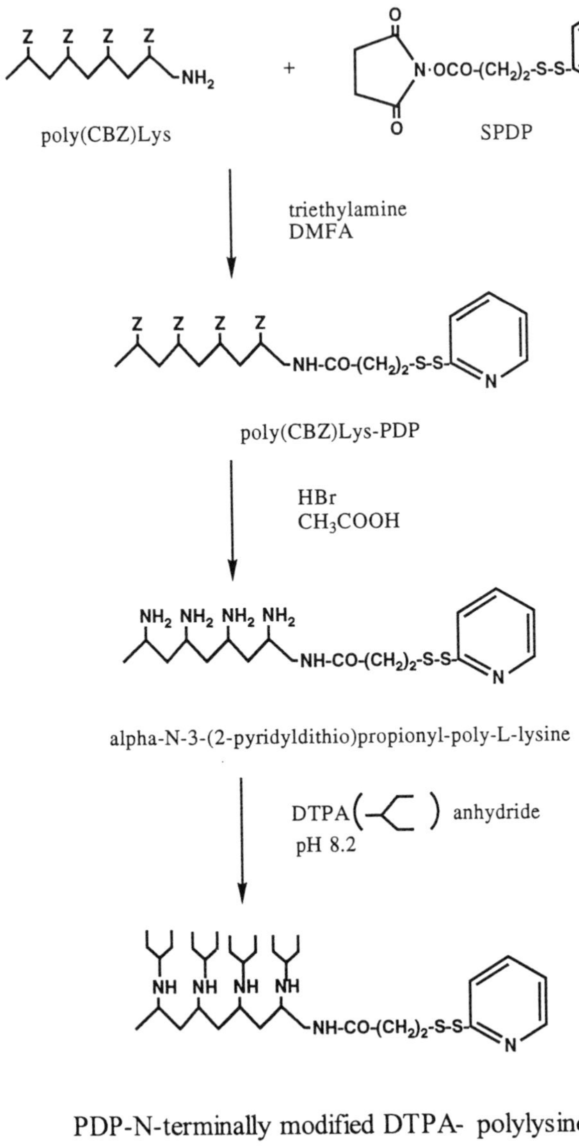

Figure 1. Synthesis of terminus-activated polychelating polymer (DTPA-PLL).

A subsequent single-point attachment of the polychelating polymer to an antibody (or antibody fragment, Fab) may be completed using various different routes via such reagents as SPDP, succinimidyl maleidomethyl cyclohexane (SMCC) or by bromoacetylation as shown in Figure 2.[9]

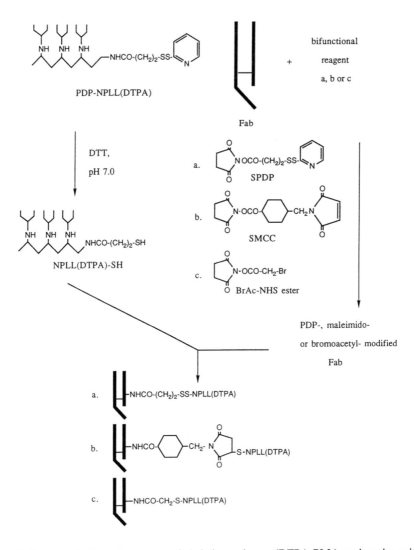

Figure 2. Scheme of alternative routes of chelating polymer (DTPA-PLL) conjugation with antibody Fab fragment via different intermediates (9).

2.2 Chelating Polymer-Modified Antibodies In Vitro and In Vivo

Single-point modification of monoclonal antibodies with chelating polymers sharply increases antibody loading with metals. Using ^{111}In as a radioactive label, under similar conditions we succeeded in binding with chelating polymer-modified Fab much higher In radioactivity than with Fab routinely modified by direct incorporation of a monomeric DTPA (Fig.3). Depending on the molecular weight of a chelating polymer used, 10 to more than 90 metal ions can be bound with a single antibody molecule, as it was proved using Gd^{3+} and $^{111}In^{3+}$, and following T_1 parameter (Gd) or gamma-radioactivity (^{111}In) of the labeled antibody.[4] Antimyosin antibody R11D10 loaded with ^{111}In via DTPA-PLL (13,000) was able to deliver to the antigen-coated surface two orders of magnitude higher radioactivity than the corresponding directly labeled DTPA-antibody, the binding kinetics being unchanged.[7]

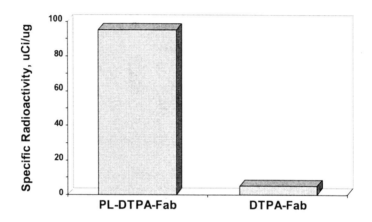

Figure 3. ^{111}In labeling of DTPA-PLL(13,000)-Fab conjugate and same quantity of Fab directly modified with DTPA (7).

The possibility of binding many metal atoms to a single antibody via a chelating polymer may be especially important for MR imaging, since for the effective contrasting of the given tissue it is necessary to achieve Gd concentration in this tissue of at least from 10 to 100 M.[10] It is also important, that a single-point antibody modification with polymer causes relatively minor decrease in antibody specific properties.[4,7,8]

Fast and effective accumulation of antibody-chelating polymer conjugate in the target area was demonstrated by us in rabbits with experimental myocardial infarction.[11] ^{111}In-Labeled conjugates between DTPA-PLL and Fab of R11D10 antimyosin antibody (that is able to accumulate specifically in the compromised myocardium) were injected intravenously after 30 to 40 min of the left coronary artery occlusion and 30 min reperfusion. As a control, the same quantity of ^{111}In-radioactivity bound with R11D10 Fab via directly coupled DTPA was used. Gamma-scintigraphy during the experiment proved fast and specific accumulation of radioactivity in the target area in case of labeled polymer-antibody conjugates (see ref. 11). We were able to achieve an effective visualization of the necrosis zone within 2 to 3 hours instead of 24 to 48 hours as in the case of antimyosin Fab traditionally loaded with ^{111}In via monomeric chelating groups.

We have also quantified the effect of specific conjugate accumulation in the target area by comparing the quantity of radioactivity accumulated in normal and infarcted myocardium tissue after 5 hours and calculating an infarct-to-normal radioactivity ratio.[12] This ratio depends on the degree of necrotization in each tissue sample and can vary between 5 and 50 for the labeled DTPA-PLL-Fab. At the same time, in case of routine DTPA-Fab preparations (direct labeling), the ratio even in the most necrotized samples is usually not higher than 1.5 to 3.

Thus, the use of metal-loaded chelating polymer-antibody conjugates permits fast and effective accumulation of the label in the target area with high target-to-normal ratios that should be sufficient for fast and contrast imaging in different imaging modalities.

3.   POLYCHELATING AMPHIPHILIC POLYMERS (PAP) AS KEY COMPONENTS OF MICROPARTICULATE DIAGNOSTIC AGENTS

To still further increase a local spatial concentration of a contrast agent for better imaging, it was a natural progression in the development of the effective contrast agents to use certain microparticulate carriers able to carry multiple contrast moieties for an efficient delivery of contrast agents to areas of interest and enhancing a signal from these areas.

3.1   Liposomes and Micelles as Carriers of Drugs and Imaging Agents

Among contrast agent carriers, liposomes, microscopic artificial phospholipid vesicles, and micelles, amphiphilic compound-formed colloidal particles with hydrophobic core and hydrophilic corona, draw a special attention because of their easily controlled properties and good pharmacological characteristics (first of all, excellent biocompatibility and biodegradability). Figure 4 demonstrates the principal schemes of micelle and liposome formation and their loading with various reporter moieties which might be covalently or non-covalently incorporated into different compartments of these particulate carriers (see further).

To increase the quantity of liposome (micelle) accumulating in the "required" areas, the use of targeted carriers (carriers with a surface-attached specific ligand) was suggested.[13] Immunoglobulins, primarily of the IgG class, are the most promising and widely used targeting moieties for various drugs and drug carriers including liposomes/micelles (in this case we are talking about immunoliposomes or immunomicelles).

To achieve long circulation and still better accumulation of diagnostic carriers in various pathological areas with leaky vasculature (tumors, infarcts, etc.), the coating of the carrier (liposome) surface with inert, biocompatible polymers, such as polyethylene glycol (PEG), was suggested.[14] These polymers form a protective layer over the liposome surface and slow down the liposome recognition by opsonins and subsequent clearance. Long-circulating liposomes and micelles are now widely used in biomedical *in vitro* and *in vivo* studies and even found their way into clinical practice.[15,16]

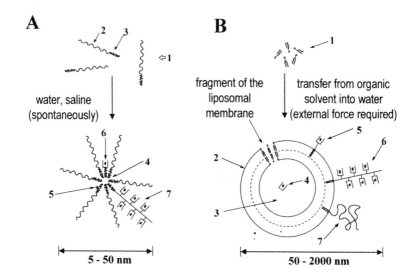

Figure 4. Schematic structures of micelle and liposome, their formation and loading with a contrast agent. **A** - micelle is formed spontaneously in aqueous media from amphiphilic compound (**1**) the molecule of which consists of distinct hydrophilic (**2**) and hydrophobic (**3**) moieties. Hydrophobic moieties form the micelle core (**4**). Contrast agent (**asterisk**; gamma- or MR-active metal-loaded chelating group, or heavy element, such as iodine or bromine) can be directly coupled to the hydrophobic moiety within the micelle core (**5**), or incorporated into the micelle as an individual monomeric (**6**) or polymeric (**7**) amphiphilic unit. **B** - liposome can be artificially prepared from individual phospholipid molecules (**1**), and consists of bilayered membrane (**2**) and internal aqueous compartment (**3**). Contrast agent (**asterisk**) can be entrapped into inner water space of liposome as a soluble entity (**4**), or incorporated into the liposome membrane as a part of monomeric (**5**) or polymeric (**6**) amphiphilic unit (similar to that in case of micelle). Additionally, liposome can be sterically protected by amphiphilic derivative of PEG or PEG-like polymer (**7**). Reproduced with permission from ref. 17.

Gamma-scintigraphy and MRI both require the sufficient quantity of radionuclide or paramagnetic metal to be associated with liposome. Metal may be chelated into a soluble chelate (such as DTPA) and then included into the water interior of a liposome.[18] Alternatively, DTPA or a similar chelating compound may be chemically derivatized by the incorporation of a hydrophobic group, which can anchor the chelating moiety on the liposome surface during or after liposome preparation,[19] see also Figure 4.

In the case of MR imaging, for a better MR signal, all reporter atoms should be freely exposed for interaction with water. This requirement makes metal encapsulation into the liposome less attractive than metal coupling with chelators exposed into the water space. In addition, low-molecular-weight water-soluble paramagnetic probes may leak from liposomes upon the contact with body fluids, which destabilizes most liposomal membranes. Membranotropic chelating agents - such as DTPA-stearylamine or DTPA-phosphatidyl ethanolamine - consist of the polar head containing chelated paramagnetic atom, and the lipid moiety which anchors the metal-chelate complex in the liposome membrane. This approach has been shown to be far more superior in terms of the relaxivity of the final preparation when compared with liposome-encapsulated paramagnetic ions.[20] Liposomes with membrane-bound paramagnetic ion demonstrate also the reduced risk of the leakage in the body. Membranotropic chelates are also suitable for micelle incorporation (they anchor in the hydrophobic micelle core), see Figure 4.

3.2 Polychelating Amphiphilic Polymers (PAP)

The idea of chelating polymers was described earlier in this chapter. We found that the concept may be slightly modified to become suitable not only for protein modification, but also for liposome and micelle loading with metal contrast agent. Since it is rather difficult to enhance the signal intensity from a given reporter metal in liposomes and micelles, one may attempt to increase the quantity of carrier-associated reporter metal (such as Gd or ^{111}In). We have tried to solve this task by using polychelating polymers additionally modified with a hydrophobic residue to assure their firm incorporation into the liposome membrane or micelle core. With this in mind, we developed PAP, a new family of co-polymers containing a hydrophilic fragment with multiple chelating groups and relatively short but highly hydrophobic phospholipid fragment on one end of a polymeric chain suitable for incorporation into liposomes and micelles. The approach is again based on the use of CBZ-protected PLL with free terminal amino-group, which is derivatized into a reactive form with subsequent deprotection and incorporation of DTPA residues.[21] In this particular case, we developed a pathway for the synthesis of amphiphilic polychelator N,α-(DTPA-polylysyl)glutaryl phosphatidyl ethanolamine (DTPA-PLL-NGPE). This polychelator easily incorporates into the liposomal membrane or micelle core in the process of liposome of micelle preparation, and sharply increases the number of chelated heavy metal atoms attached to a single lipid anchor (see Fig.1). This allows to increase the number of bound reporter metal atoms per vesicle and to decrease the dosage of an administered lipid without compromising the image signal intensity. Figure 5 describes the chemistry of PAP synthesis (a single-terminus activated chelating polymer preparation and the attachment of a hydrophobic anchor to it).

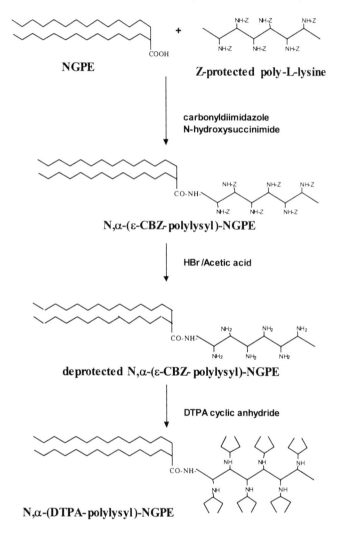

Figure 5. Synthesis of amphiphilic DTPA-PLL-NGPE consisting of hydrophilic DTPA-polylysyl (DTPA-PLL) moiety and hydrophobic N-glutaryl phosphatidyl ethanolamine (NGPE) moiety.

Figure 6 clearly demonstrates enhanced relaxivity of liposomes containing Gd-loaded PAP compared to membranothropic single chelating groups (at the same molar fraction of chelate-associated hydrophobic anchors).

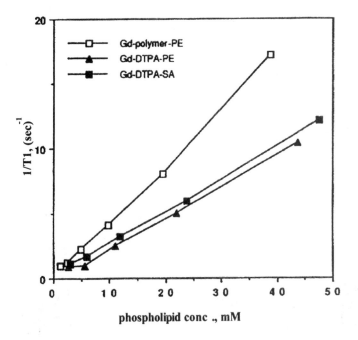

Figure 6. Molecular relaxivities of liposomes with different Gd-containing membranotropic chelators. Liposomes (egg lecithin:cholesterol:chelator = 72:25:3, size between 205 and 225 nm) were prepared by consecutive extrusion of lipid suspension in HEPES buffered saline, pH 7.4, through the set of polycarbonate filters with pore size of 0.6, 0.4 and 0.2 μm. Gd content determination was performed by Galbraith Laboratories, Inc. The relaxation parameters of all preparation were measured at room temperature using a 5-MHz RADX nuclear magnetic resonance proton spin analyzer. The relaxivity of liposomes with polymeric chelator is noticeably higher because of larger number of Gd atoms bound to a single lipid residue. Reproduced with permission from ref. (21).

Micelles formed by self-assembled amphiphilic polymers (such as PEG-phosphatidyl ethanolamine, PEG-PE) can also be loaded with amphiphilic PL-based chelates carrying diagnostically important metal ions such as ^{111}In and Gd.[22] The final preparations are quite stable *in vivo*. Upon subcutaneous administration, micelles penetrate the lymphatics and can serve as fast and efficient lymphangiographic agents for scintigraphy or MR imaging.

To illustrate the performance of contrast PAP-modified liposomes (micelles), Figure 7 shows images of lymph nodes in rabbit with Gd-polychelate-liposomes. Gd-DTPA-PLL-NGPE-liposomes with an average diameter of 215 nm were subcutaneously injected into the forepaw of sedated rabbit, and images were acquired using fat suppression mode and T_1 weighted pulse sequence on 1.5 Tesla GE Signa MRI instrument. High content of Gd in liposomes due to the use of membrane-incorporated amphiphilic chelating polymer permitted fast visualization of rabbit lymph nodes within several minutes post injection. It is important to mention here that the time between contrast agent administration and image acquisition for any other lymphotropic-imaging agent is usually in a range of several hours.

Gd-polychelate-liposomes additionally coated with PEG (to make them long circulating and suitable for the visualization of the blood pool) were used to enable experimental MR angiography in dog experiments. Gd-DTPA-PLL-NGPE / PEG-liposome preparation containing more than 30% wt of Gd, was intravenously injected into a normal dog, and the MR imaging was performed with a Siemens Expert 1.0 Tesla magnetom. Since the blood half-life for the preparation was experimentally found to be around 2 hours, the imaging session was conducted for 1 hour. Within this period of time all major vessels remained clearly delineated. The preparation enabled also the visualization of much smaller vessels such as the renal artery, the carotid arteries and jugular vein (communicated by Biostream, Inc.).

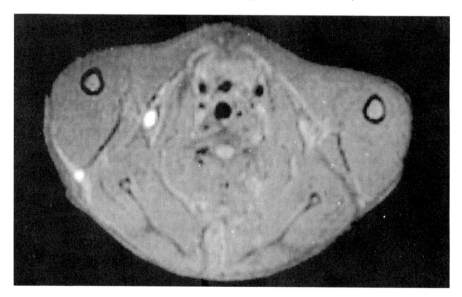

Figure 7. Transverse scan of axillary/subscapular lymph node area in rabbit 10 min after subcutaneous injection of PEG-containing Gd-polychelate liposomes. Both nodes are well visible.

4. CT-VISUALIZATION OF BLOOD POOL WITH LONG-CIRCULATING IODINE-CONTAINING MICELLES

Computed tomography (CT) represents an imaging modality with high spatial and temporal resolution. Without enhancement, all tissues are not well enough discriminated, but targeted radiopaque contrast agents restore tissue discrimination and improve diagnostic utility.[23,24] In order to formulate a material whose distribution is limited to the blood pool, three properties seem necessary: 1) a size larger than fenestrated capillaries (> 10 nm), 2) resistance to phagocytosis, and 3) the radiopaque moiety structurally incorporated within the particulate.

Amphiphilic polymers are able to spontaneously form stable and long-circulating micelles with the size of 10 to 80 nm in an aqueous media and seem to be carriers meeting all the above mentioned requirements.[25] Recently, we have described synthesis and *in vivo* properties of a block-copolymer of methoxy-poly(ethylene glycol) (MPEG) and iodine-substituted poly-L-lysine (PLL).[26,27] In the aqueous solution, this copolymer forms stable and heavily loaded with iodine (up to 30% of iodine by weight) micelles with the size below 100 nm and may serve as an efficient CT imaging agent for the blood pool visualization, see the scheme on Figure 8.

To prepare (MPEG-iodoPLL)block-copolymer, we obtained an N-hydroxysuccinimide(NHS)-activated derivative of MPEG-propionic acid, conjugated it with CBZ-protected PLL, deprotected CBZ-PLL block, and modified free amino-groups in PLL with an NHS-activated derivative of 2,3,5-triiodobenzoic acid. The iodine content of final preparation was 33.8%. Micelles prepared from the resulting amphiphilic iodine-containing co-polymer were concentrated up to ca. 25 mg I/ml. The presence and stability of iodine-containing polymeric micelles were assessed by gel permeation HPLC using Shodex KW-804 column with standards. The micelles with the size of about 75 nm were obtained, as was determined by quasi-elastic laser light scattering. Iodine concentration in the preparations obtained was determined on a Toshiba CT scanner using iodine dilutions of known concentration for comparison.[26,27]

CT blood pool imaging experiments were performed in rabbits and rats. Animals were anesthetized throughout the entire experiments, and the micelle biodistribution was studied during 3 hours following administration of 170 mg I/kg. CT images were acquired at certain times on Toshiba 900 S/xII CT scanner (Toshiba American Medical Systems, Tustin CA).

Imaging parameters were as following: 80 kV/150 mA, slice thickness 1 mm. Signal intensities of blood pool (aorta, heart), spleen and liver were determined on the images on each time point using an operator-defined region-of-interest and image processing software (DIP StationTM, Hayden Image Processing Group).

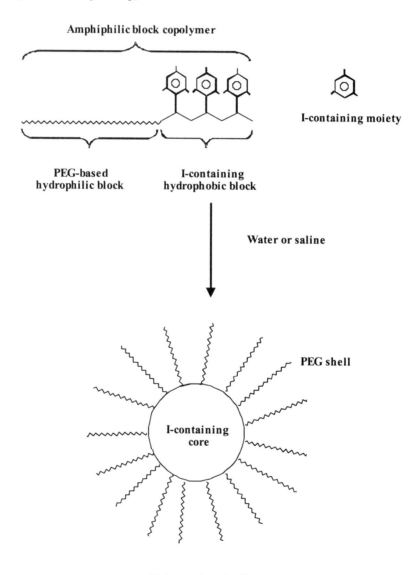

Figure 8. Scheme of micellization of iodine-containing amphiphilic co-polymers in aqueous media. Reproduced with permission from Ref. (26).

Table 3. Organ opacification (in Hounsfield units) in rat after injection of iodine-containing long-circulating micelles

Time (min)	Aorta	Liver	Spleen
0	85	92	67
60	232	155	156
120	235	145	159
180	253	156	191

A significant enhancement of blood pool (aorta and heart) by 3-fold was observed upon the injection of iodine-containing micelles, and quantitative analysis of regions of interest in blood, liver and spleen showed no sign of decrease in blood opacification during the 3 hours studied following intravenous injection (Table 1). We believe that these MPEG-iodolysine micelles are less susceptible to phagocytosis due to their small size and their PEGylated outer surfaces. The iodine containing blocks are situated in the core of the micelle and remain there until the micelle slowly dissociates into unimers.[26] Such selective blood-pool contrast agents may find an application for minimally invasive angiography, oncologic imaging of angiogenesis, ascertaining organ blood volume, and identifying hemorrhage.

REFERENCES

1. Larson, S.M., Carrasquillo, J.A., and Reynolds, J.C., 1984, *Cancer Invest.* 2:363-381.
2. 1988, *Radiolabeled Monoclonal Antibodies for Imaging and Therapy* (S.C..Srivastava, ed.), Plenum Press, New York.
3. Paik, C.H., Ebbert, M.A., Murphy, P.R., Lassman, C.R., Reba, R.C., Eckelman, W.C., Pak, K.Y., Powe, J., Steplewsky, Z., and Koprowski, H., 1983, *J. Nucl. Med.* 24:1158-1163.
4. Torchilin, V.P., Klibanov, A.L., Nossiff, N.D., Slinkin, M.A., Strauss, H.W., Haber, E., Smirnov, V.N., and Khaw, B.A., 1987, *Hybridoma* 6:229-240.
5. Manabe, Y., Longley, C., and Furmanski, P., 1986, *Biochem. Biophys. Acta* 883:460-467.
6. Torchilin, V.P., and Klibanov, A.L., 1991, *CRC Crit. Rev. Ther. Drug Carriers System* 7:275-308.
7. Slinkin, M.A., Klibanov, A.L. and Torchilin, V.P., 1991, *Bioconjugate Chem.* 2:342-348.
8. Torchilin, V.P., Trubetskoy, V.S., Narula, J., Khaw, B.A., Klibanov, A.L., and Slinkin, M.A., 1993, *J. Contr. Release* 24:111-118.
9. Trubetskoy, V.S., Narula, J., Khaw, B.A., and Torchilin, V.P., 1993, *Bioconjugate Chem.* 4:251-255.
10. Lauffer, R.B., 1990, *Magn. Reson. Q.* 6:65-84.
11. Khaw, B.A., Klibanov, A., O'Donnell, S.M., Saito, T., Nossiff, N., Slinkin, M.A., Newell, J.B., Strauss, H.W., and Torchilin, V.P., 1991, *J. Nucl. Med.* 32:1742-1751.
12. Torchilin, V.P., Klibanov, A.L., Huang, L., O'Donnel, S., Nossiff, N.D. and Khaw, B.A., 1992, *FASEB J.* 6:2716-2719.
13. Torchilin, V.P., 1985, *CRC Crit. Rev. Ther. Drug Carriers Syst.* 2:65-115.
14. Torchilin, V.P., and Trubetskoy, V.S., 1995, *Adv. Drug Deliv. Rev.* 16:141-155.
15. Torchilin, V.P., 1996, *Mol. Med. Today* 2:242-249.

16. Trubetskoy, V.S., and Torchilin, V.P., 1996, *STP Pharma Sci.* 6:79-86.
17. Torchilin, V.P., 1997, *Quat. J. Nucl. Med.* 41:141-153.
18. Tilcock, C., Unger, E., Cullis, P., and MacDougall, P., 1989, *Radiology* 171:77-80.
19. Kabalka, G., Davis, M., Holmberg, E., Maruyama, K., and Huang, L., 1991, *Magn. Res. Imaging.*, 9:373-377.
20. Tilcock, C., 1993, In *Liposome Technology* (G. Gregoriadis, ed.), 2nd ed., CRC Press: Boca Raton, FL, vol. 2, pp. 65-87.
21. Trubetskoy, V.S., and Torchilin, V.P., 1994, *J. Lip. Res.* 4: 961-980.
22. Trubetskoy, V.S., Frank-Kamenetsky, M.D., Whiteman, K.R., Wolf, G.L., and Torchilin, V.P., 1996, *Acad. Radiol.* 3: 232-238.
23. Wolf, G.L., 1995, In *Handbook of Targeted Delivery of Imaging Agents* (V.P. Torchilin, ed.), CRC Press, Boca Raton, FL, pp. 3-22.
24. Leander, P., 1996, *Acta Radiol.* 37: 63-68.
25. Trubetskoy, V.S., and Torchilin, V.P., 1995, *Adv. Drug Del. Rev.* 16: 311-320.
26. Trubetskoy, V.S., Gazelle, G.S., Wolf, G.L., and Torchilin, V.P., 1997, *J Drug Targeting* 4: 381-388.
27. Torchilin, V.P., Frank-Kamenetsky, M.D., and Wolf, G.L., 1999, *Acad. Radiol.* 6: 61-65.

CONJUGATION CHEMISTRIES TO REDUCE RENAL RADIOACTIVITY LEVELS OF ANTIBODY FRAGMENTS

Yasushi Arano
Laboratory of Radiopharmaceutical Chemistry, Faculty of Pharmaceutical Sciences, Chiba University, 1-33 Yayoi-cho, Inage-Ku, 263-8522 Chiba, Japan

1.   INTRODUCTION

Monoclonal antibodies have been used to deliver radionuclides to target tissues for diagnostic and therapeutic purposes. Although some clinical studies indicated successful target visualizations and complete recession of tumors, the use of radiolabeled intact antibodies has shown some limitations such as slow penetration into tumors, nonhomogeneous tumor distribution, high blood pool background and high incidence of immune response [1-5]. The slow elimination rates of intact antibodies from circulation cause the delay in the optimal imaging time and induce bone marrow toxicity in targeted radiotherapy. Since most of the unfavorable distribution characteristics of intact antibodies result from the presence of effector or linking regions that are not required for the specific binding with antigen, antibody fragments such as Fab and F(ab')₂ have been applied as vehicles to deliver radioactivity to target tissues. More recently, genetic engineering has provided a new class of antibody fragment that consists of antibody variable heavy and light regions linked with a short peptide or a disulfide bond to form a single molecule of low molecular weight (single chain Fv fragment)[6,7]. These classes of antibody fragments showed faster pharmacokinetics and even distribution in the tumor mass in a size-related manner[4,5]. Such characteristics render antibody fragments attractive vehicles to deliver

Biomedical Polymers and Polymer Therapeutics
Edited by Chiellini *et al.*, Kluwer Academic/Plenum Publishers, New York, 2001

radioactivity for both diagnostic and therapeutic nuclear medicine. Indeed, promising results have been reported with radiolabeled antibody fragments in animal studies and clinical studies[8, 9]. However, high and persistent localization of the radioactivity was observed in the kidney especially when antibody fragments are labeled with metallic radionuclides, which compromises tumor visualization in the kidney region and limits therapeutic potential[10-12]. Metallic radionuclides possess many advantages over radioiodine for both diagnosis and therapeutic applications, due to their superior physical characteristics (half-life, gamma or beta energy properties), availability, cost and simple radiolabeling procedures with a combination of appropriate chelating agents. Thus, radiolabeled antibody fragments would become much more useful for diagnostic and therapeutic nuclear medicine if the renal radioactivity levels could be reduced without impairing the radioactivity levels in the target tissues. In this manuscript, our chemical approaches to reduce renal radioactivity levels of antibody fragments are briefly described.

2.   RENAL HANDLING OF ANTIBODY FRAGMENTS

It has been well recognized that the kidney plays an important role in the plasma turnover of low molecular weight proteins (LMWPs)[13,14]. Many proteins with a molecular weight smaller than serum albumin are filtered in the glomerulus, and subsequently reabsorbed at the proximal tubular cells from the luminal fluid. Although only a few articles have been documented on the exact mechanisms of localization of antibody fragments in the kidney, it is most likely that antibody fragments such as Fab and Fv fragments would be accumulated in the kidney after glomerular filtration, followed by reabsorption at the proximal cells. This was supported by the recent findings that infusion of basic amino acids such as L-lysine reduced renal accumulation of radiolabeled antibody fragments[15-17].

Recent study of gallium-67 (^{67}Ga)-labeled Bz-NOTA-Fv fragment clearly demonstrated the important role of radiometabolites in the renal radioactivity levels (where Bz-NOTA represents 2-(p-(isothiocyanatobenzyl-1,4,7-tetraazacyclononane-1,4,7-triacetic acid) [18]. After administration into mice, ^{67}Ga-NOTA-Fv fragments were metabolized to ^{67}Ga-NOTA-adducts of lysine (ε-amine) and methionine (N-terminus α amine). This antibody possesses 13 lysine residues and terminal methionine groups in each chain, and isothiocyanato group binds to primary amine groups. While ^{67}Ga-NOTA-lysine retained within the renal cells for long postinjection intervals, ^{67}Ga-NOTA-methionine showed faster elimination rate from the renal cells, and eventually excreted in the urine by 24 h postinjection. These findings along with previous studies of radiolabeled polypeptides strongly implied that antibody fragments would be accumulated in the renal cells after glomerular filtration, followed by reabsorption at the proximal cells [19-23]. Then, antibody fragments are metabolized after transportation to the lysosomal compartment of the cells. The long and persistent localization of the radioactivity in the renal cells would be attributed to slow elimination rates of radiometabolites generated after lysosomal proteolysis of antibody fragments.

3. METABOLIZABLE LINKAGES TO FACILITATE EXCRETION OF RADIOMETABOLITES FROM RENAL LYSOSOMES.

One potential strategy to reduce the renal radioactivity levels of radiolabeled antibody fragments may involve an interposition of a cleavable linkage between antibody fragments and radiolabeled compounds of urinary excretion. In this design, a radiolabeled compound of urinary excretion characteristics is selectively released from a parental antibody fragment immediately after lysosomal proteolysis in the kidney. The rationale behind the chemical design is well supported by the metabolic studies of radiolabeled polypeptides and antibody fragments described above. This chemical design of radiolabeled antibodies was first reported by Haseman et al. who placed a diester bond between an intact antibody and a radiometal chelate to accelerate hepatic radioactivity levels of an intact antibody [24]. After this report, a number of radiolabeled antibodies with cleavable linkages have been synthesized and evaluated in animal models. However, contradictory results have been obtained with regard to the radiochemical design. It was thought that use of radioiodines would provide an insight in understanding the structure/distribution relationship of the chemical approach with regard to other radionuclides, and a novel radioiodination reagent was designed and evaluated in animal models.

When considering lower pH environment of the lysosomes, generation of radiolabeled compounds with carboxylic acid from parental antibody fragments would be advantageous to achieve rapid elimination rate of the radioactivity from the kidney. Ester bonds are one of the representative chemical bonds that allow specific release of carboxylic acid from covalently conjugated antibody molecules. Thus, we developed a new radioiodination reagent that conjugate radiolabeled *meta*-iodohippuric acid with polypeptides *via* an ester bond (maleimidoethyl 3-iodohippurate, MIH), as shown in Figure 1 [25].

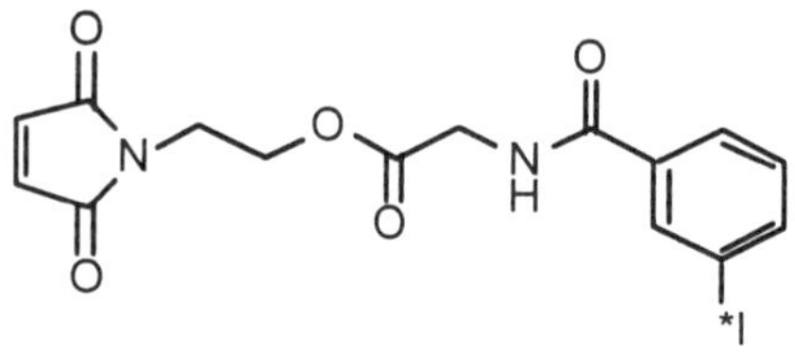

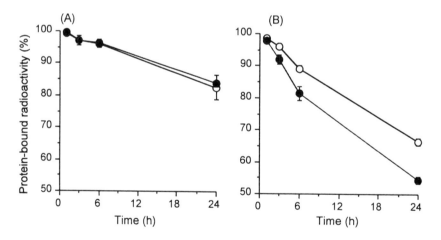

Figure 1. Chemical structure of maleimidoethyl 3-iodohippurate.

MIH was designed to liberate *meta*-iodohippuric acid in the lysosomal compartment of the liver or kidney when conjugated to an intact antibody (IgG) or a Fab fragment. In animal models, radioiodinated MIH-conjugated intact IgGs demonstrated significantly lower radioactivity levels in non-target tissues such as the liver when compared with a control antibody without a cleavable linkage. However, a small but significant decrease in the radioactivity levels in the target (tumor cells) was also indicated, due to a cleavage of the ester bond in the plasma. When MIH was conjugated to a Fab fragment, the ester bond in the conjugate possessed the stability in a buffered-solution of neutral pH identical to that in MIH-conjugated IgG. However, MIH-conjugated Fab fragment liberated significantly higher amounts of *meta*-iodohippuric acid than MIH-conjugated IgG when incubated in human serum or murine plasma (Fig.2) [26].

Figure 2. Comparative stability of ester bonds in MIH-conjugated intact IgG (O) and MIH-conjugated Fab fragment (●) in a buffered-solution of neutral pH (A) or in murine plasma (B).

These findings suggested that MIH may not be applicable to antibody fragments due to insufficient stability of the ester bond of this reagent in plasma. Furthermore, the chemical design using cleavable linkages that liberate radiometabolites of urinary excretion in lysosomal compartment would be restricted to antibodies that are not internalized into target cells. If this chemical design is applied to antibodies that internalize into target cells, reduction of the radioactivity levels would be observed not only in non-target tissues but also in target tissues.

4. METABOLIZABLE LINKAGES TO LIBERATE RADIOMETABOLITES FROM ANTIBODY FRAGMENTS BY RENAL BRUSH BORDER ENZYMES.

The renal handling of peptides has been extensively investigated. These studies indicated that some glomerularly-filtered peptides are hydrolyzed to free amino acids during a short contact time with the brush border enzymes present on the lumen of the renal proximal cells[14]. Indeed, several peptidases have been identified on the brush border membrane of the renal tubules. Thus, it was thought that the renal localization of the radioactivity may be reduced if radiolabeled compounds of urinary excretion are released from antibody fragments before they are incorporated into the renal cells by the action of the brush border enzymes. This approach is applicable to a wide variety of antibody fragments whether they are internalized to target cells or not. This approach may also advantageous to reduce the renal radioactivity levels at earlier postinjection intervals, because radiolabeled compounds are released from the antibody fragments before they are taken up and transported to and metabolized in the lysosomal compartment of the renal cells.

To estimate the hypothesis, a Fab fragment of a monoclonal antibody against osteogenic sarcoma (OST7) was used as a model. Because of its high urinary excretion from the renal tubules and high stability against *in vivo* deiodination, *meta*-iodohippuric acid was selected as the radiometabolite. Then, since glycyl-lysine sequence is a substrate for one of the brush border enzymes, carboxypeptidase M, a new radioiodination reagent that conjugate *meta*-iodohippuric acid to antibody fragments *via* the peptide bonds was developed[27]. Chemical structure of the reagent, 3'-iodohippuryl N^ε-maleoyl-L-lysine (HML), is shown in Figure 3.

When the peptide bond between glycine and lysine is cleaved, *meta*-iodohippuric acid is released from the conjugate. For comparison, two additional radioiodinated Fab fragments were prepared. *Meta*-iodohippuric acid was conjugated to Fab fragments though an amide bond to prepare MPH-Fab by procedures similar to those employed for HML-Fab.

Radioiodine was also attached directly to the Fab fragment, as shown in Figure 4.

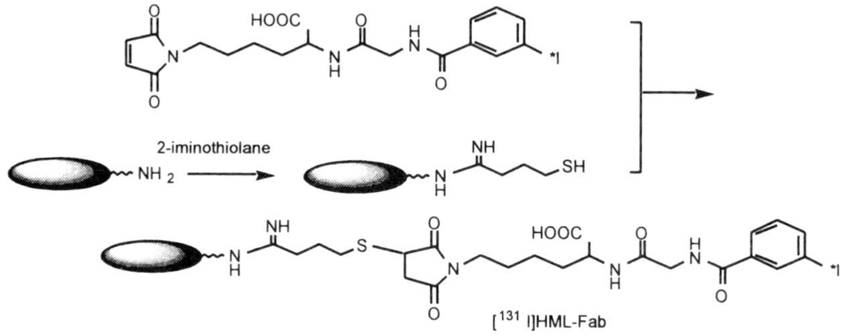

Figure 3. Conjugation of radioiodinated HML to Fab fragments using 2-iminothiolane.

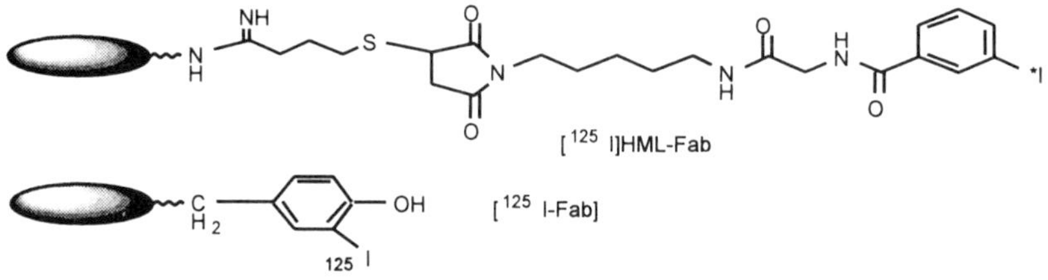

Figure *4.* Chemical structures of radioiodinated MPH-Fab and directly radioiodinated Fab fragments.

The biodistribution of the radioactivity after simultaneous administration of [131I]HML-Fab and [125I]MPH-Fab is shown in Figure 5.

No significant differences were observed in the radioactivity levels in the blood between the two. However, significant differences were observed in the kidney. MPH-Fab showed high radioactivity levels in the kidney and reached its peak at 1 h postinjection, whereas HML-Fab demonstrated significantly lower levels of the radioactivity in the kidney as early as 10 minutes postinjection. The radioactivity levels in the blood and the kidney after concomitant administration of [131I]HML-Fab and directly radioiodinated Fab (125I-Fab) are also illustrated in Figure 5.

[131I]HML-Fab showed slightly higher radioactivity levels in the blood. In the kidney, [131I]HML-Fab significantly reduced the radioactivity levels from 10 minutes postinjection onward. Furthermore, while the directly radioiodinated Fab showed the highest radioactivity levels in the kidney at 30 minutes postinjection, such an increase in the radioactivity in the kidney was not observed with HML-Fab. These findings indicated that HML decreased the renal radioactivity levels of radiolabeled Fab fragment from early postinjection times.

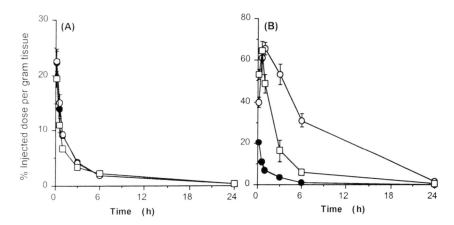

Figure 5. Biodistribution of radioactivity after concomitant administration of [^{131}I]HML-Fab (●), [^{125}I]MPH-Fab (O) or ^{125}I-Fab (□) into mice.

The radioactivity excreted in the urine for 6 hours postinjection of HML-Fab was then analyzed to assess the present chemical design of radiolabeled antibody fragments. On size-exclusion HPLC, more than 85 % of the radioactivity was eluted at the low molecular weight fractions and about 12 % of the radioactivity was eluted in fractions similar to those of the intact Fab fragment. On reversed-phase HPLC of the urine sample, over 93 % of the low molecular weight fractions had a retention time identical to that of *meta*-iodohippuric acid. These findings indicated that the low renal radioactivity levels of HML-Fab were attributed to rapid and selective release of *meta*-iodohippuric acid in the kidney.

To investigate the mechanisms of rapid and selective release of *meta*-iodohippuric acid from HML-Fab, *meta*-iodohippuric acid was conjugated to the thiolated Fab fragment through an ester bond (MIH-Fab), and the biodistribution of the radioactivity of the two conjugates were compared. In our previous study using galactosyl-neoglycoalbumin, NGA, both HML- and MIH-conjugated NGA selectively released *meta*-iodohippuric acid at similar rates after lysosomal proteolysis in hepatic parenchymal cells. Thus, if the low renal radioactivity levels of HML-Fab was caused by the rapid release of *meta*-iodohippuric acid after lysosomal proteolysis in the renal cells, then, both HML- and MIH-conjugated Fab fragments would provide similar radioactivity levels in the kidney. When [^{131}I]HML-Fab and [^{125}I]MIH-Fab were simultaneously administered to mice, [^{125}I]MIH-Fab showed faster clearance of the radioactivity from the blood, due to a partial cleavage of the ester bond in plasma, as stated before. Despite this, [^{131}I]HML-Fab demonstrated significantly lower radioactivity levels in the kidney from 10 minutes postinjection onward. The low renal radioactivity levels achieved by [131I]HML-Fab were clearly demonstrated when the kidney-to-blood ratios of the radioactivity of the three radioiodinated Fab were compared (Fig.6).

[131I]MIH-Fab showed lower kidney-to-blood ratios of the radioactivity than those of 125I-Fab. However, both two radiolabeled Fab fragments reached their peak ratios at 1 hour postinjection, suggesting that the radiometabolite derived from [125I]MIH-Fab would possess faster elimination rate from the lysosomal compartment of the renal cells than those of 125I-Fab. On the other hand, [131I]HML-Fab demonstrated almost constant radioactivity ratios of about 1 from 10 min to 3 hours postinjection, and the ratios were significantly lower than those observed with [125I]MIH-Fab. These findings suggested that HML-Fab released *meta*-iodohippuric acid before the antibody fragment was transported to and metabolized in the lysosomes of the renal cells.

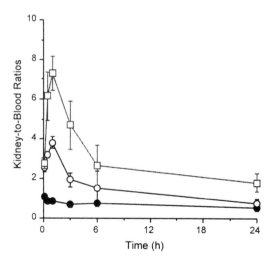

Figure 6. Kidney-to-blood ratios of the radioactivity after administration of [131I]HML-Fab (●), [125I]MIH-Fab (○) and 125I-Fab (□) in mice.

To further understand the mechanisms responsible for the rapid and selective release of *meta*-iodohippuric acid from [131I]HML-Fab fragment, subcellular localization of the radioactivity was investigated at 10 and 30 minutes postinjection of the three radioiodinated Fab fragments. Both [125I]MIH-Fab and 125I-Fab showed the major radioactivity at membrane fractions at 10 minutes postinjection. At 30 minutes postinjection, the majority of the radioactivity of the two Fab fragments migrated to the lysosomal fractions. This suggested that most of the administered Fab fragments would be transported to and metabolized in the lysosomal compartment after 30 minutes. On the other hand, most of the renal radioactivity of [^{131}I]HML-Fab was observed in the membrane fractions at both 10 and 30 minutes after injection. As shown in Figure 6, [^{131}I]HML-Fab showed kidney-to-blood ratios of the radioactivity of about 1 from 10 minutes to 3 hours postinjection. Thus, the gathered findings suggested that HML-Fab released *meta*-iodohippuric acid at the membrane fractions of the renal cells before transported to the lysosomal compartment, presumably by the action of the brush border enzymes present on the lumen of the renal tubules.

The biodistribution of the radioactivity in nude mice bearing osteogenic sarcoma at 3 hours postinjection of [^{131}I]HML-Fab and ^{125}I-Fab is shown in Table 1. No significant differences were observed in the radioactivity levels in the tumor and blood. However, [^{131}I]HML-Fab significantly reduced the renal radioactivity levels. As a result, [^{131}I]HML-Fab amplified the tumor-to-kidney ratios of the radioactivity by a factor of four.

Table 1. Biodistribution of radioactivity after concomitant administration of [^{131}I]HML-Fab and ^{125}I-Fab into nude mice bearing osteogenic sarcoma.

	[^{131}I]HML-Fab	^{125}I-Fab
Blood	3.38 (0.33)	3.61 (0.48)
Liver	0.69 (0.11)	1.06 (0.07)
Kidney	2.08 (0.25)	9.57 (2.76)
Intestine	0.54 (0.08)	1.04 (0.17)
Spleen	0.58 (0.10)	1.81 (0.28)
Stomach[b]	0.12 (0.03)	1.40 (0.75)
Tumor	10.30 (1.77)	11.06 (1.65)

[a] Expressed as % injected dose per gram tissue. Mean (SD) of six mice at 3 h postinjection.
[b] Expressed as % injected dose per organ.

5. CONCLUSION

In conclusion, the findings in this study indicated that [^{131}I]HML-Fab reduced renal radioactivity levels at early postinjection times without impairing the radioactivity levels in the target due to rapid and selective release of *meta*-iodohippuric acid in the kidney. These findings indicated that HML would be a promising reagent for targeted imaging and therapy using antibody fragment as vehicles. Although further studies are required, these studies suggested that radiochemical design of antibody fragments that liberates radiometabolites of urinary excretion by the action of brush border enzymes may constitute a new strategy for reducing the renal radioactivity levels of antibody fragments. These findings also provide a good basis for further application of this strategy to a variety of radionuclides of clinical importance.

REFERENCES

1 Fujimori, K.; Covell, D. G.; Fletcher, J. E.; Weinstein, J. N., 1989, *Tumors. Cancer Res.*, 49, 5656-5663.

2 Goldenberg, D. M.; Sharkey, R. M., 1993, *Int. J. Oncol.*, 3, 5-11.

3 Jain, R. K., 1990, *Tumors. Cancer Res.*, 50, 814s-819s.

4 Yokota, T.; Milenic, D. E.; Whitlow, M.; Schlom, J., 1992, *Cancer Res.*, 52, 3402-3408.

5 Yokota, T.; Milenic, D. E.; Whitlow, M.; Wood, J. F.; Hubert, S. L.; Schlom, J., 1993, *Cancer Res.* 53, 3776-3783.

6 Pantoliano, M. W.; Bird, R. E.; Johnson, L. S., 1991, *Biochemistry*. 30, 10117-10125.

7 Whitlow, M.; Bell, B. A.; Feng, S. L.; Filpula, D.; Hardman, K. D.; Hubert, S. L.; Rollence, M. L.; Wood, J. F.; Schott, M. E.; Milenic, D. E.; Yokota, T.; Schlom, J., 1993, *Protein Engineering.*, 6, 989-995.

8 Begent, R. H. J.; Verhaar, M. J.; Chester, K. A.; Casey, J. L.; Green, A. J.; Napier, M. P.; Hopestone, L. D.; Cushen, N.; Keep, P. A.; Johnson, C. J.; Hawkins, R. E.; Hilson, A. J. W.; Robson, L., 1996, *Nature Med.* 2, 979-984.

9 Yoo, T. M.; Chang, H. K.; Choi, C. W.; Webber, K. O.; Le, N.; Kim, I. S.; Eckelman, W. C.; Pastan, I.; Carrasquillo, J. A.; Paik, C. H., 1997, *J. Nucl. Med.* 38, 294-300.

10 Behr, T. M.; Becker, W. S.; Bair, H. J.; Klein, M. W.; Stuhler, C. M.; Cidlinsky, K. P.; Wittekind, C. W.; Scheele, J. R.; Wolf, F. G., 1995, *J. Nucl. Med.* 36, 430-441.

11 Buijs, W. C. A. M.; Massuger, L. F. A. G.; Claessens, R. A. M. J.; Kenemans, P.; Corstens, F. H. M., 1992, *J. Nucl. Med.* 33, 1113-1120.

12 Choi, C. W.; Lang, L.; Lee, J. T.; Webber, K. O.; Yoo, T. M.; Chang, H. K.; Le, N.; Jagoda, E.; Paik, C. H.; Pastan, I.; Eckelman, W. C.; Carrasquillo, J. A., 1995, *Cancer Res.* 55, 5323-5329.

13 Maack, T.; Johnson, V.; Kau, S. T.; Frgueiredo, J.; Sigulem, D., 1979, *Kidney Int.* 16, 251-270.

14 Silvernagl, S., 1988, *Physiol. Rev.* 68, 911-1007.

15 Hammond, P. J.; Wade, A. F.; Gwilliam, M. E.; Peters, A. M.; Myers, M. J.; Gilbey, S. G.; Bloom, S. R.; Calam, J., 1993, *Br. J. Cancer.* 67, 1437-1439.

16 Behr, T. M.; Becker, W. S.; Sharkey, R. M.; Juweid, M. E.; Dunn, R. M.; Bair, H.-J.; Wolf, F. G.; Goldenberg, D. M., 1996, *J. Nucl. Med.* 37, 829-833.

17 Kobayashi, H.; Yoo, T. M.; Kim, I. S.; Kim, M.-K.; Le, N.; Webber, K. O.; Pastan, I.; Paik, C. H.; Eckelman, W. C.; Carrasquillo, J. A., 1996, *Cancer Res.* 56, 3788-3795.

18 Wu, C.; Jagoda, E.; Brechbiel, M.; Webber, K. O.; Pastan, I.; Gansow, O.; Eckelman, W. C., 1997, *Bioconjugate Chem.* 8, 365-369.

19 Arano, Y.; Mukai, T.; Uezono, T.; Wakisaka, K.; Motonari, H.; Akizawa, H.; Taoka, Y.; Yokoyama, A., 1994, *J. Nucl. Med.* 35, 890-898.

20 Arano, Y.; Mukai, T.; Akizawa, H.; Uezono, T.; Motonari, H.; Wakisaka, K.; Kairiyama, C.; Yokoyama, A., 1995, *Nucl. Med. Biol.* 22, 555-564.

21 Duncan, J. R.; Welch, M. J., 1993, *J. Nucl. Med.* 34, 1728-1738.

22 Duncan, J. R.; Stephenson, M. T.; Wu, H. P.; Anderson, C. J., 1997, *Cancer Res.* 57, 659-671.

23 Franano, F. N.; Edwards, W. B.; Welch, M. J.; Duncan, J. R., 1994, *Nucl. Med. Biol.* 21, 1023-1034.

24 Haseman, M. K.; Goodwin, D. A.; Meares, C. F.; Kaminski, M. S.; Wensel, T. G.; McCall, M. J.; Levy, R., 1986, *Eur. J. Nucl. Med.* 12, 455-460.

25 Arano, Y.; Wakisaka, K.; Ohmomo, Y.; Uezono, T.; Mukai, T.; Motonari, H.; Shiono, H.; Sakahara, H.; Konishi, J.; Tanaka, C.; et al., 1994, *J. Med. Chem.* 37, 2609-2618.

26 Arano, Y.; Wakisaka, K.; Ohmono, Y.; Uezono, T.; Akizawa, H.; Nakayama, N.; Sakahara, H.; Tanaka, C.; Konishi, J.; Yokayama, A., 1996, *Bioconjugate Chem.* 7, 628-637.

27 Arano, Y.; Fujioka, Y.; Akizawa, H.; Ono, M.; Uehara, T.; Wakisaka, K.; Nakayama, M.; Sakahara, H.; Konishi, J.; Saji, H., 1999, *Cancer Res.* 59, 128-134

SYNTHETIC PEPTIDE AND THEIR POLYMERS AS VACCINES

Synthetic Peptide Polymers as the Next Generation of Vaccines

D. Jackson, E. Brandt, L. Brown, G. Deliyannis, C. Fitzmaurice, S. Ghosh, M. Good, L. Harling-McNabb, D. Dadley-Moore, J. Pagnon, K. Sadler, D. Salvatore, and W. Zeng
Cooperative Research Centre for Vaccine Technology, Department of Microbiology & Immunology, The University of Melbourne, Parkville 3052, Australia

1. INTRODUCTION

Vaccines follow the availability of clean water as the most cost effective method of achieving improved public health. Infectious diseases, autoimmune disorders and some cancers are amenable to prophylactic and therapeutic treatment by vaccines and it is the realisation of this fact that is responsible not only for the efforts being made by a large number of groups world wide in designing new vaccines but also in the efforts of public health organisations in delivering vaccines to the community. Amongst the latest technologies being applied to the development of vaccines, synthetic peptides offer a versatile approach to the problems of vaccine design. Immunisation with peptides can induce humoral (antibody) and cellular (cytotoxic T cell and helper T cell) immune responses capable of cross-reacting with intact antigen. Because of this ability to stimulate both humoral and cellular arms of the immune response, a great deal of effort has been put into evaluating peptides as vaccines not only in those situations where antibody is an important determinant in immunity but also in cases where cytotoxic T cells are required for the elimination of virus infected cells or cancer cells.

Biomedical Polymers and Polymer Therapeutics
Edited by Chiellini *et al.*, Kluwer Academic/Plenum Publishers, New York, 2001

1.1   Why are synthetic peptides promising vaccine candidates?

Important practical and commercial advantages in producing synthetic peptide-based artificial antigens and immunogens include the following:

- infectious material is not required which is an important consideration when immunising against highly dangerous infectious diseases such as AIDS, hepatitis or rabies
- cell mediated immune responses as well as antibody responses can be elicited by synthetic peptides
- deleterious sequences that are present in some proteins, for example the EBNA antigens of Epstein-Barr virus and the M protein of streptococcal bacteria both of which are believed to induce autoimmune disease, can be deleted
- non-protein groups such as lipid, carbohydrate and phosphate can be readily introduced to improve immunogenicity, stability and solubility
- quality control is often simplified because the material can be chemically defined
- peptides are stable indefinitely in freeze-dried form at ambient temperature, avoiding the requirement for maintenance of a "cold chain" during storage and distribution
- the cost of production of large amounts of synthetic peptides to GMP standards is competitive with that of traditional alternatives.

1.2   Requirements that must be met by synthetic peptide-based vaccines.

For any peptide to be able to induce an effective antibody response it must contain particular sequences of amino acids known as epitopes that are recognised by the immune system. In particular, for antibody responses, epitopes need to be recognised by specific immunoglobulin (Ig) receptors present on the surface of B lymphocytes. It is these cells which ultimately differentiate into plasma cells capable of producing antibody specific for that epitope. The epitope recognised by surface immunoglobulin receptors also needs to possess a conformation that is similar to that assumed by the same sequence within the intact antigen so that antibodies capable of cross reacting with the native antigen are produced. In addition to these B cell epitopes, the immunogen must also contain epitopes that are presented by antigen presenting cells (APC) to specific receptors present on helper T lymphocytes, the cells which are necessary to provide the signals required for the B cells to differentiate into antibody producing cells. Helper T cell epitopes are bound by molecules present on the surface of APCs that are coded by class II genes of the major histocompatibility complex (MHC). The complex of the class II molecule and peptide epitope is then recognised by specific T-cell receptors (TCR) on the surface of T helper lymphocytes. In this way the T cell, presented with an antigenic epitope in the context of an MHC molecule, can be activated and provide the necessary signals for the B lymphocyte to differentiate. Figure 1 illustrates the basic recognition events leading to the production of antibody in response to a complex antigen.

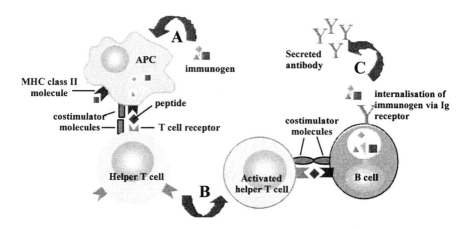

Figure 1. The first step in the generation of an antibody response is the uptake of immunogen by APC (A). Complex antigens undergo proteolysis to form peptides, some of which are bound by major histocompatibility complex class II molecules and transported to the APC surface. Helper T cells that bear receptors capable of interacting with the peptide/class II complexes can then bind to the APC, an event which enables additional interactions through co-stimulatory molecules. These recognition events result in the transmission of activation signals to the T cell, one consequence of which leads to expression by the T cell of other co-stimulatory molecules on its surface (B). The activated T cell is now poised to respond to those B cells that display similar peptide/class II complexes on their surfaces, acquired as a result of internalisation of the same immunogen through specific surface immunoglobulin receptors. It is this interaction between T cells and B cells that is termed "help," and that results in triggering of the B cell to differentiate into a plasma cell capable of secreting antibody of the same specificity as that of the immunoglobulin receptor (C). Cytokines are also produced by each cell type shown which profoundly influence the type of immune response that is elicited.

It should be noted that if antigen is provided in the form of a simple peptide rather than as part of a complex immunogen, the need for internalisation and subsequent intracellular fragmentation may be bypassed because the peptide epitopes can bind directly to class II molecules on the cell surface.

In the case of viral infections and in many cases of cancer, antibody is of limited benefit in recovery and the immune system responds with cytotoxic

T cells (CTL) which are able to kill the virus-infected or cancer cell. Like helper T cells, CTL are first activated by interaction with APC bearing their specific peptide epitope presented on the surface, this time in association with MHC class I rather than class II molecules. Once activated the CTL can engage a target cell bearing the same peptide/class I complex and cause its lysis. It is also becoming apparent that helper T cells play a role in this process; before the APC is capable of activating the CTL it must first receive signals from the helper T cell to upregulate the expression of the necessary costimulatory molecules.

In general then, an immunogen must contain epitopes capable of being recognised by helper T cells in addition to the epitopes that will be recognised by surface Ig or by the receptors present on cytotoxic T cells. It should be realised that these types of epitopes may be very different. For B cell epitopes, conformation is important as the B cell receptor binds directly to the native immunogen. In contrast, epitopes recognised by T cells are not dependent on conformational integrity of the epitope and consist of short sequences of approximately nine amino acids for CTL and slightly longer sequences, with less restriction on length, for helper T cells. The only requirements for these epitopes are that they can be accommodated in the binding cleft of the class I or class II molecule respectively and that the complex is then able to engage the T-cell receptor.

Because individuals possess different alleles of the genes that encode class I and class II MHC proteins, an epitope that is recognised by one individual may not be recognised by the MHC molecules of other individuals. Therefore, there is an issue of polymorphism associated with recognition of antigenic epitopes by the immune system. Furthermore, because many diseases are caused by organisms where the target antigens are polymorphic, multiple serotypes of target antigens exist for many important pathogens of man and animals; examples of such pathogens include influenza virus, the malaria parasite, streptococcal bacteria, human immunodeficiency virus, *Dichelobacter nodosus*, the causative agent of footrot in sheep, and *Escherichia coli* to name a few.

The polymorphism of MHC proteins which are essential recognition elements of the immune system and also the polymorphism of target proteins of pathogens presents the designer of vaccines with the problem of displaying sufficient epitopes to cover the universe of different host MHC types as well as multiple pathogen serotypes. Although attempts have been made to solve the problem of polymorphic target antigens by defining conserved epitopes within pathogens, such epitopes do not always elicit protective immune responses. It is therefore necessary to develop immunogens in which some, if not all, of the variant epitopes are represented as well as including sufficient epitopes that will be recognised by class II and, if a cellular immune response involving cytotoxic T cells is required, class I MHC molecules.

Because synthetic peptides are routinely assembled as short monomers representing individual epitopes, they may lack immunogenic activity either because they do not contain appropriate helper T-cell epitopes or conformational integrity is lacking in the B-cell epitope; in addition they may be rapidly metabolised before having a chance to initiate an immune response. In many cases one or more of these factors contribute to the poor immunogenicity often observed with synthetic peptides. The rest of this chapter will address some of the recent improvements in peptide vaccine technology that have been used to advance the development of synthetic peptide-based vaccines.

1.2.1   Providing conformational integrity for B cell epitopes

The antigenic structure of a large number of antigens has been investigated and the results of many of these experiments have led to the identification of antigenic sites of proteins. This subsequently led to the synthesis of peptides representing these putative antigenic sites in the hope that they would induce antibody capable of binding to the native protein. The results of early experiments with synthetic peptides soon demonstrated that unless a B-cell epitope was easily able to assume the conformation that it adopted in the original antigen, then antibodies elicited by it would not bind to the native protein. Outside of the context of the whole protein sequence, peptide sequences do not usually contain sufficient structural information to fold correctly and consequently one of the problems confronting the designers of peptide vaccines is to find a way to induce the correct conformation into synthetic peptide epitopes.

Successful approaches to the problem of mimicking the correct conformation have included the synthesis of antigenically active peptides of hepatitis B surface antigen[1]. The linear forms of the peptides investigated were found to be inactive whereas cyclisation through the formation of a disulphide bond led to antigenically active species. Another example of a conformation-dependent epitope is found in the M protein of group A streptococcus (GAS); the M protein is a coiled-coil alpha helical surface protein of the bacterium and requires alpha helical conformation in order to be recognised by antibody. A simple peptide representing an antigenic sequence of M protein does not assume the correct helical structure when made as a simple synthetic peptide but when flanked by sequences that possessed a propensity for folding into an alpha helix, the synthetic peptide did express antigenic activity[2]. Finally, Mutter et al.[3] have introduced the TASP (template associated synthetic peptide) technique to assemble four epitopes on a cyclic template and reported that the geometry of the arrangement brought the individual peptides into juxtaposition which allowed the correct conformation of each to be assumed (see below).

These and other solutions to the problem of correctly folding peptides illustrate not only that the location by epitope mapping of antigenic epitopes within a protein's sequence is a prerequisite for the design of synthetic peptide-based vaccines but also that some knowledge of the three dimensional structure of the antigen may be required. In many cases, unless individual epitopes are identified and appropriate structural features incorporated into them, antibody of the correct specificity will not be induced.

1.2.2   Providing T cell help for the induction of antibody

(i) the use of carrier proteins as a source of T cell helper epitopes. The most common method for inducing antibody responses against synthetic peptides that contain B cell epitopes is to couple them to carrier proteins which, by virtue of the range of epitopes available for class II presentation, act as a source of stimulus for helper T cells. Methods for chemically coupling peptides to proteins range from the use of relatively non-specific reagents such as glutaraldehyde to highly specific heterobifunctional reagents that circumvent the formation of carrier protein-carrier protein or peptide-peptide conjugates yielding only peptide-carrier protein conjugates. This approach has often been successful, particularly in experimental work aimed at the generation of antibody reagents. For example, peptide-specific antibodies have been used to demonstrate the presence of proteins not previously detected by more conventional methods[4]. When coupled to carrier proteins, antibodies to self-antigens such as human chorionic gonadotrophin[5] and the luteinising hormone releasing hormone[5,6] have been elicited and shown to have biological effects. Chemical coupling of peptide to a carrier protein can, however, cause changes in the determinant(s) of interest[7] and because it is difficult to control the coupling reaction, the production of a heterogeneous preparation of peptide-carrier protein complexes often results which complicates quality control. It has also been reported that if prior exposure to the carrier protein has occurred, and we should remember that popular carrier proteins such as tetanus toxoid are themselves used in vaccination regimens, then suppression of the B cell response to the epitope of interest can occur limiting the amount of useful antibody that is elicited [8]. Another limitation of this approach is the availability of carrier protein; if a successful vaccine based on a peptide-protein carrier protein system is developed then sufficient carrier protein must be available. As mentioned above some toxoids in current use as vaccines themselves are popular as carrier proteins due to prior registration of the material, but the availability of sufficient carrier proteins could be problematic if a peptide-based vaccine was to become too popular. Another potential problem is that, although the use of a carrier protein usually provides sufficient numbers of different T helper epitopes to be recognised by most or all individuals within an outbred population, the method does not easily lend itself to the reproducible production of a vaccine candidate consisting of multiple B cell epitopes representing different serotypes attached to a single carrier protein molecule.

(ii) the use of synthetic T cell epitopes. The assembly of vaccines containing totally synthetic defined T cell epitopes would solve some of the problems associated with the use of carrier proteins. Useful immune responses have been achieved in experimental situations using different strategies including assembly of tandem (T-cell epitope)-(B-cell epitope) constructs[9] and non-specific polymerisation of the peptide[10]. The importance of order or orientation of epitopes became apparent using these approaches; antibody of irrelevant specificity may be induced unless the immunogen is constructed with the individual epitopes in the correct orientation[11]. The use of synthetic peptides representing T cell epitopes also seems to avoid the problem of carrier protein induced epitope suppression[12] and in at least one system a totally synthetic vaccine based on LHRH and a synthetic T cell epitope has been shown to abolish fertility[13]. Using orthogonally protected lysine residues, simple branched immunogens containing B and T cell epitopes have also been investigated[14]. This study demonstrated that the choice of T cell epitope used to elicit help for antibody production was important (Fig.2) and also that the branched immunogens investigated elicit higher titres of antibody than the corresponding tandem linear arrangement of epitopes. Simple branched immunogens also displayed a greater stability to the proteolytic activity of serum (Fig.2).

Although immune responses have been achieved in experimental models by the use of totally synthetic B and T cell epitope constructs, the problems of displaying multiple epitopes to cover the universe of MHC molecules and multiple serotypes are not addressed by this approach. The discovery of "promiscuous" T helper epitopes, that is, epitopes that elicit helper T cells in multiple strains or even in different species of animal[15] seemed to offer a possible solution at least to part of the problem, but frequently the promiscuity is not universal and additional helper T cell epitopes are required.

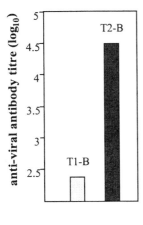
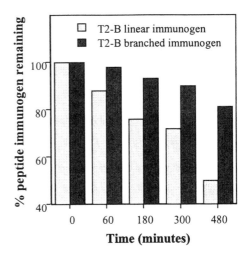

Time (minutes)

Figure 2. A comparison of the relative efficiencies of different T-cell epitopes to provide help in the induction of antibody responses and the relative stabilities of branched and linear immunogens to proteolysis. The left hand panel shows the abilities of two immunogens constructed with the same B cell epitope but with different helper T cell epitopes, T1 or T2, to elicit antibody. The right hand panel demonstrates that an immunogen composed of the T2 and B epitopes in a branched configuration is more resistant to proteolysis following exposure to serum than the same epitopes when assembled as a tandem linear arrangement.

1.2.3   The multiple antigenic peptide system (MAPS)

Each of the approaches described above poses its own problems including technical difficulties in assembling long peptides, the introduction of novel determinants at epitope-epitope junctions[16] which may redirect the immune response and the masking of epitopes by non-specific polymerisation processes. The MAPS developed by Tam[17] and extended by others particularly Hudercz[18] heralded a significant step forward allowing the assembly of many, but usually identical, peptide sequences attached to a core of branching lysine residues (Fig.3) to yield a multivalent vaccine candidate.

This technology has been adopted by many groups and many successes have been reported in experimental animals; the first MAPs consisted of a single type of peptide epitope but the availability of orthogonally protected side groups of lysine allows the assembly of MAPs from which different peptide sequences can be synthesised.

Despite the successes of this technique, a limitation of the method is the difficulty in assuring purity of the product. The process usually involves the construction of very large structures and because assembly involves the simultaneous synthesis of all epitopes, there is a high probability that steric constraints and aggregation of the growing dendrimers will reduce the fidelity of the sequences being assembled. Furthermore, the number of different epitopes that can be introduced into these MAPS is limited.

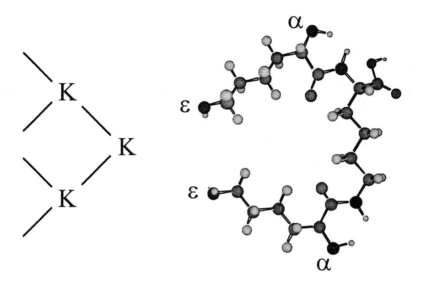

Figure 3. Schematic representation of a tetravalent MAP formed by the addition of a lysine residue to each of the two primary amine groups of a central lysine (K) residue, an arrangement that provides four primary amine groups from which separate epitopes can be synthesised. The right hand schematic is an energy minimised molecular model of the trivalent lysine core indicating the alpha and epsilon amino groups to which individual epitopes are added.

1.2.4   Use of solid phase supports to produce polyvalent epitopes

Peptide synthesis is usually carried out using solid phase chemistry on a beaded support which is derivitised with appropriate active groups that act as the synthesis points for peptide chains. Exposure of the peptide-support linkage to acidic conditions normally results in release of the peptide from the support, but use of an appropriate acid stable link between peptide and support as well as incorporation of cleavable cross-links within the beaded solid phase can be used to allow peptides to be synthesised in the usual way but remaining attached to the polymer backbone. In this way polymeric forms of the target peptide can be assembled in which the peptides are pendant from a polymer backbone. This approach has been reported to be useful for eliciting antibodies to peptide epitopes[19] and in the induction of T cell responses[20]. Such soluble polymeric supports have also been used in the preparation of immunoadsorbents which allow the preparation of antibodies from serum samples[19,21]. This approach does not, however, allow for the incorporation of large numbers of different peptide epitopes nor does it allow for purification of the epitopes because, as with assembly of MAPs, synthesis of the epitopes is carried out simultaneously and not prior to polymerisation.

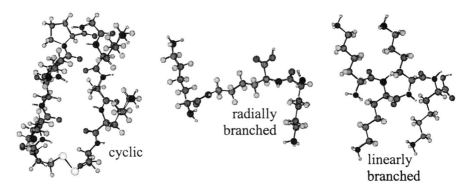

Figure 4. Molecular models of various templates to which synthetic peptide epitopes can be ligated. Each template is tetravalent providing four points to which peptide epitopes can be attached. The cyclic template is assembled by closure of a disulphide bond within the sequence, the radially branched template shown is assembled by adding a single tier of lysine residues to the alpha and epsilon amino groups of a central lysine residue and the linearly branched template is a simple linear sequence of four lysine residues.

1.2.5   Chemical ligation as a solution to purity problems

A solution to the problem of achieving homogeneity when MAPS are assembled is to synthesise and purify core and peptide epitopes separately and then chemically ligate the products. By choosing appropriate chemistries the ligation process can be specifically directed to ensure that the correct final structure is achieved. Particularly elegant approaches have included the use of oxime chemistry[22] in which the specific reaction of aldehyde groups with aminooxy groups to form an oxime bond is used.

Using a template similar to that described by Mutter etal.[3], Zeng et al.[23] successfully fabricated a pure tetramer by chemical ligation of four 23 amino acid residue peptides using oxime chemistry to the attachment points of the cyclic template. This tetramer was also shown to be more immunogenic than the individual peptide monomers[24] even though the structure of the individual peptide epitopes showed no particular features of conformational structure or stability[25] similar to those reported by Mutter et al. for the TASP that they had assembled.

By judicious use of ligation chemistries, the method also allows flexibility in the attachment of epitopes through different points of their sequence thereby determining orientation and overall structure of the molecule, variations that can perhaps be used to tailor the type of immune response that is elicited. Cyclic, linear and branching templates have all been reported for construction of synthetic peptide-based immunogens[3,23] and schematics of some of these structures are shown in Figure 4.

Despite the fact that purified products can be made using chemical ligation, a limitation of the technology is that only relatively small numbers of epitopes can be incorporated. Not only are the total numbers of epitopes small but these methods also limit the number of different epitopes that can be incorporated into a single molecule.

1.2.6   Free radical induced polymerisation of acryloyl peptides

We have recently developed a method for the assembly of large numbers of identical or different pure epitopes into polyvalent artificial proteins using free radical initiated polymerisation[26,27]. In a way similar to the polymerisation of acrylamide to form polyacrylamide, peptides acylated with the acryloyl group will polymerise to form peptide side chains that are pendant from an inert alkane backbone. In this way, large synthetic structures can be assembled while avoiding errors inherent in long sequential syntheses. Each epitope sequence is first assembled using conventional solid phase synthesis but before deprotection of the side chains and release from the solid support, the terminal amino group is acylated using acryloyl chloride (Fig.5). Individual acryloyl peptides can then be purified and then polymerised, either singly or in admixture with similarly derivatised peptides, to yield polymers with peptide epitopes attached to the alkane backbone. In practice, the size of peptide epitopes prevents polymerisation by steric hindrance and so acrylamide or a similar "spacer" is added to distance the pendant epitopes along the length of the backbone. An advantage of this approach, as with other chemical ligation strategies mentioned above, is the ability to purify individual defined epitopes prior to polymerisation. This offers advantages in quality control which are lacking in other methods available for producing multivalent immunogens.

This technology is very flexible in terms of the types of structures that can be produced and the following variations are easily investigated:

- individual peptides can be made to terminate in either a carboxyl or carboxyamide group to mimic the natural situation of the particular epitope where it may form the C-terminus of the protein or an internal sequence of amino acids
- introduction of a lysine residue possessing an orthogonally protected epsilon amino group at the C-terminus of the peptide provides the opportunity to have peptide epitopes pendant from their C-termini whereas introduction of lysine in a central position allows the peptide to be pendant from its mid-point

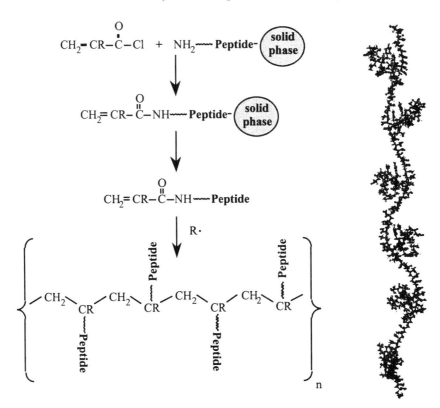

Figure 5. Schematic of the reaction sequence used to acryloylate and subsequently polymerise individual purified peptide epitopes into high molecular weight polymers. The strategy lends itself to the copolymerisation of the same or different peptides as well as the inclusion of other materials into the backbone. The molecular model on the right hand side represents a part of a copolymer composed of a single type of peptide epitope and an acrylamide backbone. In this model, a ratio of acrylamide:peptide of 10:1 is represented and six epitopes pendant from the backbone are shown.

- because the peptide epitopes are individually assembled, additional molecular features such as intra-peptide disulphide loops can be incorporated to mimic structures similar to those found in the native protein
- the physical and chemical properties of the backbone can be tailored by incorporation of, for example, hydrophilic, hydrophobic, charged or polar groups
- in the case of co-polymerisation of two or more acryloyl peptides, the distribution of the peptides along the polymer backbone will be dependent on the physical properties of the component peptides.

This will allow many different interaction possibilities to arise depending on the sequence of the epitopes. The resulting arrangements may even allow for the simulation of conformational determinants in a way similar to that reported for TASP-immunogens[3].

The immunological properties of polymeric epitopes produced by this type of free radical induced polymerisation have been reported and in all cases examined their immunogenicity is superior to that of monomeric peptides[27]. More recently we have used this approach to copolymerise 8 peptide epitopes of group A streptococci and were able to induce protective antibodies against different bacterial isolates in outbred mice (data not shown).

1.2.7 Self adjuvanting synthetic peptide-based immunogens

The multivalent forms of immunogen discussed above can clearly induce higher levels of antibody and helper T cell activity than the corresponding monomeric determinants, but in most of these cases it has been necessary to deliver these synthetic immunogens with an adjuvant (an agent that augments specific immune responses to antigens). Immunological adjuvants include oil-based formulations such as complete Freund's adjuvant (CFA), which has been used in experimental animal models for many years but is too reactogenic for human use and is currently loosing favour even as an experimental adjuvant because it can cause pathology at the site of injection. It contains heat-killed *Mycobacterium tuberculosis* in a mixture of mineral oil and emulsifier. Currently, the only adjuvants licensed for general use in humans are the calcium and aluminium salts (phosphate or hydroxide), which have a high adsorptive capacity for some proteins. These substances can enhance the response to certain antigens but in general have little if any effect in adjuvanting peptides. Despite the recent promotion of many novel adjuvant formulations and molecules, few exist that are potent and yet display sufficiently low-toxicity to be suitable for human or animal use.

The most effective experimental adjuvants have an immunostimulatory component such as killed bacteria or biologically active bacterial cell wall components such as the lipid A portion of bacterial lipopolysaccharide (LPS) or muramyl dipeptide from bacterial peptidoglycan. These components are known to activate lymphocytes and/or macrophages. Efficient synthetic adjuvants some of which are based on these materials are now available and some at least show promise for use with synthetic peptide vaccines. In 1983, Wiesmuller, Bessler and Jung[28] described a synthetic analog of the N-terminal moiety of bacterial lipoprotein from *E.coli* (S-{2,3-bis(palmitoyloxy)-(2-RS)-propyl}-N-palmitoyl-R-cysteine) termed Pam3Cys (Fig.6). When this compound is coupled to a peptide epitope, the resulting lipopeptides were found to be potent immunogens. Subsequently, many investigators have shown that chemical conjugation of this lipid component directly onto an epitope-containing peptide creates a "self-adjuvanting" immunogen that can elicit strong antibody responses when delivered in the absence of any other adjuvant.

An example from our own laboratory[29] shows that the serum antibody responses achieved following inoculation of a synthetic peptide covalently coupled to Pam3Cys are comparable to those achieved when the peptide is emulsified in Freund's complete adjuvant. In contrast, attachment of other lipid components including palmitic acid or cholesterol had little or no adjuvanting effect (Fig.6).

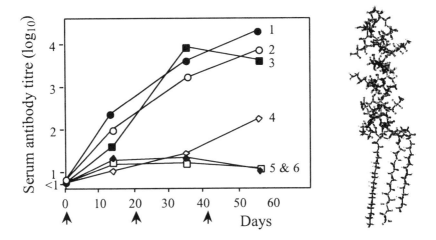

Figure 6. Comparison of the serum antibody levels at various times following inoculation with a synthetic peptide injected alone or in the presence of adjuvant. Synthetic peptide delivered (1) intranasally and covalently coupled to Pam3Cys (2) intraperitoneally and covalently coupled to Pam3Cys (3) intraperitoneally emulsified in CFA (4) intraperitoneally and covalently coupled to two palmitic acid molecules (5) intraperitoneally and covalently coupled to a cholesterol molecule (6) intraperitoneally in normal saline. The arrows indicate the times of immunisation. The right hand side of the figure is a representation of the structure of the synthetic peptide with Pam3Cys attached at the N-terminus. The three palmitic acid residues of the Pam3Cys moiety are seen at the bottom of the model.

The ability to assemble self-adjuvanting immunogens obviates the need for emulsification in traditional and often harmful oil based adjuvants and also allows different delivery routes to be accessed. The Pam3Cys peptide was equally effective when delivered by the intranasal route as when delivered parenterally (Fig.6). Furthermore, the Pam3Cys-peptide elicited IgA antibody forming cells in the lung and draining lymph nodes following intra-nasal administration[29]. Clearly the ability to induce IgA has ramifications for vaccines that are required to induce mucosal immune responses such as those to prevent respiratory diseases as well as HIV and other sexually transmitted pathogens. The ability to deliver vaccines intranasally also offers an alternative to "needle" administration. Successful delivery of Pam3Cys-containing immunogens by the oral route has also been reported[30] and offers potential for vaccination not only against gastrointestinal diseases but, because the antigenic stimulation of mucosa at one site may cause immunity at other mucosal sites, against pathogens which infect other mucosal portals of entry including the respiratory tract and urinary genital tract.

A further enhancement in the potency of immunogens is achieved by combining the strategies for producing multivalent immunogens with self-adjuvanting molecules. High titres of antibody which are long lived were induced by tetrameric constructs to which Pam3Cys was attached[29]. In this case it was found that low doses of Pam3Cys-containing tetramer induced higher levels of antibody than those elicited by the monomeric Pam3Cys-peptide. Again, high levels of antibody were achieved by the intraperitoneal or intranasal routes.

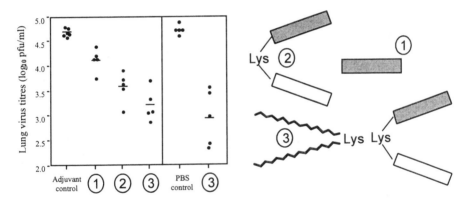

Figure 7. Viral clearing responses of mice inoculated with different synthetic peptide constructs containing the CTL epitope alone, 1; the CTL epitope covalently attached to a helper T cell epitope, 2; the CTL epitope covalently attached to a helper T cell epitope and two molecules of palmitic acid, 3. The constructs were delivered subcutaneously in Seppic Montanide ISA720 adjuvant (left panel) or in phosphate buffered saline (right panel). Mice were challenged 4 weeks later with the influenza virus from which the epitopes were derived and the titre of infectious virus in lungs was determined 5 days later. The titres are expressed as plaque forming units per ml of lung homogenate supernatant.

Strong cytotoxic T cell responses can also be induced when lipid is attached to the appropriate epitope[31]. Recovery from infection with influenza virus requires the induction of an efficient cytotoxic T cell response and we have shown that the ability of a CTL epitope from the nucleoprotein of influenza virus to elicit a viral clearing responses is improved by the incorporation of a T helper epitope sequence. The clearance of virus is further augmented by the addition of 2 palmitic acid groups (Fig.7).

These experiments were carried out by emulsifying the synthetic peptide constructs in Seppic Montanide ISA 720 adjuvant but when inoculated in saline alone, the vaccine containing the CTL epitope, T helper epitope and palmitic acid was still able to induce a response that could clear as much as 98% of the virus from the lungs of influenza-infected mice. In contrast, the non-lipidated epitope was found to be ineffective at eliciting a virus clearing response. Although we have not yet examined whether Pam3Cys is more efficient than two palmitic acid molecules in this cytotoxic T cell system the requirements for CTL induction may differ from those for antibody induction because we have found that cholesterol can replace palmitic acid in eliciting viral clearing responses.

Not only have encouraging results been obtained with models of virus infection such as the influenza virus system but they have also been reported in several cancer models[32] The destruction of cancer cells is often dependent on a vigorous CTL response and there is every reason to believe that efficient delivery of tumour-specific CTL epitopes could provide "off the shelf" cancer vaccines. A realisation of the central role that dendritic cells play in the efficient presentation of antigen to the immune system is nowhere more evident than with the development of anti-cancer vaccines[33] and it is possible that successes in this area could provide a momentum which will benefit peptide vaccines in general.

2.   CONCLUSION

As the development of synthetic peptide-based vaccines progresses, the early problems which slowed their progress are being solved and overcome by new technologies. The ultimate goal of a peptide-based vaccine is to design a formulation which will display adequate numbers of epitopes to address MHC diversity and the occurrence of multiple serotypes in a highly immunogenic, self-adjuvanting preparation.. The development of rationally designed vaccines, such as peptide-based vaccines, requires a contribution from diverse disciplines ranging from immunology and peptide and polymer chemistry to infectious disease epidemiology and bacterial physiology. Many groups from these disparate fields continue to contribute to this effort and thanks to recent advances, some of which have been described in this chapter, these goals are now being realised.

An enormous amount of progress has been made in the last few years towards the development of totally synthetic and self-adjuvanting vaccine candidates which are capable of eliciting humoral and cell-mediated immune responses. In an increasing number of cases vaccine candidates based on peptides and peptide polymers have been shown to confer protection against infectious diseases and cancer and have also been shown to successfully modify endocrine functions involved in reproduction. The first generation of peptide vaccines may not be far away.

ACKNOWLEDGMENTS

This work was supported by grants from the National Health and Medical Research Council of Australia and from the Cooperative Research Centre for Vaccine Technology.

REFERENCES

1. Ionescu-Matiu, I., Kennedy, R. C., Sparrow, J. T., et al., 1983, Epitopes associated with a synthetic hepatitis B surface antigen peptide. *J Immunol* 130: 1947-52.
2. Relf, W. A., Cooper, J., Brandt, E. R., et al., 1996, Mapping a conserved conformational epitope from the M protein of group A streptococci. *Pept Res* 9: 12-20.
3. Mutter, M., Tuchscherer, G. G., Miller, C., et al., 1992, Template-assembled synthetic proteins with four-helix-bundle topology. *J. Am. Chem. Soc.* 114: 1463-1470.
4. Sutcliffe, J. G., Shinnick, T. M., Green, N., Liu, F. T., Niman, H. L.,Lerner, R. A., 1980, Chemical synthesis of a polypeptide predicted from nucleotide sequence allows detection of a new retroviral gene product. *Nature* 287: 801-5.
5. Talwar, G. P., Sharma, N. C., Dubey, S. K., et al., 1976, Isoimmunization against human chorionic gonadotropin with conjugates of processed beta-subunit of the hormone and tetanus toxoid. *Proc Natl Acad Sci U S A* 73: 218-22.
6. Fraser, H. M., Gunn, A., Jeffcoate, S. L., Holland, D. T., 1974, Effect of active immunization to luteinizing hormone releasing hormone on serum and pituitary gonadotrophins, testes and accessory sex organs in the male rat. *J Endocrinol* 63: 399-406.
7. Briand, J. P., Muller, S., Van Regonmortel, M. H. V., 1985, Synthetic peptides as antigens: pitfalls of conjugation methods. *J. Immunol. Meth* 78: 59.
8. Schutze, P. M., Leclerc, C., Jolivet, M., Audibert, F., Chedid, L., 1985, Carrier-induced epitopic suppression, a major issue for future synthetic vaccines. *J Immunol* 135: 2319-2322.
9. Francis, M. J., Hastings, G. Z., Syred, A. D., McGinn, B., Brown, F., Rowlands, D. J., 1987, Non-responsiveness to a foot-and-mouth disease virus peptide overcome by addition of foreign helper T-cell determinants. *Nature* 330: 168-70.
10. Borras-Cuesta, F., Fedon, Y., Petit-Camurdan, A., 1988, Enhancement of peptide immunogenicity by linear polymerization. *Eur J Immunol* 18: 199-202.
11. Dyrberg, T., Oldstone, M. B., 1986, Peptides as antigens. Importance of orientation. *J Exp Med* 164: 1344-9.
12. Etlinger, H. M., Knorr, R., 1991, Model using a peptide with carrier function for vaccination against different pathogens. *Vaccine* 9: 512-4.
13. Ghosh, S., Jackson, D. C., 1999, Antigenic and immunogenic properties of totally synthetic peptide-based anti-fertility vaccines. *Int Immunol* 11: 1103-1110.
14. Fitzmaurice, C. J., Brown, L. E., McInerney, T. L., Jackson, D. C., 1996, The assembly and immunological properties of non-linear synthetic immunogens containing T-cell and B-cell determinants. *Vaccine* 14: 553-60.

15. Panina-Bordignon, P., Tan, A., Termijtelen, A., Demotz, S., Corradin, G., Lanzavecchia, A., 1989, Universally immunogenic T cell epitopes: promiscuous binding to human MHC class II and promiscuous recognition by T cells. *Eur J Immunol* 19: 2237-42.

16. Sharma, P., Kumar, A., Batni, S., Chauhan, V. S., 1993, Co-dominant and reciprocal T-helper cell activity of epitopic sequences and formation of junctional B-cell determinants in synthetic T:B chimeric immunogens. *Vaccine* 11: 1321-6.

17. Tam, J. P., 1988, Synthetic peptide vaccine design: synthesis and properties of a high-density multiple antigenic peptide system. *Proc Natl Acad Sci U S A* 85: 5409-13.

18. Hudecz, F., 1995, Design of synthetic branched-chain polypeptides as carriers for bioactive molecules. *Anticancer Drugs* 6: 171-93.

19. Butz, S., Rawer, S., Rapp, W., Birsner, U., 1994, Immunization and affinity purification of antibodies using resin-immobilized lysine-branched synthetic peptides. *Pept Res* 7: 20-3.

20. Jackson, D. C., Fitzmaurice, C. J., Sheppard, R. C., McMurray, J., Brown, L. E., 1995, The antigenic and immunogenic properties of synthetic peptide-based T-cell determinant polymers. *Biomed Pep Prot and Nucl Acids* 1: 171-176.

21. Brandt, E. R., Hayman, W. A., Currie, B., et al., 1996, Opsonic human antibodies from an endemic population specific for a conserved epitope on the M protein of group A streptococci. *Immunology* 89: 331-7.

22. Rose, K., 1994, Facile synthesis of homogeneous artificial proteins. *J. Am. Chem. Soc.* 116: 30-33.

23. Zeng, W., Rose, K. and Jackson, D.C. Polyoximes: a flexible approach for producing homogeneous artificial proteins (1995). In: "Peptides 1994", Proceedings of the Twenty-Third Peptide Symposium (September 4-10, 1994, Braga, Portugal). L.S. Maia, Ed. ESCOM, Leiden, The Netherlands 1995, pp 855-856.

24. Rose, K., Zeng, W., Brown, L. E., Jackson, D. C., 1995, A synthetic peptide-based polyoxime vaccine construct of high purity and activity. *Mol Immunol* 32: 1031-7.

25. Wilce, J. A., Zeng, W., Rose, K., Craik, D. J., Jackson, D. C., 1996, 1H NMR structural study of free and template-linked antigenic peptide representing the C-terminal region of the heavy chain of influenza virus hemagglutinin. *Biomed Pept Proteins Nucleic Acids* 2: 51-8.

26. O'Brien-Simpson, N. M., Ede, N. J., Brown, L. E., Swan, J., Jackson, D. C., 1997, Polymerisation of unprotected synthetic peptides: a view towards synthetic peptide vaccines. *J. Am. Chem. Soc.* 119: 1183-1188.

27. Jackson, D. C., O'Brien-Simpson, N., Ede, N. J., Brown, L. E., 1997, Free radical induced polymerization of synthetic peptides into polymeric immunogens. *Vaccine* 15: 1697-705.

28. Wiesmuller, K. H., Bessler, W., Jung, G., 1983, Synthesis of the mitogenic S-[2,3-bis(palmitoyloxy)propyl]-N- palmitoylpentapeptide from Escherichia coli lipoprotein. *Hoppe Seylers Z Physiol Chem* 364: 593-606.

29. Zeng, W., Jackson, D.C., Murray, J., Rose, K., and Brown, L.E. 1999, Totally synthetic lipid containing polyoxime peptide constructs are potent immunogens *Vaccine* in press.

30. Nardelli, B., Haser, P. B., Tam, J. P., 1994, Oral administration of an antigenic synthetic lipopeptide (MAP-P3C) evokes salivary antibodies and systemic humoral and cellular responses. *Vaccine* 12: 1335-9.

31. Deres, K., Schild, H., Wiesmuller, K. H., Jung, G., Rammensee, H. G., 1989, In vivo priming of virus-specific cytotoxic T lymphocytes with synthetic lipopeptide vaccine. *Nature* 342: 561-4.

32. Melief, C. J., Offringa, R., Toes, R. E., and Kast, W. M. 1996, Peptide-based cancer vaccines. *Curr Opin Immunol* 8: 651-657.

33. Minev, B. R., Chavez, F. L., and Mitchell, M. S. 1998, New trends in the development of cancer vaccines. *In Vivo* 12: 629-638

DEVELOPMENT OF FUSOGENIC LIPOSOMES AND ITS APPLICATION FOR VACCINE

Fusogenic Liposomes Efficiently Deliver Exogenous Antigen through the Cytoplasm into the Class I MHC Processing Pathway

Akira Hayashi and Tadanori Mayumi
Department of Biopharmaceutics, Graduate School of Pharmaceutical Sciences, Osaka University, 1-6, Yamadaoka, Suita, Osaka 565-0871, Japan

1.   INTRODUCTION

For patients with infectious diseases including AIDS or those with cancer, induction of protective immunity, which can eliminate the cells involved in the respective pathologic conditions, is essential. In particular, induction of cytotoxic T lymphocytes (CTL) becomes important in such patients, since CTL plays a central role in protective immunity. It was reported that endogenous antigens in the cytoplasm are generally presented to the T cells in conjunction with class I MHC molecules, while exogenous antigens are processed by endocytosis, then presented to the T cells in conjunction with class II MHC molecules. In recent years, attention has been focused on component vaccines due to their higher safety. However, since component vaccines are exogenous antigens, most component vaccines are endocytosed after administration, then presented to the T cells in conjunction with class II MHC molecules. That is, it is a major drawback of component vaccines that the efficiency of antigen presentation in conjunction with class I MHC molecules, which is essential for CTL induction, is markedly low. Moreover, when peptides that bind to the receptors of class l MHC molecules are used as vaccines, there is the risk that these vaccines may undergo degradation in endosomes during the usual pathway of antigen presentation via endocytosis even when antigen molecules extravasated from endosomes into the cytoplasm. Therefore,

Biomedical Polymers and Polymer Therapeutics
Edited by Chiellini *et al.*, Kluwer Academic/Plenum Publishers, New York, 2001

antigen presentation in conjunction with class I MHC molecules is not expected when such vaccines are administered. To design effective component vaccines, it is essential to develop a novel technique considering intracellular kinetics of antigens; that is, efficient providing of antigens in conjunction with class I MHC molecules for the antigen presentation pathway after direct induction of antigens into the cytoplasm.

To date, we have been developing fusogenic liposomes (FL) whose membrane fusion capacity was identical to that of Sendai virus. FL facilitates direct and efficient induction of inclusion substances into the cytoplasm by membrane fusion without accompanying cytotoxicity. Furthermore, using FL, when we transferred the TNF-α gene to the artery supplying a mouse tumor, TNF- α was efficiently expressed in this artery, which facilitated continuous supply of TNF- α to the region of the tumor, then resulting in therapeutic values of obtaining tumor reduction without any side effect. Thus, it was revealed that FL shows extraordinary superior endocytotic efficiency both *in vitro* and *in vivo* as well as showing superior safety.

Therefore, we applied FL to a component vaccine and attempted to solve the above problems. FL can deliver the encapsulated soluble protein directly into the cytosol of cultured cells and introduce it into the class I MHC antigen-presentation pathway. Moreover, a single immunization with ovalbumin (OVA) encapsulated in FLs but not in simple liposomes results in the potent priming of OVA-specific CTLs. Thus, FLs function as an efficient tool for the delivery of CTL vaccines.

These studies will be discussed here.

2.   FL CAN DELIVER MATERIALS INTO CYTOPLASM

To date, we have been developing FL. The FL, prepared by fusing simple liposomes with Sendai virus particles[1,2,3,4,5,6], can fuse with the cell membrane like the native virus particle and deliver their contents directly and efficiently into the cytoplasm both *in vitro* and *in vivo* as well as showing superior safety (Fig.l).

2.1   Delivery of DTA into Cytoplasm

Fragment A of diphtheria toxin (DTA) is known to kill cells by inactivating elongation factor 2 even when only one molecule of this protein is introduced into the cytoplasm[1,2,6]. Whereas it is absolutely non-toxic even

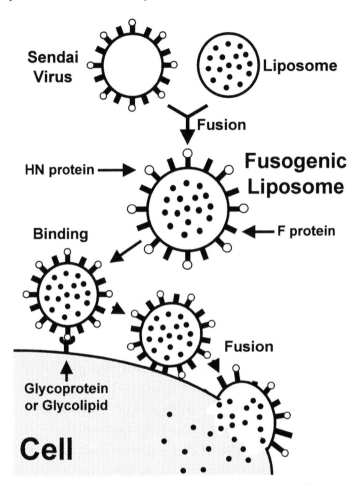

Figure 1. Fusion Mechanism of FL.

if it is taken up by endocytosis, because it cannot reach the cytoplasm owing to degradation by lysosomal enzymes[7,8]. Therefore DTA is a superior marker of the carrier's capability to delivery proteins into cytoplasm. We found that FL containing DTA killed various cultured cells quite[1,2,6]. Because Sendai virus can infect a wide variety of cells with sialic acid, a common component of glycolipid and glycoprotein, as receptors, the FL containing DTA may kill a variety of cells, including tumorigenic cells or suspension cells.

We examined whether the liposomes containing DTA could kill tumorigenic sarcoma-180 (S-180) cells in suspension *in vitro*[9] (Fig.2). FL containing DTA killed S-180 cells in a dose-dependent manner, 50μl of the liposomes with OD_{540} of 0.1 killed 1 x 10^6 S-180 cells. Neither simple liposome containing DTA nor empty FL had any influence on the viability of S-180 cells (Fig.2). These results suggested that FL containing DTA could kill S-180 cells efficiently *in vitro* by introducing DTA into the cytoplasm.

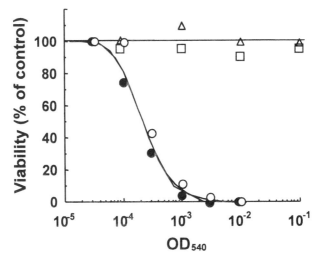

● **Fusogenic liposome containing DTA**
△ **Simple liposome containing DTA**
□ **UV-inactivated Sendai virus**
○ **Intact Sendai virus**

Figure 2. In vitro cytotoxicity of FL containing DTA against S-180.

3.   FL AS A GENE TRANSFER VECTOR

Delivery of foreign genes into animal cells plays an important role not only in *in vitro* research but also in clinical studies for gene therapy. Many viral and nonviral vectors for gene transfer have been developed for these purposes, and they have both advantages and disadvantages[10,11,12]. Most viral vectors can transfer genes efficiently and some of them can be used for gene transfer into tissue cells *in vivo*. However, the structure and the stability of transferred genes are restricted by those of the virus genome. In addition their antigenicity is graver problems to apply clinical trials. In contrast, nonviral vectors such as cationic lipid (CL)-DNA complex can transfer genes with little restriction in their structure but with lower efficiency than viral vectors.

We have clarified that FL delivers their contents directly and efficiently into cytoplasm both *in vitro* and *in vivo*[12a] therefore, we applied FL to a gene transfer vector and attempted to solve the above problems.

3.1 The Characteristics of FL as a Gene Transfer Vector

We examined the characteristics of FL as in comparison with CL-DNA complex, a common nonviral gene transfer vector.

One of the important characteristics of macromolecule transfer into the cells by FL is the active and rapid absorption onto the plasma membrane by HN glycoprotein and the succeeding efficient membrane fusion triggered by F glycoprotein[6,13](Fig.1). In contrast, the gene transfer activity of the CL-DNA complex depends on passive binding to the cell surface and on uptake through non-specific endocytosis by target cells[14]. To investigate the effects of these different characteristics on the efficiency of gene transfer, we examined the expression of the firefly luciferase gene transferred by these vectors under various conditions.

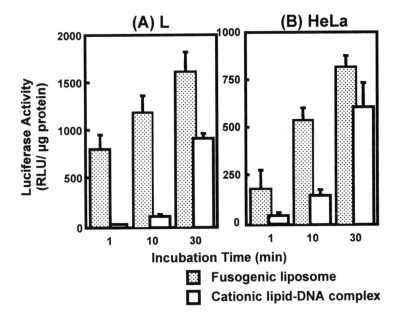

Figure 3. Effect of exposure on gene transfer activity.

First, we examined the effect of the length of the exposure of the vectors to the cells on the gene transfer[15] (Fig.3). L and HeLa cells were treated either with FL or with CL-DNA complex containing 0.5 μg of pCAL2; luciferase expressed plasmid, for 1, 10, or 30 min. Neither FL nor CL-DNA complex had any apparent cytotoxicity under these conditions. The cells treated with FL had significant luciferase activity even when they were incubated with the vector for 1-10 min, while they had to be incubated with CL-DNA complex for at least 30 min to have the same level of luciferase activity. When the cells were incubated with these vectors for 10 min, those treated with FL had 4 to 13-fold more luciferase activity than those treated with CL-DNA complex.

We also examined the effect of serum proteins in the incubation medium on the gene transfer[15] (Fig.4). FL efficiently transferred DNA in the presence of serum proteins. The cells incubated with FL in the presence of 40 % FCS still had 70 % of the luciferase activity of those incubated with the vector in the absence of serum proteins. On the contrary, serum proteins strongly inhibited the gene transfer mediated by CL-DNA complex. The cells incubated with CL-DNA complex in the presence of 5 % FCS showed no obvious luciferase activity compared with those incubated with the vector in the absence of serum proteins.

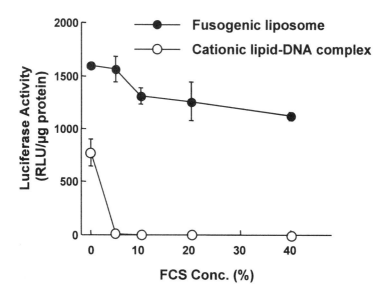

Figure 4. Effect of serum proteins in the medium on gene transfer activity.

The sensitivity of CL-DNA complex-mediated gene transfer to serum proteins may be due to the degradation of DNA by nucleases contained in the serum, because DNA formed an unstable complex with CL but was not encapsulated in the CL membrane. In contrast, DNA in FL was encapsulated in the membranous structure and protected from nucleases[3]. Another possible explanation is that the absorption to the cell surface and/or the fusogenic activity of CL may be interfered by serum lipoproteins. The resistance of FL-mediated gene transfer to FCS was surprising in another aspect, because serum contains a high concentration of sialic acid, which is a putative Sendai virus receptor, whereas saturation of gene transfer mediated by FL at a higher concentration suggested the limited availability of the receptors on cell surface. These data showed that the components containing sialic acid in serum would not compete with cell surface receptors for the binding of FL to the cell membrane.

The rapid and efficient gene transfer mediated by FL shown in Figures 4 and 5 was clearly due to the active binding and the fusion of FL with the cell membrane. In contrast, CL-DNA complex required a longer incubation period and a higher DNA concentration to facilitate gene transfer. However, the higher the concentration of CL-DNA complex was, the more the cells became damaged. These characteristics of FL-mediated gene transfer, together with resistance to serum components, suggested that FL would be superior to CL-DNA complex, especially for gene transfer into tissue cells *in vivo.*

3.2 New Cancer Therapy Using FL as a In Vivo TNF-α Gene Transfer Vector

Reinforcement of anti-tumor immune reactions is a new approach for treating cancer that escapes normal host defense. Induction of an anti-tumor effect by various cytokines is one such approach[16]; the potent activity of the cytokines serving to activate the immune system. However, although systemic administration of cytokines was reported to be effective in animal models, severe side effects make this approach impractical[17,18].

The regional supplementation of cytokines by gene transfer enables a higher local concentration of cytokines to be sustained near the target cells for a longer period than with systemic administration. These approaches include the transplantation of tumor cells genetically modified with cytokine genes *in vitro* [19] and the direct transfer of cytokine genes into tumor cells *in vivo*[20,21]. In therapy, the latter approach has significant advantages in that it minimizes delays in treatment and the cost and labor to establish the genetically-modified tumor cell lines from each patient. However, this method seems to be less effective, probably because the number of tumor cells expressing cytokines and the amount of cytokines produced are insufficient to induce an effective immune reaction.

In this chapter, we examined, using FL as a prominent *in vivo* gene transfer vector; the effect of TNF-α produced in the arteries leading to tumors on tumor growth *in vivo*. The use of arteries leading to tumors as the site to produce cytokines for inducing anti-tumor immune reaction has several advantages over the use of solid tumor cells. First, cytokines are produced more efficiently in the arteries, because the endothelial cells are more accessible to gene transfer vectors than solid tumor. Second, cytokines produced in the artery can be delivered to a large number of tumor cells, which derive nutrition from the artery vessel. Third, the production of cytokines in the artery or tumor-artery endothelial cells may alter the characteristics of their own cells, thus allowing easier access to the tumor cells by immunocompetent cells. We chose TNF-α as the effector molecule, because it has the ability to modulate the function of tumor-vascular endothelial cells[22,23] and enhance the immune reaction[24,25].

We examined whether TNF-α produced in arteries leading to tumors and tumor tissue could suppress the growth of S-180 using FL as a outstanding *in vivo* gene transfer vector[26] (Fig.5). Administration of TNF-α genes into the femoral artery leading to tumors resulted in significant regression of the tumor, with 4 out of 11 mice cured of tumors. On the other hand, administration of luciferase genes had no effect on tumor suppression. Furthermore, injection of TNF-α genes into the left femoral artery opposite the site in which S-180 cells were inoculated did not suppress tumor growth. Intratumor injection of TNF-α genes did not result in anti-tumor activity either. These results suggest that administration of TNF-α genes into the artery leading to the tumors using FL is more effective than intratumor injection, and local production of TNF-α is essential to tumor suppression.

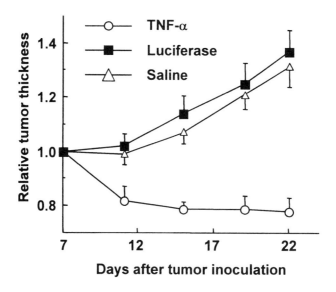

Figure 5. Tumor growth after direct gene transfer into the artery leading to the tumors.

Our results demonstrate that *in vivo* transfection of TNF-α genes using FL to the artery leading to tumors and tumor-vessels could efficiently suppress tumor growth. An extremely low dose of TNF-α was sufficient to induce tumor regression. In contrast, when TNF-α genes were directly injected into the tumor, no anti-tumor effect was observed (Fig.5). This is probably because intratumoral injection transduces only some of the cells in the tumor and the concentration of TNF-α produced *in vivo* is too low to induce a reaction with most of the tumor cells, resulting in failure to activate anti-tumor immunity. With the approach described in this study, TNF-α produced *in vivo* can react with the majority of the tumor cells and stimulate anti-tumor immunity more efficiently.

By using FL as the gene transfer vector, DNA was transferred only into the femoral artery and its downstream tissue, the site for S-180 solid tumors. FL mediate rapid and efficient gene transfer[15], but are inactivated after entering the blood stream by the complement system. Thus, DNA would be introduced only into local tissues at the site of inoculation. No TNF-α was detected in the serum of mice injected with TNF-α gene. TNF-α would be produced only in local tissues.

In summary, transfection of TNF-α gene into the endothelial cells located upstream of the tumors and the tumor tissues specifically enhanced the TNF-α concentration at the tumor sites and was found to be effective in tumor regression. This system using FL, which targets not only most tumor cells, but also tumor-endothelial cells, is a novel and effective approach to *in vivo* gene therapy of solid cancer.

4.   FL AS A VACCINE CARRIER

As mentioned above, CTL play a critical role in the protection against viral infections[27,28] and malignancy[29]. CTL responses have been generated in vivo using replicating vaccines such as plasmid DNA[30], retroviral vectors or other modified pathogens[31,32,33] expressing foreign proteins. However, these approaches may pose safety risks, especially in immuno-compromised individuals[34]. Exogenous proteins and peptides used as non-replicating vaccines, although they are safer, fail to efficiently elicit CTL responses in vivo. The responses against these proteins/peptides are generated when short peptides intracellularly processed are presented to CD8+ CTL precursors in association with class I MHC[35,36]. An essential event in this process is the intracytosolic presence of antigens, as has been observed for endogenously synthesized viral proteins. Exogenous proteins and peptides are internalized by endocytosis/phagocytosis and are processed within the vacuolar compartment for presentation by class II MHC molecule[37]. They fail to enter the cytosolic compartment and therefore, are not efficiently presented via the class I MHC pathway[37]. Several immunization approaches using liposomes have elicited detectable levels of in vivo priming of class I MHC-restricted CTL to protein antigens[38,39]. However, it has not been determined whether this results from the direct delivery of protein antigens to the cytosol.

We have clarified that FL delivers their contents directly and efficiently into cytoplasm both *in vitro* and *in vivo* [9,12a,26,40]. This is a unique delivery system because most vectors thus far utilized are assumed to induce the priming of class I MHC-restricted CTL based on the leakage of only a small amount of protein molecules from endosomes/phagosomes to cytoplasm. In this chapter, we demonstrate that FL can deliver the encapsulated soluble protein directly into the class I MHC antigen-presentation pathway through membrane fusion but not through endocytosis/phagocytosis (Fig.6). Consequently, a single immunization with ovalbumin (OVA) in FL results in high level of priming of OVA-specific CTL. These results suggest that the delivery system using FL is an effective strategy for eliciting class I MHC-restricted CTL.

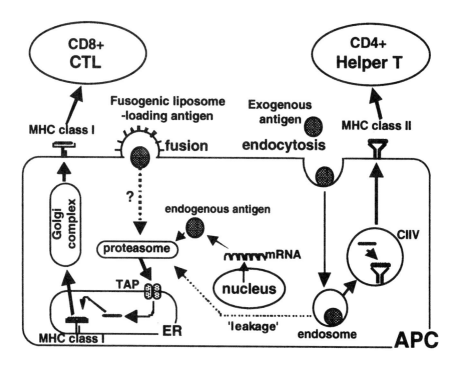

Figure 6. Class I and class II MHC antigen presentation pathway

4.1 Directly Delivery of FL Exogenous Antigen into the Class I MHC Presentation Pathway

To investigate whether FL-mediated delivery of soluble protein antigen leads to class I MHC antigen presentation, FL encapsulating OVA as a model of soluble protein antigen (OVA-FL) were prepared.

In addition, to examine the efficacy of OVA-FL in the cytosolic delivery of the content (OVA) and in the surface expression of OVA-derived peptides in association with class I MHC, EL4 lymphoma cells were incubated with OVA-FL. The resultant EL4 cells designated OVA-FL/EL4 cells were then subjected to cytotoxic assays by OVA-specific CTL. OVA-specific CTL were generated by immunizing C57BL/6 mice with OVA emulsified in complete Freund's adjuvant (CFA) and subsequently stimulating spleen cells from these mice with OVA gene-transfected EL4 cells (EG7)[41]. The results show that OVA-specific CTLs elicit comparable levels of cytotoxicity on OVA-FL/EL4 cells to those on EG7 targets, but significant cytotoxicity was not observed on EL4 targets prepared by incubation with either empty FL or OVA encapsulated in simple liposomes (Fig.7). The efficacy of OVA-FL/EL4 cells as targets were confirmed in OVA-FL (Fig.7). The cytotoxicity of OVA-FL/EL4 cells as well as EG7 cells was blocked in the presence of anti-class I MHC-common monoclonal antibody. These results indicate that FL can efficiently deliver the encapsulated protein into the cytoplasm of fused cells, resulting in the incorporation of the protein fragment into the class I MHC complex. And we make doubly sure the protein antigen encapsulated in FL is internalized by cells not through endocytosis but through the fusion with FL[42].

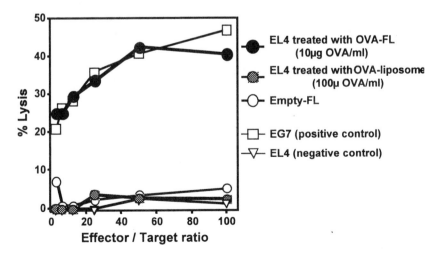

Figure 7. FL deliver OVA into the class I MHC processing pathway.

4.2   Efficient Priming of OVA-Specific CTL By Immunization With FL Containing

The remarkable efficiency by which OVA encapsulated in FL is processed and presented with class I MHC molecules *in vitro* led us to examine whether this vector (FL) might target OVA into the class I MHC presenting pathway in vivo and prime the CTL responses. We immunized C57BL/6 mice s.c. with various doses of OVA encapsulated in FL or simple liposomes or emulsified in CFA (Fig.8). Spleen cells from these *in vivo* primed mice were stimulated *in vitro* with EG7 cells, and the effectors generated were subjected to cytotoxicity assays using EG7 target cells. Generation of CTL under these conditions is dependent on priming *in vivo*[43]. The priming with OVA in simple liposomes led to only marginal levels of anti-OVA CTL generation. In contrast, the injection of OVA-FL resulted in high or moderate levels of anti-OVA CTL priming, respectively. CTL were also primed by injection of OVA in. However, there was a considerable difference in the generation of CTL between FL and CFA priming with the same amounts of OVA.

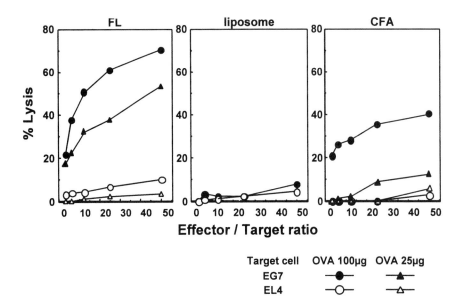

Figure 8. Efficient priming of OVA-specific CTL by immunization with FL containing OVA.

Our results demonstrate that FL can deliver exogenous protein antigens directly into the cytosol through the FL-cell fusion and direct the protein antigens encapsulated in FL into the class I MHC antigen-presentation pathway. Several approaches using various carriers for exogenous protein antigens have been shown to induce the in vivo priming of CTL against these antigens[38,39,41,44]. Moreover, some strategies[39,45,46], such as conjugation with iron oxide beads[44], succeeded in directing exogenous antigens into the class I MHC presentation pathway. However, these approaches did not deliver exogenous proteins directly into the cytosol and the class I MHC presentation was based on the internalization of antigens by macrophages through endocytic/phagocytic processes and the random escape from the vacuolar compartment to the cytosol. Thus, FL, which can deliver entrapped exogenous antigens directly into the cytosol, are more desirable as a carriers for CTL vaccine.

A similar approach to that used here was reported by Miller et al.[47]. These authors prepared proteoliposomes (virosomes) by reconstituting biologically active Sendai virus glycoprotein into the liposomes[48] and showed that rhesus monkeys developed an antigen-specific, class I MHC-restricted CTL response after vaccination with proteoliposomes containing antigen[47]. However, it is unclear whether these proteoliposomes carried the encapsulated contents directly into the cytosol to the class I MHC processing pathway. We also prepared proteoliposomes encapsulating OVA (OVA-proteoliposomes) by the dialysis method and compared the efficacy of OVA-proteoliposomes in the delivery of the encapsulated protein into the cytosol to that of OVA-FL. Both hemagglutinating and hemolytic activities are required for fusion with cells, but OVA-proteoliposomes exhibited these activities only weakly. In particular, a hemolytic activity was only marginal. In contrast too high levels of CTL susceptibility of OVA-FL/EL4 (Fig.7), EL4 cells incubated in vitro with OVA-proteoliposomes were found to exhibit only marginal cytotoxicity by OVA-specific CTL. These observations indicate that FL act as a much more efficient carrier of the encapsulated protein into the cytosol than proteoliposomes reconstituted of Sendai virus spikes.

A means of delivering exogenous protein antigens into the cytosol would potentially be useful for developing CTL vaccines. While such a delivery may be achieved using replicating vectors, this method poses risks for recipients based on the pathogenicity of the vectors used. Non-replicating vectors, although producing safer vaccines, generally fail to efficiently induce class I MHC-restricted immune responses. In this study, we showed that FL, a non-replicating vector, are capable of delivering CTL antigens directly into the cytosol and thereby introducing these into the class I MHC antigen-presentation pathway. Furthermore, FL would be a useful probe for studying class I MHC-restricted antigen processing in the cytoplasm. Thus, the present approach could contribute not only to developing non-replicating safe CTL vaccines but also to a better understanding of intracytoplasmic antigen processing that leads to the class I MHC antigen presentation pathway.

ACKNOWLEDGMENTS

We thank Dainippon Pharmaceutical Co., Ltd., Osaka, Japan and Dr. J. Miyazaki (Osaka university, School of Medicine, Osaka, Japan) for providing the plasmid pHT13 and pCAGGS, respectively. We thank the Cancer Cell Repository, Institute of Development, Aging and Cancer (Tohoku University, Sendai) for EL4 cells. This work was supported in part by a Grant-in-Aid for Scientific Research from the Ministry of Education, Science and Culture of Japan and by a grant from Japan Health Science Foundation.

REFERENCES

1. Uchida, T., Kim, J., Yamaizumi, M., Miyake, Y. and Okada, Y., 1979,. *J. Cell Biol.*, 80: 10-20.
2. Nakanishi, M., Uchida, T., Sugawa, H., Ishlura, M. and Okada, Y., 1985, *Exp. Cell Res.*, 159: 399-409.
3. Kato, K., Nakanishi, M., Kaneda, Y., Uchida, T. and Okada, Y., 1991, *J. Biol. Chem.*, 266: 3361-3364.
4. Kato, K., Kaneda, Y., Sakurai, M., Nakanishi, M. and Okada, Y., 1991, *J. Biol. Chem.*, 266: 22071-22074.
5. Nakanishi, M. and Okada, Y., 1993, *Liposome Technology* (Gregoriadis G., eds.) CRC Press: Florida, pp. 249-260
6. Nakanishi, M., Ashihara, K., Senda, T., Konda, T., Kato, K. and Mayumi, T., 1995, *Peptide and Protein Drug Delivery* (Lee, VHL., Hashida, M. and Mizushima, Y., eds.) Harwood Academic Publishers: The Netherlands. pp. 337-349.
7. Yamaizumi, M., Mekada, E., Uchida, T. and Okada, Y., 1978, *Cell*, 15: 245-250.
8. Uchida, T., 1982, *Molecular Action of Toxins and Viruses* (Cohen, P. and Heyningen, SV. Eds.) Elsevier Biomedical Press: Amsterdam. pp. 1-31.
9. Mizuguchi, H., Nakanishi, M., Nakanishi, T., Nakagawa, T., Nakagawa, S. and Mayumi, T., 1996, *Br. J. Cancer*, 73: 472-476.
10. Friedmann, T., 1989, *Science*, 244: 1275-1281.
11. Miller, A. D., 1992, *Nature*, 357: 455-460.
12. Mulligan, R. C., 1993, *Science*, 260: 926-932.
12a. Nakanishi, T., Hayshi, A., Kunisawa, J., Tsutsumi, Y., Tanaka, K., Yashiro-Ohtani, Y., Nakanishi, M., Fujiwara, H., Hamaoka, T., Mayumi, T., 2000, *Eur.J.Immnol.*, 30: 1740-1747.
13. Bagai, S., and Sarkar, D. P., 1994, *FEBS Lett.* 353: 332-336.
14. Legendre, J-Y., and Szoka Jr., F. C., 1992, *Pharm. Res.* 9: 1235-1242
15. Mizuguchi, H. Nakagawa, T., Nakanishi, M., Imazu, S., Nakagawa, S. and Mayumi, T., 1996, *Biochem. Biophys. Res. Commun.*, 218: 402-407
16. Rosenberg, S.A., 1988, Today, 9: 58-62.

17. Siegel, J.P., and Puri, R.K.,1991, *J. Clin. Oncol.*, 9: 694-704.

18. Moritz, T., Niederle, N., Baumann, J., May, D., Kurschel, E., Osieka, R., Kempeni, J., Schlick, E., and Schmidt, C.G., 1989, *Cancer Immunol. Immunother.*, 29: 144-150.

19. Hock, H., Dorsch, M., Richter, G., Kunnzendorf, U., Kruger-krasagakes, S., Blankenstein, T., Qin, Z., and Diamantstein, T., 1994, *Nat. Immunol.*, 13: 85-92.

20. Addison, C.L., Braciak, T., Ralston, R., Muller, W.J., Gauldie, J., and Graham, F.L., *Proc. Natl. Acad. Sci.* USA, 92: 8522-8526.

21. Cordier, L., Duffour, M.T., Sabourin, J.C., Lee, M.G., Cabannes, J., Ragot, T., Perricaudet, M., and Haddada, H., 1995, *Gene Ther.*, 2: 16-21.

22. Nawroth, P.P., and Stern, D.M., 1986, *J. Exp. Med.*, 163: 740-745.

23. Shimomura, K., Manda, T., Mukumoto, S., Kobayashi, K., Nakano, K., and Mori, J., 1988, *Int. J. Cancer*, 41: 243-247.

24. Philip, R., and Epstein, L., 1986, *Nature*, 323: 86-89.

25. Blankenstein, T., Qin, Z., Uberla, K., Muller, W., Rosen, H., Volk, H-D., and Diamantstein, T., 1991, *J. Exp. Med.*, 173: 1047-1052.

26. Mizuguchi, H., Nakagawa, T., Toyosawa, S., Nakanishi, M., Imazu, S., Nakanishi, T., Tutsumi, Y., Nakagawa, S., Hayakawa, T., Ijuhin, N. and Mayumi, T., 1998, *Cancer Res*, 58: 5725-5730.

27. Rouse, B. T., Norley, S. and Martin, S., 1988, *Rev. Infect. Dis.*, 10: 16-33.

28. Lukacher, A. E., Braciale, V. L. and Braciale, T. J., 1984, *J. Exp. Med.*, 160: 814-826.

29. Cerundolo, V., Lahaye, T., Horvath, C., Zanovello, P., Collavo, D. and Engers, H. D., 1987, *Eur. J. Immunol.*, 17: 173-178.

30. Ulmer, J. B., Donnelly, J. J., Parker, S. E., Rhodes, G. H., Felgner, P. L., Dwarki, V. J., Gromkowski, S. H., Deck, R. R., DeWitt, C. M., Friedman, A., Hawe, L. A., Leander, K. R., Martinez, D., Perry, H. C., Shiver, J. W., Montgomery, D. L. and Liu, M. A., 1993, *Science*, 259: 1745-1749.

31. Aldovini, A. and Young, R. A., 1991, *Nature*, 351: 479-482.

32. Schafer, R., Portnoy, D. A., Brassell, S. A. and Paterson, Y., 1992, *J. Immunol.*, 149: 53-59.

33. Hahn, C. S., Hahn, Y. S., Braciale, T. J. and Rice, C. M., 1992, *Proc. Natl. Acad. Sci.* USA., 89: 2679-2683.

34. Guillaume, J. C., Saiag, P., Wechsler, J., Lescs, M. C. and Roujeau, J. C., 1991, *Lancet* 337: 1034-1035.

35. Townsend, A. R., Rothbard, J., Gotch, F. M., Bahadur, G., Wraith, D., and McMichael, A. J., 1986, *Cell*, 44: 959-968.

36. Van-Bleek, G. M., and Nathenson, S. G., 1990, *Nature*, 348: 213-216.

37. Germain, R. N., 1986, *Nature*, 322: 687-689.

38. Reddy, R., Zhou, F., Nair, S., Huang, L. and Rouse, B. T., 1992, *J. Immunol.*, 148: 1585-1589.

39. Jondal, M., Schirmbeck, R. and Reimann, J., 1996, *Immunity*, 5: 295-302.

40. Nakanishi, M., Mizuguchi, H., Ashihara, K., Senda, T., Akuta, T., Okabe, J., Nagoshi, E., Masago, A., Eguchi, A., Suzuki, Y., Inokuchi, H., Watabe, A., Ueda, S., Hayakawa, T. and Mayumi, T., 1998, *J. Control. Release*, 54:61-68.

41. Ke, Y., Li, Y. and Kapp, J. A., 1995, *Eur. J. Immunol.*, 25: 549-553.

42. Nakanishi,T. et al., in preparation.

43. Moore, M. W., Carbone, F. R. and Bevan, M. J., 1988, *Cell*, 54: 777-785.

44. Kovacsovics-Bankowski, M., Clark, K., Benacerraf, B. and Rock, K. L., 1993, *Proc. Natl Acad. Sci.* USA, 90: 4942-4946.

45. Lee, K. D., Oh, Y. K., Portnoy, D. A. and Swanson, J. A., 1996, *J. Biol. Chem.*, 271: 7249-7952.

46. Pfeifer, J. D., Wick, M. J. Roberts, R. L. Findlay, K. Normark, S. J. and Harding, C. V., 1993, *Nature*, 361: 359-362.
47. Miller, M. D., Gould-Fogerite, S., Shen, L., Woods, R. M. Koenig, S., Mannino, R. J. and Letvin, N. L., 1992, *J. Exp. Med.*, 176: 1739-1744.
48. Gould-Fogerite, S., and Mannino, R. J., 1985, *Anal. Biochem.*, 148: 15-25.

A NOVEL HYDROPHOBIZED POLYSACCHARIDE/ONCOPROTEIN COMPLEX VACCINE FOR HER2 GENE EXPRESSING CANCER

[1]Hiroshi Shiku, [1]Lijie Wang, [2]Kazunari Akiyoshi, and [2]Junzo Sunamoto
[1]*Second Department of Internal Medicine, Mie University School of Medicine, 2-174 Edobashi, Tsu-city, Mie514-8507, Japan,* [2]*Department of Synthetic Chemistry & Biological Chemistry, Graduate School of Engineering, Kyoto University, Kyoto 606-5801, Japan*

1. INTRODUCTION

Lysis of tumor cells by $CD8^+$ CTLs has been demonstrated to be an important effector mechanism of tumor defense and rejection. They destroy target cells by recognizing epitope peptides in a MHC class I restricted manner. In certain murine tumor systems, epitope peptides recognized by $CD8^+$ CTLs have been identified, and the feasibility of vaccination with those peptides either for the prevention or the treatment of tumors has been demonstrated.

However, immunization with a single epitope containing one anchor motif occasionally resulted in weak and limited immunological responses. Soluble proteins with epitope peptides may be alternative candidates for vaccines. By more stimulation from various types of potential CTL and T helper epitopes with different anchor motifs, protein antigens may have more potential to elicit immunological responses in a wider range than single peptides, and therefore, the protein induced immunity may be stronger than the immunity induced by a single epitope peptide. Although the use of recombinant soluble proteins containing epitope peptide sequences is a promising approach, failure to induce specific $CD8^+$ CTL activity by

Biomedical Polymers and Polymer Therapeutics
Edited by Chiellini *et al.*, Kluwer Academic/Plenum Publishers, New York, 2001

immunizing hosts with such exogenous protein molecules has been experienced repeatedly. Investigators have been looking for ways to overcome these problems by novel approaches to induce MHC class I restricted CTL activity by the use of recombinant proteins.

Earlier, we reported, that a soluble truncated hybrid protein of *gag* and *env* of the human T lymphotropic virus type 1 could induce specific CD8[+] T cell dependent immunity, when reconstituted into mannan-derivative-coated liposomes[1]. Considering the increasing evidence for receptors which can specifically bind polysaccharides, we considered the coating by polysaccharides as more important than liposomes for the delivery of proteins. Therefore, we designed a novel and simple protein delivery system for cancer immuno-therapy: cholesteryl group-bearing polysaccharides, mannan and pullulan (CHM and CHP), complexed with the truncated protein encoded by the HER2 proto-oncogene, which is overexpressed with high frequency in breast, stomach, ovarian and bladder cancer as well as in other kinds of cancer. In this study, we investigated whether or not the HER2 protein alone or complexed with CHM or CHP can induce responses by CD8[+] CTLs and tumor rejection specific for HER2 expressing tumor cells. The humoral immune response against HER2 generated by this novel vaccine was investigated. Its usefulness for the prevention against tumor growth and the therapy of established tumors was evaluated.

2.   RESULTS AND DISCUSSION

2.1   Induction of HER2 specific CD8[+] CTLs by *in vivo* priming with CHM- or CHP-HER2 complexes

We have recently demonstrated that HER2 can be a target for tumor rejecting immune responses against syngeneic murine HER2[+] tumor cells[2]. We defined two different peptides, HER2p63-71 and HER2p780-788, binding with K^d, one of murine MHC class I, that can induce CD8[+] CTLs. The growth of HER2[+] syngeneic tumors was suppressed in mice immunized with HER2p63-71 or p780-788. Base on these findings, we prepared a truncated HER2 protein with 147 N-terminal amino acids which contain are of murine K^d restricted CTL epitopes HER2p63-71. BALB/c female mice were subcutaneously immunized twice at a one-week interval with 20ƒÊg HER2 protein complexed with cholesteryl group-bearing polysaccharides, or controls[3,4,5]. Spleen cells were obtained from animals one week after the second immunization, and were sensitized *in vitro* with mitomycin-C treated CMS17HE as described in Materials and Methods. Killer cell activity specific for HER2 expressing target cells was induced from spleen cells of the animals immunized with complexes of cholesteryl group-bearing polysaccharide (CHM or CHP), and HER2 protein. No HER2 specific killer activity was induced from spleen cells derived from the mice immunized with complexes of cholesterol bearing polysaccharides and control protein CAB, or CHM, CHP alone, HER2 protein alone or CAB protein alone[3]. The specific killer activity induced in animals immunized with the CHM-HER2 complex was blocked *in vitro* with monoclonal antibodies against CD3, CD8 or K^d monoclonal antibodies. In contrast, monoclonal antibodies against CD4, I-A^d, L^d or D^d did not show any significant blocking activity. The results suggest that the induced killer cells are $CD8^+$ CTLs with K^d restriction.

We asked whether CTLs induced in BALB/c mice primed with the CHM-HER2 complex can also recognize HER2p63-71 peptide since the truncated HER2 protein consists of 147 amino acids from the N-terminus. P1.HTR cells pulsed with peptides were subjected to a ^{51}Cr release assay using CTLs derived from mice immunized with CHM-HER2. The CTLs specifically lysed P1.HTR pulsed with a HER2 derived peptide p63-71, but not P1.HTR pulsed with other peptides with K^d binding motives, or non-pulsed target cells[3].

2.2   Immunization with the CHP-HER2 complex is therapeutically effective against HER2 expressing tumors

BALB/c mice inoculated with 2 ∞ 10^6 HER2 expressing CMS7HE tumor cells were given weekly immunization of 20μg protein of CHP-HER2 complex starting on the day of the challenge or 3, 7, or 14 days after the tumor challenge, respectively. Complete tumor rejection was observed when the immunization was initiated either on the day of tumor challenge or on day 3 after primary tumor challenge (Fig.1). When the immunization was started 7 days or 14 days after the tumor inoculation, only marginal suppression of tumor growth was observed without complete rejection.

2.3 $CD8^+$ T cells are the major effector cells in vivo for the rejection of $HER2^+$ tumors

We questioned whether $CD8^+$ T cells were also the effector cells *in vivo* for the prevention of HER2 expressing tumor cells. After the second immunization with CHM-HER2, the mice were administered monoclonal antibodies against CD4, CD8, or their combination one day before the tumor challenge. The suppression of the tumor growth was abolished by the administration of monoclonal antibodies against CD8, or the combination of monoclonal antibodies against CD4 and CD8, but not by the administration of monoclonal antibodies against CD4 alone.

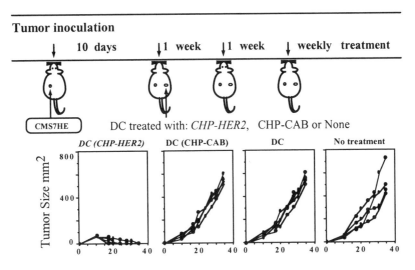

Figure 1. Therapeutic effect of the CHP-HER2 vaccine. BALB/c mice were challenged with $2\infty10^{6}$ CMS7HE subcutaneously and weekly given CHP-HER2 complex containing 20µg of protein starting on the day of challenge, or 3, 7 or 14 days later. Each group consisted of four mice, a line represents a single mouse.

2.4   DCs can incorporate CHP-HER2 complex and specifically stimulate CD8^{+} T cells and CD4^{+} T cells.

We questioned whether bone marrow-derived DCs could incorporate CHP-HER2 complex and stimulate T cells by providing the cognate target peptides. BALB/c mice were immunized twice with CHP-HER2 complex at a weekly interval. One week after the last immunization, CD4^{+} T cell and CD8^{+} T cell subpopulations were prepared. DCs cultured with CHP-HER2 complex or a control CHP-CAB complex, for 3 hours, or untreated DCs were used as antigen presenting cells to stimulate T cells. Both CD4^{+} T cells and CD8^{+} T cells showed a significantly stronger response to DCs treated with CHP-HER2 complex than to DCs treated with CHP-CAB, or to DCs without prior treatment (Fig.2). The result shows that DCs can incorporate CHP-HER2 complex and present cognate peptides to both CD4^{+} T cells and CD8^{+} T cells after appropriate processing.

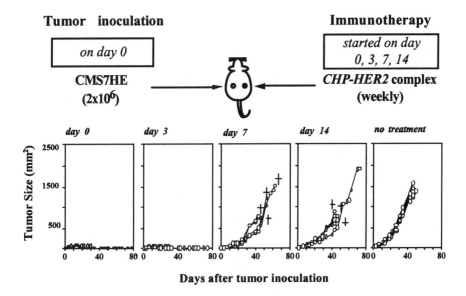

Figure 2. Bone marrow-derived DCs demonstrate a potent APC function. DCs pretreated with CHP-HER2 complex, control CHP-CAB complex, or untreated DCs were used as stimulator cells. Responder CD4$^+$ T cells and CD8$^+$ T cells were obtained from nylon fiber-purified spleen cells of mice immunized with CHP-HER2 complex. ^{3}H-TdR proliferation assay was performed. Both CD4$^+$ T cells and CD8$^+$ T cells showed significantly responded only to CHP-HER2 complex pretreated DCs.

2.5   Experimental cell therapy using DCs pretreated with CHP-HER2 complex

We further examined whether DCs treated ex vivo with CHP-HER2 complex could be used as vaccine against HER2 expressing tumor cells. 2 ∞ 10^6 CMS7HE cells were inoculated subcutaneously into BALB/c mice. 10 days after inoculation, vaccination with 4 ∞ 10^5 DCs pretreated with CHP-HER2 complex or CHP-CAB control complex, or DCs without treatment was started subcutaneously at a weekly basis. In the group of mice vaccinated with CHP-HER2 complex pretreated DCs, complete eradication of tumor was observed in all of 4 mice (Fig.3). In contrast, tumor growth in mice of groups either treated with DCs pretreated with CHP-CAB control complex or DCs alone was similar to the tumor growth observed in mice without vaccination.

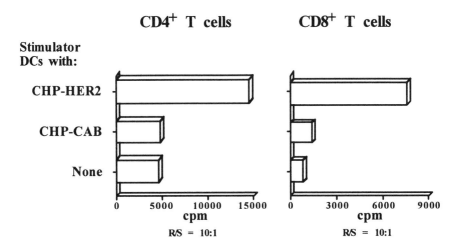

Figure 3. The therapeutic effect of DCs pretreated with CHP-HER2 complex. BALB/c mice were challenged with 2∞10⁶ CMS7HE subcutaneously. Vaccination with 410⁵ DCs pretreated with CHP-HER2 complex, CHP-CAB control complex or without prior treatment was started on day 10 after tumor challenge and continued on a weekly basis. Strong tumor suppression was only observed in the group of mice vaccinated with CHP-HER2 complex pretreated DCs. Four mice were used for each experimental group, a line represents a single mouse.

3.   CONCLUSION

In our present study, the HER2 protein required hydrophobized polysaccharides for the induction of a CTL response and the rejection of HER2 expressing tumor cells *in vivo*. These CTLs were K^d restricted CD8$^+$ T lymphocytes specifically recognizing the HER2 derived peptide p63-71, a part of the HER2 protein used for the vaccine, indicating that a soluble oncoprotein could enter the class I pathway by the help of CHM or CHP.

The CHP-HER2 complex revealed to be as therapeutically potent. When mice were immunized with either CHP-HER2 complex they could generate both CD4$^+$ T cells and CD8$^+$ T cells specifically reactive with DCs pretreated with CHP-HER2 complex. It is interesting that for these animals either immunized with CHP-HER2 complex or HER2 protein alone, DCs pretreated with CHP-HER2 complex but not HER2 protein alone could strongly stimulate specific CD8$^+$ T cells in. These clearly show that DCs, whether pretreated with CHP-HER2 complex or HER2 protein alone can incorporate and process the antigen peptides and finally present them to CD4$^+$ T cells.

Having established that bone marrow-derived DCs can efficiently stimulate both CD4$^+$ T cells and CD8$^+$ T cells, we examined their usefulness for immunotherapy of HER2 expressing tumors. In fact treatment of mice inoculated with CMS7HE 10 days prior to immunization, obvious suppression of tumor growth was observed in the group utilizing DCs pretreated with CHP-HER2 complex. Non-specific adjuvant effect of CHP seems to be unlikely because DCs treated with CHP-CAB control complex showed no effect for tumor suppression when compared with mice without any immunization. It is of particular interest that in mice immunized with CHP-HER2 complex pretreated DCs, there was complete tumor eradication observed in all of 4 mice. In our experience, immunization either with CHP-HER2 complex or CHM-HER2 complex, complete tumor suppression was possible only when we initiated it sooner than 4 days after tumor inoculation. The present data strongly suggests that CHP-HER2 complex can be effectively used as a cancer vaccine in concert with bone marrow-derived DCs for immunological cell therapy. Since murine K^d and human HLA-A24 share a resembling anchor motif for peptides, HER2p63-71 and HER2p780-788 were examined for induction of CTLs in HLA-A24$^+$ individuals. Human CD8$^+$ CTL clones specific for these peptides were established and they lysed HER2$^+$ tumor cells in HLA-A24-restricted manner. These results strongly indicate immediate usefulness of CHP-HER2 complex for clinical cancer immunotherapy since an identical target antigen peptide HER2p63, can be recognized by both murine and human CTL.

REFERENCES

1. Noguchi Y., Noguchi T., Sato T., Yokoo Y., Itoh S., Yoshida M., Yoshiki T., Akiyoshi K., Sunamoto J., Nakayama E. and Shiku H.; 1991, *J. Immunol.,* 146: 3599-360,
2. Nagata Y., Furugen R., Hiasa A., Ikeda H., Ohta N., Furukawa K., Nakamura H., Furukawa K., Kanematsu T. and Shiku H.; 1997, *J. Immunol.,* 159:1336-1343
3. Gu Xiao-G., Schmitt M., Hiasa A., Nagata Y., Ikeda H., Sasaki Y., Akiyoshi K., Sunamoto J., Nakamura H., Kuribayashi K. and Shiku H.; 1998, *Cancer Res.,* 58: 3385-3390
4. Akiyoshi K. and Sunamoto J.; 1993, *Macromolecules,* 26: 3062-3068
5. Nishikawa T., Akiyoshi K. and Sunamoto J.; 1996, *J. Am. Chem. Soc.,* 18: 6110-6115

PART   3

POLYMERS IN GENE THERAPY

PHARMACEUTICAL ASPECTS OF GENE THERAPY

Philip R. Dash and Leonard W. Seymour
CRC Institute for Cancer Studies, University of Birmingham, B15 2TA, U.K.

1. INTRODUCTION

The concept of gene therapy as a treatment for human disease is almost as old as molecular biology itself. One of the first proposals for the use of genes in the treatment of human diseases was made by Tatum in his 1958 Nobel Prize acceptance speech, while the term gene therapy was coined by Szybalski in the early 1960s[1]. However, it wasn't until the early 1990s that the science of gene therapy had progressed enough to enter clinical trials[2].

The idea underlying gene therapy is that human disease might be treated by the transfer of genetic material into specific cells of a patient in order either to replace a defective gene or to introduce a new function to the cell[3,4]. The potential of gene therapy to treat a wide range of diseases has led to very rapid advances in the science of gene transfer and in understanding the molecular basis of many diseases. Despite this, and a growing commercial interest in gene therapy, it is a relatively recent therapeutic approach to the treatment of disease and as such much work is still to be done to realise its full potential.

This chapter will review the potential of gene therapy to treat a range of diseases as well as the technology that is currently available to achieve this.

Biomedical Polymers and Polymer Therapeutics
Edited by Chiellini *et al.*, Kluwer Academic/Plenum Publishers, New York, 2001

1.1   Gene therapy strategies

There are a wide variety of approaches to the use of gene therapy in the treatment of disease. Firstly, gene transfer can take place either *ex vivo* or *in vivo*. In the *ex vivo* technique the patient's cells are harvested, treated, genetically modified and re-administered to the patient. In the *in vivo* approach, a gene, normal or modified, is introduced directly into the body or in one specifically targeted organ of the patient such as muscle, bronchial epithelium or cancer tumour cells[5,6].

Secondly, gene therapy can be used to replace a defective gene or to insert a novel gene. Replacement gene therapy is generally used for the treatment of inherited diseases such as cystic fibrosis or muscular dystrophy, where the disease pathology is caused by an incorrectly functioning gene. The introduction of a novel gene into a patient's cells to produce a new function in that cell can be used in the treatment of more complex diseases such as cancer, AIDS or cardiovascular disease. For example, in the treatment of cancer the introduction of a gene encoding a cytokine can be used to try to provoke an immune response to the tumour[7].

1.2   Commercial potential of gene therapy

Gene therapy has the potential to be one of the most important developments in medicine over the next few decades. Progress has been so rapid that, coupled with knowledge of the molecular biology of diseases gained from the Human Genome Project, many analysts have predicted a billion dollar market for gene therapy products within the next 5 years[1].

Most of the research on gene therapy has been conducted in the United States, where over 20 dedicated gene therapy companies have been formed with an estimated combined market valuation of over $1 billion. In total it has been estimated that the United States spends approximately $300-400 million a year, from government and industrial sources, in support of gene therapy research.

In Europe, only 9 dedicated gene therapy companies have been established, and major pharmaceutical companies have been slow to get involved in gene therapy research [1] and as a result investment in gene therapy is much lower than in the United States. In the UK, for example, annual expenditure on gene therapy research from government, industrial and charitable sources is thought to be less than $38 million dollars[1].

2. POTENTIAL DISEASE TARGETS FOR GENE THERAPY

2.1 Gene therapy of inherited genetic diseases

The replacement of inherited defective genes is, perhaps, the most obvious use of gene therapy. The aim of replacement gene therapy is to replace a missing or defective gene that is responsible for the disease in order to restore normal function. The ultimate aim is the permanent correction of the disorder, possibly by integration of the gene into stem cells. A major limitation to this type of treatment is the problem of gaining access to the relevant tissues. For this reason, the early gene replacement strategies have focused on readily accessible tissue such as muscle, lung, blood and bone marrow[3].

Blood cells and bone marrow are associated with a number of relatively common genetic diseases and as such are a major target for gene therapy. Gene transfer into lymphocytes or haemopoetic stem cells potentially allows the treatment of a variety of diseases including immune deficiency disorders such as adenosine deaminase deficiency (ADA-SCID); storage disorders such as Gaucher's disease and haemoglobinopathies such as sickle cell anaemia and the thalassemias. The only clinical trials involving gene transfer into lymphocytes have been concerned with the correction of ADA-SCID, a very rare immune deficiency syndrome, and most work has concentrated on transfer of genes into stem cells[1].

Cystic fibrosis is the most common inherited genetic disease in Europe and the United States. A mutation in the cystic fibrosis transmembrane conductance regulator (CFTR) gene results in the production of an abnormal protein that does not properly control the flow of chloride ions across membranes. The major clinical manifestation of this disease is in the airway epithelia, resulting in chronic inflammation with subsequent damage to the lung tissue. The average life expectancy of patients with CF is around 30 years and this has stimulated great interest in its treatment with gene therapy. The non-functioning protein results in a build-up of mucus in the lungs with subsequent inhibition of gas transfer and a deterioration in the function of the epithelial cilia. There have been several clinical trials of the transfer of the CFTR gene into airway epithelial cells of cystic fibrosis patients, and the results of these are discussed later.

Another, relatively common, genetic disease that has the potential to be treated by gene therapy is Duchenne muscular dystrophy (DMD). Most effort to replace the defective DMD gene has concerned an *ex vivo* strategy to transfect isolated muscle cells with multiple copies of the DMD gene and re-implant the cells back into the patient's muscle. There has so far been little work on the *in vivo* delivery of the DMD gene as there are few vectors suitable for the efficient delivery of genes into muscle cells *in vivo* and ensure tissue specific expression[1].

2.2   Cancer gene therapy

Cancer is the second commonest cause of death in the industrialised world. In the UK over a third of people will develop cancer at some point in their lifetime, and around half of cancers are not curable using the traditional treatments of surgery, radiotherapy or chemotherapy. Gene therapy has emerged as a potential alternative to the existing treatments, and its potential is such that protocols for the treatment of cancer now account for over 80% of the gene therapy clinical trials[3].

There are numerous approaches to using gene therapy to treat cancer, and some of these are outlined below.

Immunotherapy. The immunotherapy approach to cancer gene therapy is one in which the immunogenicity of tumours is increased by causing the local production of immunomodulatory agents such as cytokines (interleukin-2 or granulocyte macrophage colony stimulating factor for example) or by increasing the level of MHC antigen expression[7]. Increasing the immunogenicity of tumours may then lead to an antitumour response. Because immunity is a systemic reaction, this immune reaction could potentially eliminate all the tumour cells in the body, including sites of metastatic deposit.

Gene Directed Enzyme Pro-drug Therapy (GDEPT). The introduction of genes that encode enzymes capable of converting pro-drugs to cytotoxic drugs is the basis of the GDEPT approach to cancer gene therapy. A relatively harmless pro-drug can be administered to a patient following the transfection of some tumour cells with genes encoding enzymes that will activate the pro-drug *in situ* to form a cytotoxic drug that will kill the tumour cell. This approach may be considered as using gene therapy to improve upon conventional chemotherapy. The local expression of an activating enzyme ensures that the peripheral toxicity often associated with chemotherapy is reduced. The use of a relatively harmless pro-drug ensures that high doses can be administered to the patient, resulting in high concentrations of the cytotoxic drug being produced in vicinity of the tumour.

There are several GDEPT systems available for use in gene therapy. The enzymes nitroreductase, thymidine kinase and adenosine deaminase can be used to convert the pro-drugs CB1954, ganciclovir and 5-FC respectively into cytotoxic drugs[8]. Following the death of the tumour cell, the cytotoxic drug may be able diffuse into neighbouring cells and kill them, a phenomenon known as the bystander effect. The bystander effect ensures that it is not necessary to transfect all of the cells in the tumour, indeed it has been shown that 100% cell death can be achieved *in vitro* following transfection of less than 10% of the cells[9].

Insertion of tumour suppresser genes. Many tumours lack functioning tumour suppresser genes that might be able to prevent further tumour growth, and even promote apoptosis of the tumour cells leading to tumour regression. Insertion of a gene encoding a tumour suppresser protein such as p53 may be able to restore the ability of the cells to keep the tumour in check. The p53 protein is responsible for arresting cell cycle following DNA damage via upregulation of p21, coding for an inhibitor of cyclin-dependent kinases, upregulation of *bax* and down-regulation of *bcl*-2. It is also involved in apoptosis. It has been shown that efficient delivery and expression of the wild-type p53 gene can cause regression in established human tumours, prevent the growth of human cancer cells in culture or render malignant cells from human biopsies non-tumourigenic in nude mice[7].

2.3　Other disease targets for gene therapy

Gene therapy has the potential to be used in the treatment of almost any disease with a genetic component. There are a number of cardiovascular diseases that are under investigation for possible treatment with gene therapy. For example, familial hypercholesterolaemia might be treated by transferring the gene encoding for the low density lipoprotein receptor into hepatocytes in order to reduce the level of serum cholesterol[10-12]. Other strategies for the gene therapy of cardiovascular disease include inhibiting smooth muscle cell proliferation, that might lead to stenosis, following balloon angioplasty[13,14] and stimulation of vascular growth to overcome ischaemia[10].

Other diseases that have been investigated for possible treatment with gene therapy include AIDS[1,15], neurodegenerative diseases such as Alzheimer's and Parkinson's[16], traumatic nerve injury[17] and autoimmune diseases[1,18].

3.   CLINICAL TRIALS OF GENE THERAPY

The first gene therapy trial took place in 1990 at the National Institutes of Health (NIH) in Bethesda, USA[2]. This was related to a single gene defect in a four year old girl with adenosine deaminase (ADA) deficiency. Since then, there have been over 200 clinical trials of gene therapy protocols.

The majority of the clinical trials with gene therapy have been involved in the treatment of cancer. In Europe over 80% of the trials are concerned with cancer therapy, while in the United States the figure is 72%. Protocols for the treatment of genetic diseases (mainly cystic fibrosis) account for around 20% of clinical trials, while in the US there are a small number of trials for the use of gene therapy in the treatment of AIDS. There are four main strategies for the gene therapy of cancer that have been taken to clinical trial. Various forms of immunotherapy account for around 29% of the trials in the United States, while the GDEPT approach accounts for around 10% of all trials. Chemoprotection of stem cells by insertion of drug resistance genes is a procedure that forms the basis of around 5% of all US gene therapy trials, while the insertion of tumour suppresser genes accounts for around 4% of trials. There are also two main approaches to the treatment of inherited genetic diseases that have been taken to clinical trial. The application of the CFTR gene to the respiratory epithelium accounts for around 10% of trials, while the insertion of genes into lymphocytes or haemopoetic stem cells has been conducted in 7% of trials[3].

The vast majority of gene therapy clinical trials have been phase I clinical trials. At this stage the main purpose of a clinical trial is not to assess therapeutic efficacy, but to test the tolerance, the activity and the safety of the gene transfer protocol (including the choice of gene delivery vector). Phase I clinical trials have also been performed where there is no available animal model. By the end of 1996 only two gene therapy trials had progressed to a Phase II stage[1]. Phase II trials differ from phase I trials in that they generally involve a larger number of patients and are designed to evaluate therapeutic efficacy as well further investigation of the side-effects and potential risks. To date no one has been cured through the use of gene therapy and there has been little evidence of therapeutic efficacy in most gene therapy protocols. However, the aims of most gene therapy clinical trials have been less ambitious and have sought mainly to demonstrate detectable gene expression in the target tissues.

The results of the first gene therapy clinical trial were published in 1995[2]. This trial involved the *ex vivo* retroviral treatment of T-cells from patients suffering from adenosine deaminase deficiency with severe combined immuno-deficiency (ADA-SCID). Two patients with ADA-SCID were chosen who had not fully responded to conventional treatment (PEG-ADA) for the trial. T-cells were isolated from their blood and stimulated to proliferate *in vitro*, before being transduced with a retroviral vector containing the ADA gene. The cells were then culture-expanded and re-infused into the patients. Five to six months after beginning the gene therapy the peripheral blood T cell counts in patient 1 rapidly increased until they reached the normal range where they have remained. The levels of the ADA enzyme were also found to have increased significantly. Patient 2 also showed an increase in the number of T cells, but no significant increase in ADA levels could be detected. Despite these improvements both patients remained on PEG-ADA treatment during and after the gene therapy treatment, complicating the interpretation of the clinical efficacy of this treatment. The trial did demonstrate, however, that gene expression could be detected long term, that this expression has some detectable effect in the patients and that there were no detectable safety problems with this protocol.

Gene expression in patients has also been demonstrated in clinical trials of gene therapy for cystic fibrosis. In one such trial 15 cystic fibrosis patients were enrolled in a double-blind, placebo controlled phase I clinical trial of liposome mediated transfer of the gene encoding CFTR into nasal airway epithelial cells. Expression of the plasmid DNA was detectable in all 9 patients who received liposome treatment (through detection of RNA) and there was partial restoration of the electrophysiological defect. In most cases the response to low chloride perfusion was restored to about 20% of that seen in non-CF subjects, although in one patient the response to low chloride perfusion was within the normal range for non-CF subjects. The maximal effect occurred 3 days after exposure to the liposome/DNA complexes., but reverted to pre-treatment levels 7 days after exposure[19]. As well as demonstrating detectable gene expression in patients, this study also showed that there were no adverse clinical effects and nasal biopsies showed no adverse histological or immunological changes.

Cystic fibrosis trials with adenoviral vectors also demonstrated detectable gene expression and some temporary restoration of CFTR function. However, while initial trials demonstrated that there were no adverse effects from these trials, later studies demonstrated dose and time dependent inflammation[20].

Most clinical trials have managed to demonstrate the feasibility of gene transfer techniques *in vivo*, although much work is required to improve the efficiency of gene expression and the safety of the vectors used. In a report published in 1995, the National Institutes of Health (NIH) in the United States sponsored a review of the current state of gene therapy development[21].

This report criticised the rush towards clinical trials of gene therapy and suggested that the field required more time to mature scientifically before significant therapeutic advances could be expected[3]. As a result gene therapy may be expected to remain largely in the academic sector for longer than traditional pharmaceutics. The report went on to suggest that gene therapy research should be more focussed on the basic aspects of gene transfer and gene expression. Furthermore it was suggested that special support should be given to research on vector development.

The main challenge for the success of gene therapy is to find safe vectors capable of transporting genes efficiently into target cells and getting the genes expressed once they are inside the cell. The current lack of clinically efficient gene transfer techniques severely limits the potential application of gene therapy and justifies the recommendation of the NIH that major efforts should be made to develop new vectors[3].

4.   VIRAL GENE DELIVERY SYSTEMS

4.1   Retroviruses

Retroviruses are RNA viruses which replicate through a DNA intermediate, where the viral DNA integrates randomly into the host genome[1]. Most recombinant retroviral vectors are based on the murine leukaemia virus[22] and were one of the first vectors developed for use in gene therapy. They are also the most popular vector for gene transfer in clinical trials, being used in over in 60-70% of human gene therapy trials[3].

The capacity of retroviral vectors to carry therapeutic genes is relatively small, with a maximum insertion size of around 8 kb. They are able to target dividing cells with a high degree of efficiency of gene transfer and a moderate level of gene expression. They have a wide host range and, as a result of their integration into the host genome, they allow long term expression of the transgene[1].

The main limitations of the use of retroviruses concerns the issue of safety. Recombinant retroviral vectors are generally replication incompetent. However, it has been shown that recombination can occur in producer cell lines leading to the production of replication competent viruses and potential infection in the host[23]. Another potential safety concern is the possibility that the random integration of the retrovirus genome into the host genome might lead to insertional mutagenesis of a cellular oncogene leading to tumour formation. This appears to be a small risk, and as yet no evidence of this has been detected[1].

There are other potential drawbacks that may limit the usefulness of retroviruses in gene therapy. For example, retroviruses are generally inactivated by the complement cascade, although recently the development of new packaging cell lines has overcome this problem[24]. The small size of the retrovirus genome also limits their usefulness for gene therapy involving the transfer of genes greater than around 8 kb in size.

One of the biggest disadvantages to using murine leukaemia retroviral vectors, however, is the fact that their infectivity is limited to dividing cells. For this reason retroviruses are generally used in an *ex vivo* gene transfer capacity where the target cells can be removed from the patient, stimulated to divide *in vitro* then transfected with the retrovirus and re-administered to the patient. There are a number of techniques that can be used to try to get retroviruses to infect outside their normal host range. One such method involves the co-infection with an adenovirus, which has been shown to allow retroviral vectors to transduce cells that were normally resistant to infection[25]. A second method involves the use of an HIV-based vector to produce retroviral vectors capable of transducing non-dividing cells. The Human Immunodeficiency Virus (HIV) is able to infect non-dividing cells such as lymphocytes and as such a number of attempts have been made to produce gene transfer vectors using the HIV *gag*, *pol* and *env* genes[22]. Despite the potential advantages of an HIV-based vector these types of vectors would have to be used with caution as there is some evidence that the HIV *Env* protein itself may be neurotoxic and even immunosuppressive[26].

4.2   Adenoviruses

Adenoviruses represent the second most popular choice for gene delivery vector for gene therapy clinical trials after the retroviral vectors[3]. Adenoviral vectors are based on a family of viruses that cause benign respiratory tract infections in humans and there are 42 serotypes of adenovirus known to infect humans[27]. They are non-enveloped isometric particles approximately 45-50 nm in diameter with an icosahedral surface (capsid) and a DNA-containing core[28]. Viral replication occurs without integration into the host genome, leading to only transient expression of the transgene.

Adenoviral vectors for use in gene therapy are typically based on serotype 5, with the majority of the E1a and E1b regions deleted to prevent virus replication. The E3 region can also be deleted to provide additional space for the insertion of up to 7.5 kb of exogenous DNA[27]. This is larger than the amount that can be inserted into retroviral vectors, but is still too small to be of use in all gene therapy applications.

The main advantages of adenoviral vectors is that the efficiency of transduction is high, as is the level of gene expression, although this is only transient and can deteriorate rapidly within a few weeks[1].

One of the main disadvantages of adenoviral vectors is that cell-specific targeting is difficult to achieve as the virus has no envelope to attach cell-specific ligands to, as can be achieved with retroviruses[22], although adenoviral tropisms have been altered through modifications of the fiber knob[29]. Furthermore, the adenovirus receptor is virtually ubiquitous and consequently systemic administration is likely to lead to adenoviral uptake in cell types other than the target cell thereby reducing the specificity of the gene therapy.

There are also important safety concerns over the use of adenoviral vectors for gene therapy. Aside from the potential generation of replication competent virus, there is also the possibility of provoking an inflammatory response in the patient, particularly when repeated administrations are given. It has been demonstrated that although repeat administration of adenovirus is possible, the gene transfer becomes progressively less efficient. Furthermore, antibodies are produced against the viral and transgene products and a relatively mild inflammatory response was observed in the lungs of CD-1 mice[30]. Studies in non-human primates have also revealed evidence of an inflammatory response following exposure of lungs to adenoviral vectors. It was demonstrated that lungs treated with a high-dose (10^{10} plaque forming units) showed inflammatory cells present in the peribronchial and perivascular regions as well as alveolar accumulation of neutrophils and macrophages. An increase in the bronchoalveolar levels of the cytokine interleukin 8 was also detected. These inflammatory responses were not observed in the low dose studies which correlate to the proposed dose for phase I clinical trials in humans. However, the effect of repeat administrations of this dose of adenovirus was not investigated[31]. Adenoviruses have also been implicated in causing cardiotoxicity and brain damage, as well as causing neurogenic and pulmonary inflammation at high doses and over time[20].

Despite these apparent drawbacks, the adenovirus remains a popular vector for gene therapy due to its high gene transfer efficiency and high level of expression in a wide variety of cell types. Some effort has been made to modify the inflammatory and immunogenic properties of the adenovirus capsid, but so far little progress has been made.

4.3   Adeno-associated and other viral vectors

The adeno-associated virus (AAV) is a vector that combines some of the advantages of both the retroviral and adenoviral vectors. It is a single stranded DNA parvovirus that is able to integrate into the host genome during replication, thereby producing stable transduction of the target cell. The virus can also infect a wide range of cell types, including both dividing and non-dividing cells. AAV vectors are also not associated with any known human disease[1] and show high efficiency transduction. They can, however, only carry a fairly small therapeutic gene insert (around 5 kb).

AAV vectors have not been studied to the same extent as adenoviral or retroviral systems, however they appear to be associated with fewer safety risks than the other viral systems. This is due to the elimination of all sequences coding for viral proteins, thereby greatly reducing the risk of an immune reaction against the vector. There remain, however, the potential problems of insertional mutagenesis and the generation of replication competent virus.

Other viral vectors have also been developed for use in gene therapy. These include the herpes simplex virus, the vaccinia virus and Sinbis virus. However, these vectors have not been widely studied and it is not clear what advantages they may hold over retroviral, adenoviral or AAV vectors. Recently, limited-replicating viral vectors, such as the Onyx-015 virus, have been developed. The Onyx virus is an E1B deleted adenovirus that can only replicate in p53 deficient cells[32]. The advantage of these vectors is that their replication is limited mainly to tumour cells and that permitting the replication of the virus produces more efficient transfection of the tumour cells. Studies are ongoing to demonstrate the usefulness and safety of replicating vectors.

5. NON-VIRAL GENE DELIVERY SYSTEMS

The limitations of viral vectors in particular their relatively small capacity for therapeutic DNA, safety concerns and difficulty in targeting to specific cell types have led to the evaluation and development of alternative vectors based on synthetic, non-viral systems. Examples of some non-viral approaches to gene delivery are discussed below.

5.1 Direct injection of naked DNA

The simplest non-viral gene delivery system simply uses naked expression vector DNA. Direct injection of free DNA into certain tissues, particularly muscle, has been shown to produce surprisingly high levels of gene expression [33], and the simplicity of this approach has led to its adoption in a number of clinical protocols. In particular, this approach has been applied to the gene therapy of cancer where the DNA can be injected either directly into the tumour or can be injected into muscle cells in order to express tumour antigens that might function as a cancer vaccine [34].

The direct injection of naked DNA can also be used to treat genetic diseases, particularly of tissues that are available for direct injection such as the skin[35]. Direct injection of skin from patients suffering from the genetic skin condition lamellar ichthyosis caused by a loss of transglutaminase 1 (TGase1) expression was performed in a study by Choate and Khavari (1997)[35]. It was shown that some skin regeneration was possible following repeat injections of plasmid DNA encoding the TGase1 gene, furthermore the restoration of TGase1 expression occurred in the correct location in the suprabasal epidermis. Further analysis, however, revealed that the pattern of expression was non-uniform and failed to correct the underlying histological and functional abnormalities of the disease.

Although direct injection of plasmid DNA has been shown to lead to high levels of gene expression in some tissues, the level of expression in many other tissues is much lower than with either viral or liposomal vectors. Naked DNA is also unsuitable for systemic administration due to the presence of serum nucleases. As a result direct injection of plasmid DNA seems destined to be limited to only a few applications involving tissues that are easily accessible to direct injection such as skin and muscle cells.

5.2   Liposomal gene delivery

Negatively charged, or classical, liposomes have been used to deliver encapsulated drugs for some time[36] and have also been used as vehicles for gene transfer into cells in culture[37]. Problems with the efficiency of nucleic acid encapsulation, coupled with a requirement to separate the DNA-liposome complexes from "ghost" vesicles has lead to the development of positively charged liposomes.   Cationic lipids are able to interact spontaneously with negatively charged DNA to form clusters of aggregated vesicles along the nucleic acid. At a critical liposome density the DNA is condensed and becomes encapsulated within a lipid bilayer[37], although there is also some evidence that cationic liposomes do not actually encapsulate the DNA, but instead bind along the surface of the DNA, maintaining its original size and shape [38].

Cationic liposomes are also able to interact with negatively charged cell membranes more readily than classical liposomes. Fusion between cationic vesicles and cell surfaces might result in delivery of the DNA directly across the plasma membrane[38]. This process bypasses the endosomal-lysosomal route which leads to degradation of anionic liposome formulations. Cationic liposomes can be formed from a variety of cationic lipids, and they usually incorporate a neutral lipid such as DOPE (dioleoylphosphatidyl-ethanolamine) into the formulation in order to facilitate membrane fusion[39,40]. A variety of cationic lipids have been developed to interact with

DNA, but perhaps the best known are DOTAP (N-1(-(2,3-dioleoyloxy)-propyl)-N,N,N-trimethylammo-niumethyl sulphate) and DOTMA (N-(1-(2,3-dioleoyloxy)propyl)-N,N,N-trimethylammonium chloride). These are commercially available lipids that are sold as *in vitro* transfecting agents, with the latter sold as Lipofectin.

There have been several studies on the *in vivo*, systemic use of liposome/DNA complexes[41-43]. The factors controlling the transfection efficiency of liposome/DNA complexes following intravenous administration are still poorly understood. Complexes formed between the cationic lipid DOTMA and DNA are rapidly cleared from the bloodstream and were found to be widely distributed in the body and expression was detected mainly in the lungs but also in the liver, spleen, heart and kidneys[41]. Similar results were found when DOTAP-based liposomes were used and it was found that the main factors controlling transfection efficiency were the structure of the cationic lipid and the ratio of the cationic lipid to DNA. The type of helper lipid used was also important as the addition of DOPE was found to reduce the *in vivo* transfection efficiency of DOTAP/DNA complexes[42]. The transfection efficiency of liposome/DNA complexes *in vivo* has been shown to be relatively low, especially when compared to viral vectors[44]. One study has suggested that the *in vivo* transfection efficiency of adenoviruses is around 200 times greater than that observed with liposomes[45]. One explanation for the relatively poor transfection efficiency of liposome/DNA complexes is that they are susceptible to disruption by serum proteins. A variety of proteins are known to bind to liposomes *in vitro* and *in vivo* and may mediate membrane destabilisation[46]. There are now serious efforts being made to develop liposomal vectors that are resistant to serum disruption[47-49]. Novel cationic lipids are also being developed to try to improve the transfection efficiency of liposome/DNA complexes [40,50]. Targeting of the liposomes to specific cell types has also been investigated as a means of improving the transfection efficiency [51,22].

There have been several clinical trials of liposome/DNA complexes, although almost all of these have been involved in the treatment of cystic fibrosis[3]. Most protocols involve the use of DC-chol/DOPE liposomes directly instilled onto the nasal epithelium of CF patients. The effect of gene expression on CFTR function was determined and the presence of the gene in the target cells was determined by PCR (polymerase chain reaction) and Southern blot analysis. The results from these trials are described in Section 3. Initial clinical studies found no evidence of any safety problems with the use of liposome/DNA complexes. This is surprising as it is well documented that the liposome/DNA complexes used in clinical trials are directly cytotoxic *in vitro*[20]. Furthermore, studies in mice and macaques have demonstrated that exposure to high doses or to repeat doses of liposome/DNA complexes results in histopathology and gross lung pathology, suggesting that these vectors may not be as clinically safe as previously thought [20].

5.3   Other non-viral gene delivery systems

5.3.1   Liposome / polycation / DNA (LPD) complexes

The limitations of liposome mediated gene delivery have led to the development of novel lipids in an effort to improve these vectors. A different approach to improving the transfection efficiency of cationic liposomes has been developed by Leaf Huang at the University of Pittsburgh[44]. This approach involves the use of polycations, such as polylysine, in the formation of liposome/DNA complexes. Polylysine and the liposome preparation DC-chol/DOPE were mixed with DNA in order to form liposome/polycation/DNA or LPD complexes. These complexes showed higher transfection efficiency than the corresponding liposome/DNA complexes without the presence of polycations and also showed enhanced resistance to degradation by nucleases. A further advantage of the use of polylysine was that the resulting complexes were much smaller than complexes formed with liposomes alone. Complexes formed between DNA and DC-chol/DOPE were found to have an average diameter of 1.2 μm, possibly due to an aggregation of the liposome/DNA complexes. When the polycations polylysine or protamine were incorporated into the complexes the average diameter was reduced to around 100 nm, a size that may allow more efficient cellular internalisation.

5.3.2   Peptide mediated gene delivery

The use of cationic polymers as gene delivery vectors has been investigated by several groups and will be described in more detail in the next section. A related approach is to use naturally occurring or synthetic peptides as gene delivery systems. This approach is based upon the observation that the functionally active regions of proteins such as enzymes, receptors and antibodies are relatively small, typically consisting of around 10-20 amino acids. Synthesising peptides based upon functional regions of DNA binding proteins or a variety of viral proteins is an approach that has been used to replace the use of whole proteins (such as histone H1) or large, polydispersed polymers (such as polylysine) as gene delivery vectors[52].

A number of peptide sequences have been shown to be able to bind to and condense DNA. One such sequence is the tetra-peptide serine-proline-lysine-lysine located in the C-terminus of the histone H1 protein[53]. Rational design of peptide sequence has also been used to develop completely synthetic DNA binding peptides. For example, peptide sequences based on the tetra-amine spermine were developed by Gottschalk *et al* (1996)[52] who showed that the peptide tyrosine-lysine-alanine-(lysine)$_8$-tryptophan-lysine was very effective at forming complexes with DNA.

DNA binding peptides can also be synthesised that can be coupled to cell specific ligands, thereby allowing receptor mediated targeting of the peptide/DNA complexes to specific cell types. One such approach has been to synthesise a cationic peptide based on 16 lysines and an RGD peptide sequence[54]. The RGD sequence has been shown to facilitate binding of the peptide/DNA complex to the integrin receptor on Caco-2 cells *in vitro*.

Synthetic peptides have also been developed that enhance the release of the peptide/DNA complexes from the endosome following endocytosis. Without a means of escaping the endosome, the endosomal-lysosomal pathway presents a major barrier to transfection via receptor mediated endocytosis. The use of endosomal lysis agents, such as membrane active lipids or peptides, is important in non-viral gene delivery systems.

6. POLYMER-BASED NON-VIRAL GENE DELIVERY SYSTEMS

The two classes of non-viral vector that are being investigated most actively are the cationic lipid-based vectors and the cationic polymer-based vectors[55]. The cationic polymer-based vectors represent the newest class of non-viral delivery system to be developed and there are a number of groups working with these vectors. As this is such a new area of research there has yet to be any standardised definition of the terms involved in this field. Complexes formed between cationic polymers and DNA have been referred to as interpolyelectrolyte complexes, molecular conjugates, polylysine-DNA complexes, DNA-polylysine complexes, polyplexes and so on[55]. In this review complexes formed between poly-(L-lysine) (pLL) and DNA are referred to as polylysine/DNA complexes or abbreviated to pLL/DNA complexes. Where polymers other than poly(L-lysine) have been used the term polymer/DNA complex is used. The forward slash (/) is used to denote a non-covalent interaction, while a hyphen (-) is used to denote a covalent link.

The main advantage of using cationic polymers as gene delivery vectors is that multi-functional polymers can be developed to produce vectors with a range of properties. An example of this is shown in Figure 1.

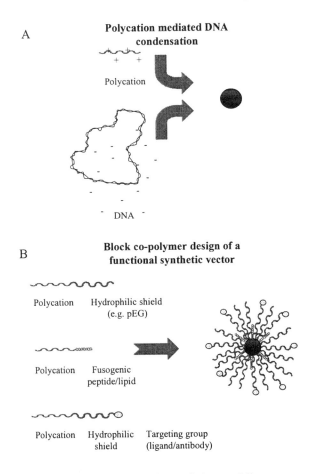

Figure 1. Potential design of a polymer-based non-viral gene delivery vector. A=polycation mediated DNA condensation, B=Development of a functional polymer-based vector through the use of block co-polymers. The rational design of vectors in this way allows the development of an "ideal" gene therapy vector. An ideal gene delivery vector would be one that is safe and non-toxic, it would be well defined and easily produced in scale. It would also be suitable for systemic administration and be able to circulate for extended periods in the bloodstream. It should also be small enough to extravasate from the vasculature into the target tissue.

The basis of most cationic polymer-based vectors is the condensation of DNA by a polycation such as pLL to a much smaller size than free DNA. Different functions can then be incorporated into the polymer/DNA complex. A great deal of interest has been generated by the possibility of attaching targeting moieties to the complex which might allow receptor mediated targeting of the complex to specific cell types. Other functional groups can also be incorporated into the complex, for example, membrane active peptides or lipids can be used to promote more efficient endosomal release of the complexes. Alternatively nuclear targeting sequences can be used to promote nuclear targeting, while hydrophilic groups can be incorporated to try to produce "stealthy" complexes that will avoid provoking an inflammatory response.

The ideal vector would also be one that could be targeted to a specific cell type, thereby reducing the transfection of peripheral cells. Upon reaching the target cell the vector should be small enough to be internalised through receptor mediated endocytosis. Once inside the endosome the vector needs some mechanism to promote its release into the cytoplasm. The vector will then need to translocate to, and enter, the nucleus where the therapeutic DNA can be expressed. The steps that such an ideal vector should be able to overcome following systemic administration are illustrated in Figure 2 and are discussed individually in more detail later in this chapter.

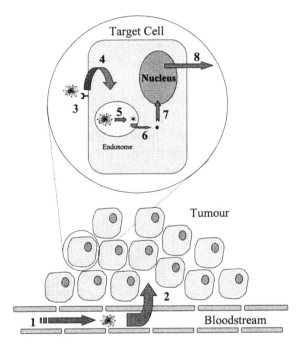

Figure 2. Diagrammatic representation of the requirements of an ideal vector for gene therapy. An ideal vector would be one that could be administered systemically (1). The vector would need to be small enough to extravasate from the vasculature into the target tissue (2). When targeting tumours tissue accumulation can be increased by the enhanced permeability and retention effect (2). Once the vector reaches the target cell it must be able to bind to cell surface receptors (3) and be internalised through endocytosis (4). Once inside the endosome, the outer layer of the vector may be removed following hydrolysis at the endosomal pH of 5.5 (5). The removal of the outer layer may then expose membrane permeabilising groups, such as lipids or fusogenic peptides, that can promote cytoplasmic delivery of the polymer/DNA core (6). The vector should then translocate to the nucleus (7), possibly through the use of nuclear targeting moieties linked to the core. The DNA should then be expressed by the cell and therapeutic protein produced (8).

6.1   Condensation of DNA by cationic polymers

DNA is an extremely large molecule. The chromatin of a typical human chromosome would be about 1 mm in length if it were extended and would therefore span the nucleus more than 100 times[56]. In its normal state it is condensed by various cellular proteins, such as the histone proteins, that allow such a large molecule to fit inside the cell nucleus. Even a plasmid DNA molecule of around 6 kb in size has an extended structure several hundred nanometers in diameter. In order to efficiently   deliver plasmid DNA into cells it is necessary to condense them to a much smaller size.

Endocytosis of plasmid DNA can be achieved if the DNA can be condensed to 100 nm or less in diameter[57], while access to tumour tissue from the vasculature is limited to particles less than 70 nm in diameter[58]. It is, therefore, important to condense plasmid DNA to as small a size as possible to facilitate gene transfer, and indeed the size of the condensed DNA complex may be one of the most important factors for successful *in vivo* gene delivery.

Polylysine was used to condense DNA as early as 1969[59] and was initially used merely as a model for the interaction of biopolymers such as DNA and histone proteins. DNA is a highly negatively charged polymer due to the repeating phosphate groups along the polymer backbone. The interaction with cationic polymers such as polylysine is, therefore, an electrostatic one. The exact parameters that govern the ability of cationic polymers to condense DNA are still under study, although it is generally accepted that neutralisation of the charges on the DNA molecule, followed by hydrophobic collapse as water is displaced from the DNA structure, may play an important role[60].

In 1987, Wu & Wu proposed using polylysine-condensed DNA as a gene delivery vector[61]. Since then it has been used as the basis of most cationic polymer-based vectors. The size of the polylysine/DNA complexes varies according to the molecular weight of the polymer used and the formation conditions such as the charge ratio between the polymer and DNA, the salt concentration and the temperature. However, complex sizes of 50-300 nm diameter can generally be achieved, while careful choice of the polymer molecular weight can produce complexes with an average diameter of less than 30 nm[57]. Careful control of mixing conditions and salt concentrations has been reported to produce complexes as small as 12 nm in diameter[62], although these results have proven difficult to reproduce.

One of the major advantages of using cationic polymers to condense DNA is that very large genes can be used. Viral vectors are limited by the amount of genetic material that can be inserted into the viral genome. This limitation does not apply to polymer-based vectors, and even DNA molecules as large as 45 Kb have been successfully condensed[63]. This permits the incorporation of gene regulatory regions such as locus control regions that may afford better control of gene expression.

6.2 Targeting strategies for polymer-based vectors

The ability to target polymer/DNA complexes to specific cell types, through a relatively simple attachment of cell specific ligands or antibodies, is one of the main attractions to using polymer-based vectors. Attempts have been made to target retroviruses, adenoviruses and liposomes to specific cell surface receptors [22,51,64,65], however the ease with which targeting moieties can be attached to cationic polymers makes these systems much more versatile.

The first ligands to be used to target polymer/DNA complexes to specific cell surface receptors were asialoglycoprotein[61] and transferrin[66,67]. Since then many other targeting moieties have been used to achieve specific receptor-mediated uptake of polymer/DNA complexes into cells. Gene delivery has been demonstrated with complexes targeted with mannose[68], fucose[69], anti-CD3 antibody[70] and other antibodies[71,72], invasin[73] and folic acid[74] amongst other ligands.

The transferrin receptor was one of the first to be exploited for receptor-mediated gene delivery. All actively metabolising cells require iron that is internalised by the cell as a transferrin-iron complex by means of receptor mediated endocytosis. To exploit this ubiquitous and efficient transport mechanism for introducing DNA into cells, conjugates of chicken or human transferrin with polycations (polylysine or protamine) were synthesised and used to form complexes with plasmid DNA. The number of transferrin molecules attached to each polylysine molecule varies according to the molecular weight of the polymer, but is generally around 1 transferrin molecule for every 50 lysine residues[66].

The transferrin-polylysine/DNA complexes were used to deliver the gene encoding the luciferase enzyme into avian erythroblasts transformed with the erbB oncogene, which is known to have a particularly active transferrin cycle[66] or into the human leukaemic cell line K-562[67] which is known to have a relatively high number of transferrin receptors on the cell surface. Efficient transfection of both these cell lines was demonstrated by the transferrin-polycation/DNA complexes and it was found that between 10 and 20 transferrin molecules per complex were required for transfection[75]. Gene expression was enhanced when an endosomolytic agent such as chloroquine was used. Endosomolytic agents are used to promote endosomal release and subsequent cytoplasmic delivery of complexes internalised through a receptor-mediated endocytosis mechanism. Their use in synthetic gene delivery vectors will be discussed later. The fact that chloroquine enhances transferrin-mediated transfection (transferrinfection) indicates that the complexes are internalised through a receptor-mediated mechanism. Further evidence that the complexes were indeed internalised by a receptor-mediated mechanism comes from the observation that transfection efficiency can be enhanced by the increase in density of the transferrin receptors on the surface of the cell that occurs following treatment of the cells with the iron chelator desferrioxamine. This treatment resulted in a 4-fold increase in receptor number and a 5-fold increase in gene expression[67].

Another well studied ligand for potential use in receptor-mediated gene delivery is asialoglycoprotein. This was the first ligand to be conjugated to polylysine and used to specifically target cells through a receptor[61]. It was demonstrated that HepG2 cells could be efficiently transfected *in vitro* with asialo-orosomucoid-pLL/DNA complexes, with a 2-fold increase in activity compared to the calcium phosphate precipitation protocol[76]. The ability to transfect hepatocytes *in vivo* was also demonstrated, although the efficiency and duration of expression were relatively low[77]. However, it was found that expression could be sustained for up to 11 weeks if the animals were subjected to a partial (66%) hepatectomy 30 minutes following gene delivery[78]. Other groups subsequently demonstrated gene expression *in vivo* up to 140 days after gene delivery by targeting the same receptor[62].

6.3   Systemic delivery of polymer-based vectors

There is now great interest in the potential of *in vivo* gene delivery systems to act as a general means of delivering therapeutic proteins either systemically or locally. The main advantage of systemically deliverable vectors is that the vectors would be taken up and expressed only in the target tissue, allowing highly specific local delivery of the therapeutic protein. Cancer protocols in particular may require systemic delivery with some degree of targeting[79]. Long circulation times allow the vector to passively accumulate in the tumour through a phenomenon known as the Enhanced Permeability and Retention (EPR) effect. This phenomenon is due to the increased permeability of tumour vasculature that permits large macromolecules to extravasate into the tumour tissue, where they are retained due to the high hydrostatic pressure in the tumour. The EPR effect has been exploited in the delivery of macromolecular pro-drugs to tumours[58] and it may have a role to play in the targeting of gene delivery vectors to tumours.

The factors that are important in the circulation of polymer/DNA complexes have yet to be fully identified or studied, however information obtained from studying the fate of macromolecules following systemic administration may provide some information on the obstacles to successful systemic delivery of DNA. The processes that affect the disposition of macromolecules before they reach their target site involve interaction with blood components, interaction with vascular endothelial cells and uptake by the reticuloendothelial system. Furthermore, the degradation of the therapeutic DNA by serum nucleases is also a potential obstacle to the delivery of functional delivery to the target cell.

The biodistribution and elimination patterns of macromolecular systems are dictated mainly by their physicochemical properties. Perhaps the two most important properties are particle size and surface charge. After intravenous injection particles greater than 5-7 μm in diameter are cleared by capillary filtration mainly in the lungs, while particles with a diameter of less than 5 μm are generally cleared from the circulation by cells of the reticuloendothelial system[80,81]. The surface charge is also well known to affect the biodistribution and blood clearance of particles, for example negatively charged particles are known to be cleared from the bloodstream than positively charged particles[82], while neutral particles exhibit an increased circulation half-life. Highly charged particles (positive or negative) have a tendency to accumulate in the liver. Cationic particles or macromolecules accumulate in the liver due to its large surface area permitting adhesion of the cationic particles with the negatively charged cell surfaces[83]. By comparison, negatively charged particles or macromolecules are taken up by liver macrophages (Kupffer cells) by scavenger receptor-mediated endocytosis[84].

The physicochemical properties of particles and macromolecules have a major effect on their biodistribution and blood clearance as these properties govern how the particles will interact with serum proteins (including serum nucleases), endothelium and the reticuloendothelial system. Some of these systems are described below.

6.3.1   Serum nuclease degradation of therapeutic DNA

Free DNA is known to be rapidly degraded by nucleases in serum, with a half-life varying from a few minutes to several hours[85]. A systemic gene delivery system should be able to prevent this degradation by serum nucleases in order to delivery functional DNA to the target cell. There remains uncertainty over whether DNA contained within liposomes is susceptible to nuclease degradation. Some groups maintain that there is some degradation of the DNA[44] while others have indicated that the DNA is completely resistant to the effects of serum nucleases[86]. By comparison, complexation of DNA with polylysine has been shown to protect the DNA from degradation by serum nucleases[44,85], suggesting that they can deliver intact therapeutic DNA to the target cells.

6.3.2   Interaction of polymer/DNA complexes with serum proteins

There have been few studies on the interaction of polymer/DNA complexes with serum proteins, yet these interactions may have a major effect on the circulation of these complexes. One of the few studies that has been performed is an analysis of the ability of various polymer/DNA complexes to interact with complement components and activate the complement cascade[87]. This study showed that the complement system may be a major barrier to the systemic administration of polycation/DNA complexes. Complexes formed between DNA and polylysine, polyethyleneimine and other cationic polymers were shown to be potent activators of the complement cascade, which may limit their *in vivo* usefulness.

The importance of serum protein interactions on circulation and transfection activity has been established with cationic liposomes. It is well known that serum has an inhibitory effect on the transfection efficiency of cationic liposome/DNA complexes[47-49]. Binding of the liposome/DNA complexes to negatively charged serum proteins, leading to a decrease in cell association has been implicated as a mechanism for the inhibitory effect of serum on transfection activity[49]. It has since been demonstrated that this inhibition can, at least partly, be overcome by increasing the charge ratio or dose of the liposome/DNA complex[48,49]. The development of novel liposome formulations can also lead to an improvement in the transfection activity in the presence of serum[47]. The *in vivo* effects of protein binding have also been investigated and it was determined that the interaction with serum proteins leads to very a rapid clearance of the liposome/DNA complexes from the circulation[46]. Furthermore, it was determined that the amount of protein bound to the liposome/DNA complexes is directly proportional to the clearance time of the complexes from the bloodstream. The incorporation of cholesterol into the liposome/DNA formulation, which increases the packing densities of the phospholipid molecules, drastically reduces the amount of bound protein and produces a corresponding improvement in circulation half-life. The incorporation of 30 mol% cholesterol increases the half-life of liposome/DNA complexes from seconds to over 5 hours[46]. The amount of serum protein binding to liposome/DNA complexes can also be reduced by increasing the dose of the liposome/DNA complex administered intravenously. This reduction in protein binding produced a corresponding increase in the circulation times, from around 4 minutes to over 80 minutes[48].

6.3.3   The Reticulo-endothelial system

The reticuloendothelial system (RES) is a broad system of cells which include reticular cells, endothelial cells, fibroblasts, histiocytes and monocytes. The phagocytic cells within this system compose the mononuclear phagocyte system (MPS), and the macrophage is the major differentiated cell of this system. The MPS also consists of bone marrow monoblasts and pro-monocytes, peripheral blood monocytes and tissue macrophages[88]. Cells of the RES and MPS, particularly the liver macrophage (Kupffer cell), are known to be important in the clearance of particles from the bloodstream. Negatively charged particles, in particular, are avidly scavenged by macrophages. The Kupffer cell is one of the most important cells in the MPS. These highly phagocytic cells are located in the sinusoid wall, usually on the endothelial surface, of the liver[89]. These cells have a number of functions, but one of the most important is their ability to endocytose and remove from the blood potentially harmful materials and particulate matter such as bacterial endotoxins, micro-organisms, immune-complexes and tumour cells. The recognition of these materials is mediated by an array of cell surface receptors that include receptors for IgG, IgA, complement components, galactose, mannose, CD4 and the carcinoembryonic antigen. Other receptors include the scavenger receptor, transferrin and DNA-binding receptors[90].

6.4   Cytoplasmic delivery of internalised polymer-based vectors

Receptor mediated endocytosis of targeted polymer/DNA complexes will result in the complexes being processed in the endosomal-lysosomal pathway, leading to degradation of the complex and the therapeutic DNA in the lysosome. Some mechanism to promote endosomal release and cytoplasmic delivery is necessary for efficient gene transfer to occur[67].

There are several strategies that can be used to try to increase the efficiency of cytoplasmic delivery. One of the most common methods for increasing the *in vitro* transfection efficiency of polymer/DNA complexes is to treat the cells with chloroquine. Chloroquine is thought to have a buffering capacity that prevents endosomal acidification, leading to swelling and bursting of the endosomes[91] and has been shown to enhance the transfection activity of polycation/DNA complexes[92,93]. This approach is limited to *in vitro* applications since the concentrations of chloroquine required to enhance transfection are likely to be toxic *in vivo*.

Cytoplasmic delivery of endocytosed polymer/DNA complexes has also been enhanced through the attachment of inactivated adenoviruses to the complex or through co-incubation of the cells with adenovirus particles and polymer/DNA complexes[94-96]. The use of adenovirus particles to enhance transfection of polymer/DNA complexes is unlikely to be widely used *in vivo* as the adenovirus particles may provoke inflammatory responses, while it has also been demonstrated that the transfection efficiency of these systems is much lower *in vivo* than *in vitro*[97].

As an alternative to the use of whole virus particles to enhance cytoplasmic delivery, fusogenic peptides have been used. Many viruses use membrane destabilising proteins to promote endosomal release, and one of the best studied of these systems is the influenza virus haemaglutin. The fusion domain of this protein is located at the N-terminus of subunit HA 2 and this peptide sequence, and modifications of it, have been shown to significantly enhance transfection efficiency in a number of polymer/DNA systems[98,99].

Another alternative is to condense DNA with a cationic polymer that possesses an intrinsic endosomolytic ability. One such polymer that has been extensively studied, and found to be very efficient at transfecting cells *in vitro*, is polyethyleneimine[100-102]. Polyethyleneimine (pEI) is a highly branched polymer composed of units which have two units of carbon per nitrogen with these units randomly distributed in the approximate ratios of one primary amino nitrogen/two secondary amino nitrogens/one tertiary amino nitrogen. When pEI is dissolved in water some of the amino groups react with the solvent to form positively charged nitrogens. Not all of the amino groups will become protonated, however, and the addition of acid will increase the number of charged amino groups. It has been suggested that only around 60% of the primary amino groups are charged in water, leaving 40% that can become protonated at lower pH, similar to which exists in the endosomes. It has been suggested that this protonation in the endosome leads to a buffering effect resulting in endosomal swelling and bursting with subsequent enhancement of transfection activity[100].

6.5   Nuclear targeting of polymer-based vectors

Delivery of DNA into the cytoplasm of the target cell does not represent the final stage of gene delivery. The DNA needs to be transported into the nucleus and expressed. It is thought that less than 1% of the plasmid molecules that reach the cytoplasm eventually reach the nucleus[103,104]. To efficiently deliver complexed DNA to the nucleus it may be necessary to incorporate nuclear targeting or nuclear localisation signals into the vector. It has been suggested that certain components of polymer/DNA complexes may possess intrinsic nuclear homing capabilities, for example polylysine[44] and the fusogenic peptide INF7[104] have both been implicated as having a possible role in increasing the efficiency of nuclear delivery of the complexed DNA. It has also been shown that certain sequences of the plasmid DNA may lead to import of the DNA into the nucleus[105].

In order to improve the efficiency of nuclear delivery it may be necessary to incorporate specific nuclear localisation signal (NLS) peptides into the polymer/DNA complex. All proteins are synthesised in the cytoplasm and those proteins that are required in the nucleus must be imported from the cytoplasm, through the nuclear pores and into the nucleus. To achieve this, nuclear proteins contain short peptide sequences termed nuclear localisation signals that allow selective transport of these proteins to the nucleus[106]. It is possible that the attachment of these peptide sequences to the polymer/DNA complexes may result in much greater transfection efficiency. There are several NLS peptides that might be suitable for incorporation into a gene delivery vector, and many of these peptides contain lysine and arginine motifs. The NLS from the SV40 large T antigen was one of the first NLS peptides to be discovered and has an amino acid sequence of PKKKRKV[106]. Since then many other sequences have been identified that may prove useful additions to a synthetic, polymer-based gene delivery vector[107-109]

REFERENCES

1. Martin, P. and Thomas, S. M., 1996, *European Union*, Brussels.
2. Blaese, R. M., Culver, K. W., Miller, A. D., Carter, C. S., Fleisher, T., Clerici, M., Shearer, G., Chang, L., Chiang, Y., Tolstoshev, P., Greenblatt, J. J., Rosenberg, S. A., Klein, H., Berger, M., Mullen, C. A., Ramsey, W. J., Muul, L., Morgan, R. A. and Anderson, W. F., 1995, *Nature* 270: 475-480.
3. Velu, T., 1996, *Gene Transfer and Therapy in Europe: research, devlopment and clinical trials*, European Union, Brussels.
4. Mulligan, R. C., 1993, *Science* 260: 926-931.
5. Boulikas, T., 1996, *Int.J.Oncol.* 9: 1239-1251.
6. Hanania, E. G., Kavanagh, J., Hortobagyi, G., Giles, R. E., Champlin, R. and Deisseroth, A. B., 1995, *Am.J.Med.* 99: 537-552.
7. Boulikas, T., 1996, *Int.J.Oncology* 9: 941-954.
8. Sugaya, S., Fujita, K., Kikuchi, A., Ueda, H., Takakuwa, K., Kodama, S. and Tanaka, K., 1996, *Human Gene Therapy* 7: 223-230.
9. Huber, B. E., Austin, E. A., Richards, C. A., Davis, S. T. and Good, S. S., 1994, *Proc.Natl.Acad.Sci.USA* 91: 8302-8306.
10. Lafont, A., Guerot, C. and Lemarchand, P., 1996, *Eur.Heart.J.* 17: 1312-1317.
11. Grossman, M., Raper, S. E. and Kozarsky, K., 1994, *Nature Genetics* 6: 335-341.
12. Von der Leyen, H. E., Gibbons, G. H. and Morishita, R., 1995, *Proc.Natl.Acad.Sci.USA* 92: 1137-1141.
13. Nabel, E. G., 1995, *Circulation* 91: 541-548.
14. Chang, M. W., Barr, E. and Seltzer, J., 1995, *Science* 267: 518-522.
15. Friedmann, T., 1997, *Scientific American* 80-85.
16. Friedmann, T., 1994, *Trends in Genetics* 10: 210-214.

17. GarciaValenzuela, E., Rayanade, R., Perales, J. C., Davidson, C. A., Hanson, R. W. and Sharma, S. C., 1997, *Journal of Neurobiology* 32: 111-122.

18. Kolodka, T. M., Finegold, M., Moss, L. and Woo, S. L. C., 1995, *Proc.Natl.Acad.Sci.USA* 92: 3293-3297.

19. Geddes, D. M., 1995, *Gene Therapy* 2: 586.

20. Malone, R. W., *Artificial Self-assembling Systems for Gene Therapy*, Vol. 3, Felgner, P., Ed., Cambridge Healthtech, San Diego, CA (1996).

21. Marshall, E., 1995, *Science* 270: 1751.

22. Miller, N. and Vile, R., 1995, *FASEB* 9: 190-199.

23. Jolly, D., 1994, *Cancer Gene Therapy* 1: 51-64.

24. Pensiero, M. N., Wysocki, C. A., Nader, K. and Kikuchi, G. E., 1996, *Hum.Gen.Ther.* 7: 1095-1105.

25. Adams, R. M., Wang, M., Steffen, D. and Ledley, F. D., 1995, *J.Virol.* 69: 1887-1894.

26. Toggas, S., Masliah, E., Rockenstein, E., Rall, G. F., Abraham, C. R. and Mucke, L., 1994, *Nature* 367: 188-193.

27. Brody, S. L. and Crystal, R. G., 1994, *Annals N.Y. Acad.Sci.* 716: 90-102.

28. Nermut, M. V., *Adenoviridae*, Elsevier Science Publications (1987).

29. Rogers, B. E., Douglas, J. T., Ahlem, C., Buchsbaum, D. J., Frincke, J. and Curiel, D. T., 1997, *Gene Therapy* 4: 1387-1392.

30. Dong, J. Y., Wang, D., Van Ginkel, F. W., Pascual, D. W. and Frizzell, R. A., 1996, *Hum.Gene Ther.* 7: 319-331.

31. Wilmott, R. W., Amin, R. S., Perez, C. R., Wert, S. E., Keller, G., Boivin, G. P., Hirsch, R., De Inocencio, J., Lu, P., Reising, S. F., Yei, S., Whitsett, J. A. and Trapnell, B. C., 1996, *Hum.Gene Ther.* 7: 301-318.

32. Heise, C., Sampson, J. A., Williams, A., McCormick, F., Von Hoff, D. D. and Kirn, D. H., 1997, *Nature Medicine* 3: 639-645.

33. Zhang, G. F., Vargo, D., Budker, V., Armstrong, N., Knechtle, S. and Wolff, J. A., 1997, *Hum.Gene Therapy* 8: 1763-1772.

34. Hart, I. R., 1995, *Gene Therapy* 2: 572.

35. Choate, K. A. and Khavari, P. A., 1997, *Hum.Gene Ther.* 8: 1659-1665.

36. Allen, T. M., 1994, *Trends Pharmacol. Sci.* 15: 215-220.

37. Gershon, H., Ghirlando, R., Guttman, S. B. and Minsky, A., 1993, *Biochemistry* 32: 7143-7151.

38. Felgner, P. and Ringwold, G. M., 1989, *Nature* 337: 387-388.

39. Jaaskelainen, I., Monkkonen, J. and Urtti, A., 1994, *Biochem. Biophys. Acta* 1195: 115-123.

40. Reimer, D. L., Zhang, Y. P., Kong, S., Wheeler, J. J., Graham, R. W. and Bally, M. B., 1995, *Biochemistry* 34: 12877-12883.

41. Liu, F., Qi, H., Huang, L. and Liu, D., 1997, *Gene Therapy* 4: 517-523.

42. Song, Y. K., Liu, F., Chu, S. and Liu, D., 1997, *Hum.Gene Ther.* 8: 1585-1594.

43. Thierry, A. R., Rabinovich, P., Peng, B., Mahan, L. C., Bryant, J. L. and Gallo, R. C., 1997, *Gene Therapy* 4: 226-237.

44. Gao, X. and Huang, L., 1996, *Biochemistry* 35: 1027-1036.

45. Laitinen, M., Pakkanen, T., Donetti, E., Baetta, R., Luoma, J., Lehtolainen, Viita, H., Agrawal, R., Miyanohara, A., Friedmann, T., Risau, W., Martin, J., F., Soma, M. and Yla-Herttuala, S., 1997, *Hum.Gene Ther.* 8: 1645-1650.

46. Semple, S. C., Chonn, A. and Cullis, P. R., 1996, *Biochemistry* 35: 2521-2525.

47. Lewis, J. G., Lin, K. Y., Kothavale, A., Flanagan, W. M., Matteucci, M. D., DePrince, R. B., Mook, R. A., Hendren, R. W. and Wagner, R. W., 1996, *Proc.Natl.Acad.Sci.USA* 93: 3176-3181.

48. Oja, C. D., Semple, S. C., Chonn, A. and Cullis, P. R., 1996, *Biochim.Biophys.Acta* 1281: 31-37.

49. Yang, J. P. and Huang, L., 1997, *Gene Therapy* 4: 950-960.

50. Felgner, P. L., 1995, *Gene Therapy* 2: 573.

51. Cheng, P. W., 1996, *Hum.Gene Ther.* 7: 275-282.

52. Gottschalk, S., Sparrow, J. T., Hauer, J., Mims, M. P., Leland, F. E., Woo, S. L. C. and Smith, L. C., 1996, *Gene Therapy* 3: 448-457.

53. Khadake, J. R. and Rao, M. R. S., 1997, *Biochemistry* 36: 1041-1051.

54. Hart, S. L., Harbottle, R. P., Cooper, R., Miller, A., Williamson, R. and Coutelle, C., 1995, *Gene Therapy* 2: 552-554.

55. Felgner, P. L., Barenholz, Y., Behr, J. P., Cheng, S. H., Cullis, P., Huang, L., Jessee, J. A., Seymour, L., Szoka, F., Thierry, A. R., Wagner, E. and Wu, G., 1997, *Hum.Gene.Therapy* 8: 511-512.

56. Alberts, B., Bray, D., Lewis, J., Raff, M., Roberts, K. and Watson, J. D., *Molecular Biology of The Cell*, Garland Publishing, Inc., New York (1989).

57. Wolfert, M. A. and Seymour, L. W., 1996, *GeneTherapy* 3: 269-273.

58. Seymour, L. W., 1992, *Crit.Rev.Ther.Drug Carrier Syst.* 9: 135-187.

59. Shapiro, J. T., Leng, M. and Felsenfeld, G., 1969, *Biochemistry* 8: 3219-3232.

60. Kabanov, A. V. and Kabanov, V. A., 1995, *Bioconjugate Chem.* 6: 7-20.

61. Wu, G. Y. and Wu, C. H., 1987, *J.Biol.Chem* 262: 4429-4432.

62. Perales, J. C., Ferkol, T., Beegen, H., Ratnoff, O. D. and Hanson, R. W., 1994, *Proc. Natl. Acad. Sci. USA* 91: 4086-4090.

63. Cotten, M., Wagner, E., Zatloukal, K., Phillips, S., Curiel, D. T. and Birnstiel, M. L., 1992, *Proc.Natl.Acad.Sci. U.S.A* 89: 6094-6098.

64. Cosset, F. L. and Russel, S. J., 1996, *Gene Therapy* 3: 946-956.

65. Lee, R. J. and Low, P. S., 1994, *J.Biol.Chem* 269: 3198-3204.

66. Wagner, E., Zenke, M., Cotten, M., Beug, H. and Birnstiel, M. L., 1990, *Proc. Natl. Acad. Sci. USA* 87: 3410-3414.

67. Cotten, M., Langle-Rouault, F., Kirlappos, H., Wagner, E., Mechtler, K., Zenke, M., Beug, H. and Birnstiel, M. L., 1990, *Proc. Natl. Acad. Sci. USA* 87: 4033-4037.

68. Liang, W. W., Shi, X., Deshpande, D., Malanga, C. J. and Rojanasakul, Y., 1996, *Biochim. Biophys. Acta* 1279: 227-234.

69. Biessen, E. A. L., Bakkeren, H. F., Kuiper, J. and Van Berkel, T. J. C., 1994, *Biochem. J.* 299: 291-296.

70. Buschle, M., Cotten, M., Kirlappos, H., Mechtler, K., Schaffner, G., Zauner, W., Birnstiel, M. L. and Wagner, E., 1995, *Human Gene Therapy* 6: 753-761.

71. Trubetskoy, V. S., Torchilin, V. P., Kennel, S. J. and Huang, L., 1992, *Bioconjugate Chem.* 3: 323-327.

72. Coll, J. L., Wagner, E., Combaret, V., Metchler, K., Amstutz, H., Iacano-di-Cacito, I., Simon, N. and Favrot, M. C., 1997, *Gene Therapy* 4: 156-161.

73. Paul, R. W., Weisser, K. E., Loomis, A., Sloane, D. L., LaFoe, D., Atkinson, E. M. and Overell, R. W., 1997, *Human Gene Therapy* 8: 1253-1262.

74. Mislick, K. A., Baldeschwieler, J. D., Kayyem, J. F. and Meade, T. J., 1995, *Bioconjugate Chem.* 6: 512-515.

75. Wagner, E., Cotten, M., Foisner, R. and Birnstiel, M. L., 1991, *Proc. Natl. Acad. Sci. USA* 88: 4255-4259.

76. Wu, G. Y. and Wu, C. H., 1988, *Biochemistry* 27: 887-892.

77. Wu, G. Y. and Wu, C. H., 1988, *J.Biol.Chem.* 263: 14621-14624.

78.Wu, C. H., Wilson, J. M. and Wu, G. Y., 1989, Targeting genes: Delivery and persistent expression of a foreign gene driven by mammalian regulatory elements in vivo. *J.Biol.Chem.* 264: 16985-16987.

79.Sikora, K., 1995, Clinical gene delivery. *Gene Therapy* 2: 562.

80.Senior, J., Crawley, J. C. W. and Gregoriadis, G., 1985, Tissue distributions of liposomes exhibiting long half-lives in the circulation after intravenous injection. *Biochim.Biophys.Acta* 839: 1-8.

81.Illum, L. and Davis, S. S., 1984, the organ uptake of intravenously administered colloidal particles can be altered using a non-ionic surfactant. *FEBS Lett.* 167: 79-82.

82.Juliano, R. L. and Stamp, D., 1975, The effect of particle size and charge on the clearance rates of liposomes and liposome encapsulated drugs. *Biochim.Biophys.Res.Commun.* 63: 651-658.

83.Nishida, K., Mihara, K., Takino, T., Nakane, S., Takakura, Y., Hashida, M. and Sezaki, H., 1991, Hepatic disposition characteristics of electrically charged macromolecules in rat in vivo and in perfused liver. *Pharm.Res.* 8: 437-444.

84.Kawabata, K., Takakura, Y. and Hashida, M., 1995, The fate of plasmid DNA after intravenous injection in mice: involvement of scavenger receptors in its hepatic uptake. *Pharm.Res.* 13: 1595-1614.

85.Chiou, H. C., M.V., T., Levine, S. M., Robertson, D., Kormis, K., Wu, C. H. and Wu, G. Y., 1994, Enhanced resistance to nuclease degradation of nucleic acids complexed to asialoglycoprotein-polylysine carriers. *Nuc.Acids.Res.* 22: 5439-5446.

86.Crook, K., McLachlan, G., Stevenson, B. J. and Porteous, D. J., 1996, Plasmid DNA-molecules complexed with cationic liposomes are protected from degradation by nucleases and shearing by aerosolisation. *Gene Therapy* 3: 834-839.

87.Plank, C., Mechtler, K., Szoka, F. C. and Wagner, E., 1996, Activation of the complement system by synthetic DNA complexes: a Potential barrier for intravenous gene delivery. *Human Gene Therapy* 7: 1437-1446.

88.Auger, M. J. and Ross, J. A., The Biology of the Macrophage, in *The Macrophage*, Lewis, C. E., and McGee, J. O. D., Eds., IRL Press, Oxford (1992).

89.Motta, P., 1975, A scanning electron microscope study of the rat liver sinusoid endothelial and Kupffer cell. *Cancer Res.* 164: 371-385.

90.Becker, S., 1988, Function of the human mononuclear phagocyte system. *Adv.Drug.Deliv.Rev.* 2: 1-29.

91.Maxfield, F. R., 1982, Weak bases and ionophores rapidly and reversibly raise the pH of endocytic vesicles in cultured mouse fibroblasts. *J.Cell Biol.* 95: 676-681.

92.Wolfert, M. A., Schacht, E. H., Toncheva, V., Ulbrich, K., Nazarova, O. and Seymour, L. W., 1996, Characterization of Vectors for Gene Therapy Formed by Self-Assembly of DNA with Synthetic Block Co-Polymers. *Human Gene Therapy* 7: 2123-2133.

93.Midoux, P., Mendes, C., Legrand, A., Raimond, J., Mayer, R., Monsigny, M. and Roche, A. C., 1993, Specific gene transfer mediated by lactosylated poly-L-lysine into heptoma cells. *Nucleic Acids Res.* 21: 871-878.

94.Curiel, D. T., High efficiency gene transfer mediated by adenovirus-polylysine-DNA complexes, in *Gene Therapy for Neoplastic Diseases*, Vol. 716, Huber, B. E. L., J.S., Ed., New York Academy of Sciences, pp. 36-59 (1994).

95.Wagner, E., Cotten, M., Curiel, D., Berger, M., Schmidt, W., Buschle, M., Schweighoffer, T. and Birnstiel, M. L., 1995, Adenovirus enhanced receptor mediated transferrinfection (AVET) applied to the generation of tumour vaccines. *Gene Therapy* 2: 574.

96.Cotten, M., Wagner, E., Zatloukal, K. and Birnstiel, M. L., 1993, *J. Virology* 67: 3777-3785.

97.Gao, L., Wagner, E., Cotten, M., Agarwal, S., Harris, C., Romer, M., Miller, L., Hu, P. C. and Curiel, D., 1993, *Human Gene Therapy* 4: 17-24.

98.Plank, C., Oberhauser, B., Mechtler, K., Koch, C. and Wagner, E., 1994, *J. Biol. Chem* 269: 12918-12924.

99.Wagner, E., Plank, C., Zatloukal, K., Cotten, M. and Birnsteil, M. L., 1992, *Proc.Natl.Acad.Sci. U.S.A.* 89: 7934-7938.

100.Boussif, O., Zanta, M. A. and Behr, J. P., 1995, *Proc.Natl.Acad.Sci.USA* 92: 7297-7301.

101. Abdallah, B., Goula, A., Benoist, C., Behr, J. P. and Demenieix, B. A., 1996, *Hum.Gen.Ther.* 7: 1947-1954.

102. Baker, A., Saltik, M., Lehrmann, H., Killisch, I., Mautner, V., Lamm, G., Christofori, G. and Cotten, M., 1997, *Gene Therapy* 4: 773-782.

103. Boulikas, T., 1997, *Int.J.Oncol.* 10: 301-309.

104. Wolfert, M. A. and Seymour, L. W., 1998, *Gene Therapy* 5: 409-414.

105. Dean, D. A., 1997, *Exp.Cell.Res.* 230: 293-302.

106. Kalderon, D., Roberts, B. L., Richardson, W. D. and Smith, A. E., 1984, *Cell* 39: 499-509.

107. Yondea, Y., Imamoto-Sonobe, N., Matsuoka, Y., Iwamoto, R., Kiho, Y. and Uchida, T., 1988, *Science* 242: 275-278.

108. Picard, D. and Yamamoto, K. R., 1987, *EMBO J.* 6: 3333-3340.

109. Standiford, D. M. and Richter, J. D., 1992, *J.Cell Biol.* 118: 991-1002.

HIGH-MOLECULAR WEIGHT POLYETHYLENE GLYCOLS CONJUGATED TO ANTISENSE OLIGONUCLEOTIDES
Synthesis and Applications

Gian Maria Bonora

Chemical Sciences Department, University of Trieste, Trieste, Italy

1.   INTRODUCTION

The oligonucleotides are short- to medium-size sequences of nucleic acids. As such they have a polianionic nature. Repeating monomer units, nucleosides, in which a ribose, or a deoxyribose unit, is joined via an N-glycosidic bond with an heterobase, purine- or pyrimidine-like, makes them. These units are connected each other by a phosphodiester bridge through the 3'- and 5'-OH functionalities of the sugars.

Usually, due to the size required for the expression of the biological activity, the average dimension of the synthetic oligonucleotides ranges between 10/12 and 18/20 nucleosides.

Taking advantage from the specificity of Watson-Crick base pairing, as originated by the formation of a regular net of hydrogen bonds between opposite heterocyclic bases of complementary nucleic acid sequences, these molecules have been tested as biological tools for elucidating gene function and as human therapeutics. It is easy to conceive that a properly designed oligonucleotide could be able to inhibit a gene by interaction with a specific part of its sequence. This inhibition of the gene expression was proposed more than 20 years ago by Zamecnik and Stephenson[1], with the purpose of blocking the effect of an exogenous viral genome inside hosting cells. This approach will be equally effective in suppressing unwanted pathological

Biomedical Polymers and Polymer Therapeutics
Edited by Chiellini *et al.*, Kluwer Academic/Plenum Publishers, New York, 2001   371

effects due to nucleic acid sequences, as those, for example, of some oncogene, or of any other gene-related pathology.

Since then, this so-called antisense strategy has been widely investigated, and two main mechanism of action can be described. The first one, the true antisense effect, is due to the arrest of the translation process, i.e. the production of proteins as expressed by the genetic message encoded in a RNA chain, the messenger or mRNA. (Fig.1A). In this case, the complex between the antisense oligonucleotide and the target RNA yields a stable duplex; as a consequence, the protein production is avoided, mainly trough the activation of an ubiquitary enzyme, the Rnase H, that destroys the RNA part of that duplex.

In the other process, the antigene mechanism is generated by the direct complexion of the original code inside the double helix of DNA. Some special oligonucleotide sequences are able to make a triple complex, a triplex, with the double helix of the DNA target: in such a way the transcription of the message in a new mRNA chain is escaped, and the following translation in the pathological proteins is fully prevented (Fig.1B).

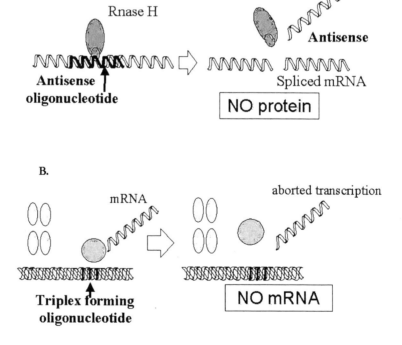

Figure 1. Antisense oligonucleotide: A. scheme of the true antisense action; B. scheme of the antigene action

Nature itself utilises this system to down-regulate several gene products, and antisense sequences have been found to be expressed inside the cells[2]. However, while the antisense approach is a convincing model for the inhibition of gene expression, several challenges have to be met to allow the use of synthetic oligonucleotides as therapeutic agents.

Among the prerequisites for a successful implementation of an oligonucleotide-based drug design strategy[3], their metabolic stability, i.e. the resistance to in vivo degradation is one of the most fundamental. Moreover, such antisense molecules, large and highly charged, present an unfavourable behaviour toward their cellular uptake. This hampers their penetration inside the cells to reach cytoplasmatic or nuclear targets. In addition, these compounds must display a high bioavailability, with slow body clearance and limited overall degradation.

As a consequence, the chemical modification of the oligonucleotide structure plays an important role in antisense research and represents the key element on the way of their adoption in therapy.

First attempts have been focused on the modification of the phosphate backbone of the chain as, for example, the introduction of a sulphur atom[4], a methyl group[5], or various amines[6]. These replacements conferred a higher nuclease stability, but are accompanied by altered RNA binding affinity, cellular uptake and Rnase H reactivity. Other modifications have been introduced at the sugar level, with an effective enhancement of stability and binding affinity to the target sequence. Every portion of the ribose, as of the deoxyribose, has been investigated, by altering both the chemical structures of the sugar[7], as well as its stereochemical property[8]. Even the heterocyclic bases have been modified at the pyrimidine[9], and at the purine level[10], mainly to enhance the binding without altering the RNase H recognition.

Alternatively, to improve the bioavailability of oligonucleotides, the attachment of macromolecules or reporter groups at the 5'- or 3'-end has been pursued.

An overall sketch of the main chemical modifications is described in Figure 2.

2. POLYMER-CONJUGATED OLIGONUCLEOTIDES

The synthesis of these conjugates has been widely investigated[11]. Among the different molecules introduced, it must be recalled the intercalating groups as acridine[12] or phenazinium[13], to increase the affinity of binding through a π-π interaction with the nucleosides of the target. Hydrophobic groups as cholesterol[14] has been also proposed to enhance the cellular association in attempt to increase the uptake. In a different approach, ligands for cellular receptors, as peptides[15] or sugars[16], have been introduced to facilitate the specific uptake of these oligonucleotides.

Besides these low-molecular weight molecules, high-molecular mass polymers have been investigated as new conjugating moieties able to endow the bound molecules with their peculiar features. In this way it is possible for example to add more hydrophobic character to the antisense derivative, to increase their cellular penetration. Besides, the presence of larger molecules will reduce their clearance from the body, increasing the time life useful to exercise a therapeutic action.

Polymers used to modify the oligonucleotides and enhance their biological efficacy are essentially policationic or biodegradating polymers. Few example of amphiphilic, synthetic macromolecules are also reported.

The first class, able to form condensed particle by electrostatic interactions with the oligonucleotides, have been widely associated with some simple ligand, as transferrin[17] or folic acid[18] to increase the efficient internalisation of the antisense molecules. The polycationic polymers used are polyamines that become cationic at physiological conditions.

CHEMICAL MODIFICATIONS:
BACKBONE AND NUCLEOSIDES

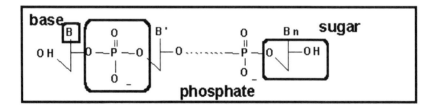

CHEMICAL CONJUGATIONS:
REPORTER GROUPS AND POLYMERS

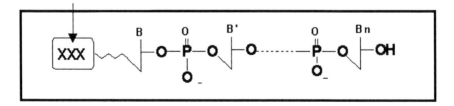

Figure 2. Scheme of the main chemical modifications of an oligonucleotide sequence

Poly (L-Lysine) (PLL) has been used to facilitate uptake of oligonucleotides[19]. This polymer interacts with the negatively charged molecules on the cell membrane and it is internalised along with the joined unit. The rate and the amount of oligonucleotide uptake are increased when conjugated with PLL, but, unfortunately, its toxicity and lack of specificity limit its usefulness.

Starbust PAMAM (polyamidoamine) are spheroidal polycation polymers, obtained by stepwise synthesis from methacrylate and ethylendiammine, leading to branched structures. The surface primary amino groups enable the dendrimers to interact with polyanions, and these complexes can mediate the transfer of antisense oligonucleotides inside the cell[20]. Furthermore, this binding extends the half- time life of the unmodified oligonucleotides, and is not citotoxic at the concentration effective for the transfer.

Polyethyleneimine (PEI) is an organic macromolecule with a very high cation charge density, since every third atom a protonated nitrogen can be present inside the aliphatic backbone. It is a versatile vector, with a very low level of citotoxicity[21].

As biodegradable polymers, these delivering vehicles for oligonucleotides show an enhanced protection from the nuclease degradation. Delivering is site specific and avoids any potential systemic toxicity. Among the few examples, poloxanes Pluronic F-127 gel and ethylenevinylacetate copolymer must be recalled[22]. Other biodegradable macromolecules usefully explored have been the copolymer DL-lactic/glycolic acid (PLGA)[23], that incorporates the oligonucleotides in microparticles, or polyanhydrides as poly (fatty acid dimer-sebacic acid) [P (FAD-SA)][24]. One promising macromolecular carrier is the synthetic N- (2-hydroxypropyl) methacrylamide (HPMA) polymer[25]. It exhibits very little binding to cell surfaces, and a very low level of immunogenicity. As expected, the molecular mass of this polymer influences the in vivo behaviour, such as the rate of excretion and the pharmacological activity, Its modification with an active thiol containing moiety has allowed its coupling with a sulfurized oligonucleotide through a disulfide bond, expected to be cleaved intracellularly. This conjugate has been demonstrated to be efficiently internalised by cultured cells. The structures of some of the above-mentioned polymers are described in Figure 3.

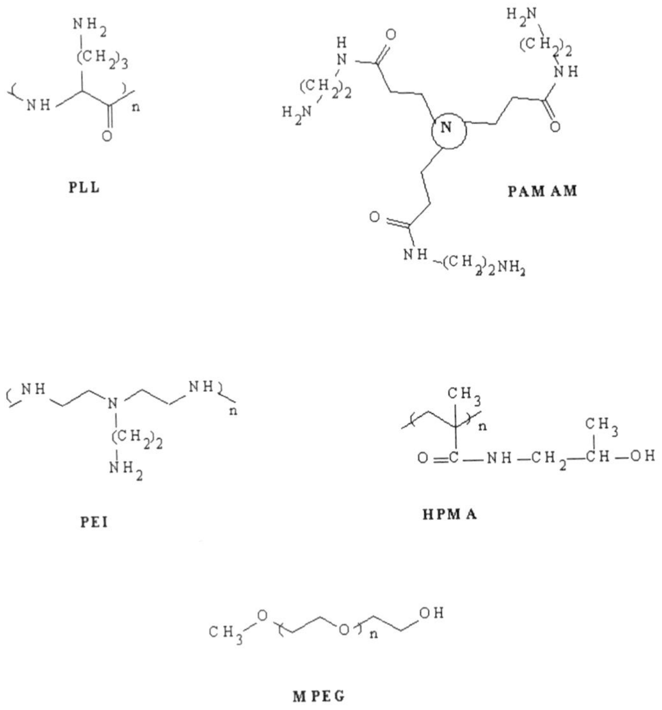

Figure 3. Structure of some of the polymers used as conjugating or complexing units of oligonucleotides

2.1   PEG-Conjugated Oligonucleotides

It must be underlined that in most of the reported applications, the polymeric unit is simply physically complexed with the antisense oligonucleotide, and not joined by a covalent bond. As a consequence the ratio between the oligonucleotide and the polymeric chain cannot be exactly predetermined, but only its average value can be estimated.

The chemical conjugation can grant a better design of the overall properties of the complex molecule, and confer more convenient properties from the point of view of its production, manipulation and storage. However, one of the main reasons of the lack of proper chemical procedures is due to the limited availability of useful functional groups on the oligonucleotide chain, and to the lower reactivity of high-molecular weight polymer chains in the standard synthetic processes.

From all these considerations, only one polymer has so far revealed a unique capacity for the chemical conjugation of oligonucleotide chain, namely the polyethylene glycol (PEG).

PEG has been successfully investigated for its applications in many biological researches, as for example a precipitating agent of macromolecules[26], or to facilitate the cell fusion in cell hybridisation processes[27]. One of its most important applications, since the description of covalent PEG-protein adducts[28], have been the study of conjugate complexes, both with small drugs, as well as with bioactive macromolecules, in which proteins constitute the largest group[29].

The great popularity of PEG-conjugates is driven by the unique combination of biological and chemical properties of this polymer. A wide description of the properties and applications of PEG has been recently reported[30]. Among its qualities, it must be recalled its large solubility in water and in most of the organic solvents needed for synthetical purposes[31]. Moreover, favourable pharmacokinetics, time distribution, and lack of toxicity and immunogenicity are strongly connected with the safety of its conjugate[32]. Last, but not least, this polymer is commercially available in a wide range of molecular weights and end-functionalities[33-34], and it is completely inert from the synthetic point of view.

The PEG-conjugates usually exhibit a high protection against the enzymatic degradations, an increased ability to penetrate the cell membranes and a reduced capability to give aspecific interactions with endogeneous proteins[30].

Despite these favourable properties, the conjugation between polyethylene glycol and oligonucleotides have received attention only recently, as a consequence of the development of new therapeutic strategies involving these molecules as novel drugs. PEG moiety can be joined to these nucleic acid derivatives by using their natural functionality or through the specific introduction of some reactive group. Generally, the conjugation can be performed during chemical as well as enzymatic synthesis, both pre- or post-synthetically, as extensively described[35]. The most convenient approach to join together an oligonucleotide and a PEG chain is the incorporation of the polymeric moiety during the synthesis of the nucleic acid derivative. In this way, side reactions are reduced, purification is easier to be performed, and it can be taken advantage from the widely employed solid-phase approaches

As far as short oligoethylene glycol chains concerns, they have been introduced as non-nucleosidic replacements in loop forming sequences[36] or in hairpin structures[37], and as linker in triple-helix forming oligonucleotides, both intra-molecularly[38] as well as inter-molecularly[39]. Moreover, penta- and hexa-ethylene glycols have been used in linking circular-forming oligonucleotides[40]. Properly tailored-mixed sequences have been recently investigated[41]. The substitution of the natural sequences inside the ribozymes has been also performed by this short oligomeric chains[42]. In any case, their incorporation has been commonly obtained by reaction of the phosphoramidite derivatives during the normal oligonucleotide synthesis, generally starting from their 5'-position. The incorporation at the 3'-terminal position is also possible by utilising a PEG-derivatized solid support[43].

Shorter oligoethylene glycol-conjugates are monodisperse samples, but the introduction of a polyethylene glycol chain implies the work with a polydisperse sample. Moreover, when such a larger polymeric chain has to be chemically incorporated on a newly synthesised oligonucleotide chains further problems are emerging. The viscosity of their solution hampers the condensation reaction, especially when the nucleic acid part is still bound to the heterogeneous support, and, owing to the reduced capacity of the solid-phase process, and its difficult upscaling, a limited amount of product can be obtained. On the other hand, a specific linker group has to be introduced to enhance the conjugation properties of the oligonucleotide counterpart, as reported[44-46]. Also, in this case, when a 3'-derivatisation is required, a properly derivatized supporting polymer must be realised[47].

Another drawback derives from the use of a large excess of the modifying polymer to increase the condensation yield. As a consequence, a difficult final separation of the PEG-conjugated derivatives from its PEG-activated reagent is observed when the conjugating reaction is performed in solution, as a consequence of avoiding the large demand of high-cost reagent during solid-phase syntheses. In any case, a non-automated solution synthesis, even if possible, requires a very laborious procedure and a quite skilful expertise.

3. LIQUID-PHASE SYNTHESIS OF PEG-CONJUGATED OLIGONUCLEOTIDES

To solve these problems a new procedure, based on a liquid-phase method has been recently proposed. In this process, called HELP or High Efficiency Liquid Phase[48], the PEG has been used as a soluble support for the oligonucleotide synthesis. To obtain a stable conjugate, introducing an additional phosphodiester bond between the polymer and the growing chain of the oligonucleotide has modified the original procedure. PEG was then utilised both as a synthetic handle, during this liquid-phase synthesis, as well as the final biological carrier in the conjugate.

Many advantages arise from this synthetic process. First of all, the oligonucleotide synthesis is carried out in homogeneous media since the chain to be synthesised is linked to a soluble supporting polymer. The polymer-bound product is recovered from the reaction mixture by a simple precipitation that allows rapid and easy elimination of excess reagents and soluble by-products. The molecular weight of the PEG ranges from 5,000 to 20,000 Da, depending from the final size of the desired oligonucleotide. The scheme of the process is reported in Figure 4.

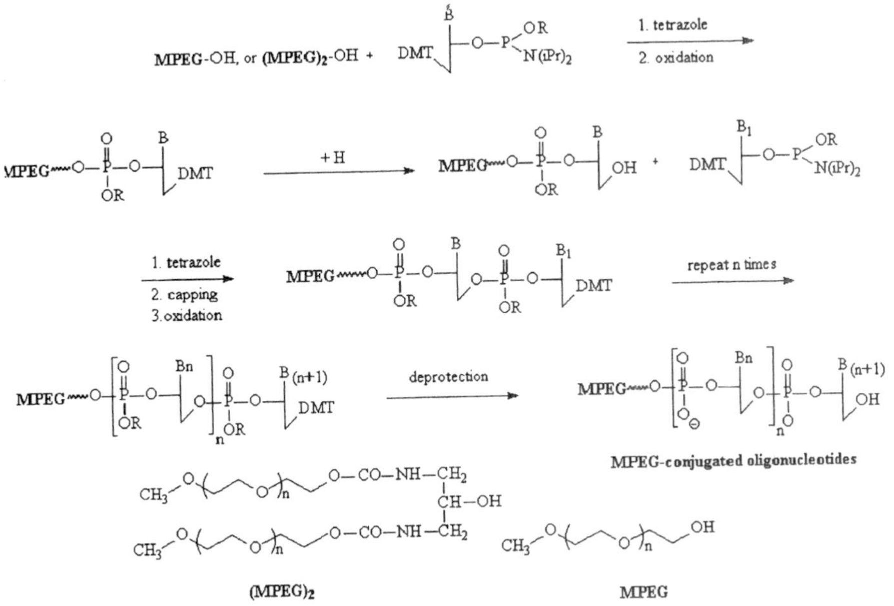

Figure 4. Scheme of the HELP (High Efficiency Liquid Phase) synthesis of the MPEG-conjugated oligonucleotides.

The same phosphoramidite-based chemistry, widely employed in the solid-phase procedures, has been successfully employed[49]. As a general observation, the phase homogeneity allows for an easy scale-up of the process, and the absence of diffusion problems inside the insoluble resin beads, in principle demands for a lower amount of reagents. It is also worth noting that the spectral transparency and the high solubility of the conjugate, allows for a rapid and non-destructive spectrometric analysis of any synthetic step. The easy evaluation of the reaction products during the purification processes is clearly emerging form this view.

As an example of an antisense oligonucleotide conjugated to a high-molecular weight polyethylene glycol we investigated a 12mer active against HIV-1[50]. A MPEG, a monomethoxy polyethylene glycol chain, of M.W. = 10,000 Da has been employed, with a very low polydispersity level. The synthetic procedure recently reported on a new Internet journal has been utilised[51]. The product has been obtained in good yield, and an appreciable amount of final product has been collected (60% of the maximum expected), if one considers the high number of precipitate-and-filtrate steps executed, as demanded during the intermediate purification.

In Figure 5 the chromatographic patterns from the ionic exchange chromatographic purification of the crude MPEG-conjugated 12mers are reported. In this case the commonly adopted reverse-phase chromatography does not work efficiently. In fact the separation of the final conjugated oligonucleotide from the shorter, failed sequences is not allowed, since a single peak is always observed, despite the presence of mixture of products.

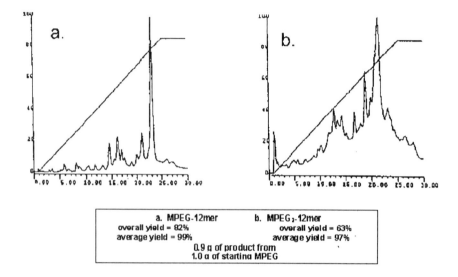

a. MPEG-12mer	b. MPEG₂-12mer
overall yield = 82%	overall yield = 63%
average yield = 99%	average yield = 97%

0.9 g of product from
1.0 g of starting MPEG

Figure 5. Anion-exchange chromatographic patterns of MPEG-conjugated anti-HIV 12mers. Main synthetic data are reported.

The presence of a large MPEG chain makes all the conjugated compounds similar toward the separation properties of this chromatographic support. On the contrary, an anion-exchange process can successfully achieve the separation; in this case the charge balance of different length-chain oligonucleotides can not be suppressed by the PEG moiety.

To verify the effect of the different shape of the PEG unit, the same oligonucleotide has been conjugated to a new branched polyethylene glycol[52]. The synthetic data are quite similar to those of the linear MPEG-conjugate, save some reduced yield during the first synthetic steps, likely due to some steric hindrance of the reactive functionality of the branched polymer.

A first biological characterisation of these high-molecular mass MPEG-conjugated oligonucleotides has been performed by their thermal denaturation studies. In other words, the dissociation of the duplexes formed with the complementary nucleic acid sequence has been examined. As shown in Figure 6, only a marginal reduction of the duplex stability has been noticed, as revealed by the difference in the melting temperatures T_m, when a high-molecular mass MPEG is present. The effect of the two MPEGs is comparable.

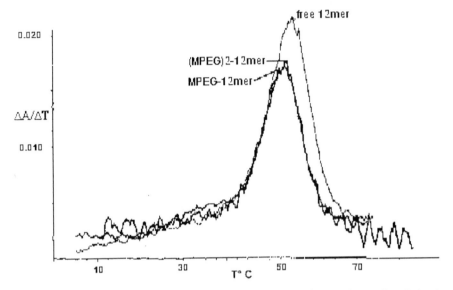

Figure 6. Thermal denaturation studies of the MPEG-conjugated HIV-12mers: first derivative of UV absorption at 260 nm as function of temperature.

A further, important feature of the presence of a polyethylene glycol chain is usually the protection given against the enzymatic digestion of the conjugated molecules. It is clearly highlighted in Figure 7 that the conjugated polymer increases the half-life of the oligonucleotide in the presence of digesting enzymes, always present in physiological media, with an almost equal effect between the linear and the branched polymeric molecules.

This later effect is mainly due to the screening action of the polyether backbone, and of its water molecules cage, against the interaction with proteins. Then, an interesting investigation has involved the behaviour of the MPEG-conjugated oligonucleotide toward the action of the Rnase H. This ubiquitary enzyme specifically degrades the phosphodiester bond of the RNA strand of a DNA.RNA hybrid, as that produced from the antisense action of an oligonucleotide. An evidence for an RNase dependent mode of action has recently been reported in these processes[53].

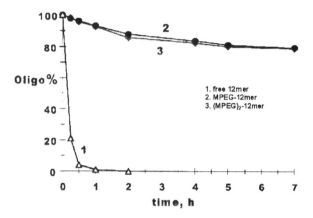

Figure 7. Nuclease stability studies of the MPEG-conjugated anti-HIV 12mers in presence of a 1:1 mixture of diesterase and 5'-nucleotidase: percent of intact oligonucleotides as function of time.

In Figure 8 it is clearly demonstrated that the conjugation with the polyethylene glycol does not affect the structure of the hybrid, and the consequent RNase activity.

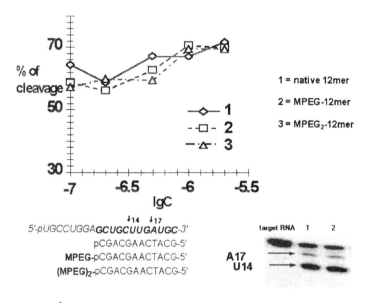

Figure 8. RNase H activity studies of MPEG-conjugated anti-HIV 12mers: percent of cleavage against increasing concentration of the oligonucletides. Bottom: scheme of the points of attack of the Rnase H on the duplexes and autoradiography of the gel-electrophoresis of products.

Again it is not possible to recognise from these data any significant difference between the linear and the branched MPEGs. This confirms that size, shape and position of the conjugating moiety does not hamper in this case the formation of a regular duplex, allowing a regular recognition from the RNase H enzyme.

Between all the biological studies, the demonstration of a direct effect on the antisense activity of the conjugated oligonucleotides is of paramount importance. It has been previously demonstrated that both the entire sequence of a 20mer complementary to part of the DNA from the HIV-1 genome, as well as the tandem complex of the shorter sequences, made by the two terminal 8mers and the central 4mer, are active in inhibiting the HIV infection[50]. The two octamers were modified at their extremities with a phenazinium (Phn) group to increase the duplex stability through an intercalating effect[13]. On this basis, the MPEG-conjugated 12mer, in tandem with the Phn-derived 8mer, is expected to give some useful information on the effect of the presence of a large MPEG chain on the overall antisense activity. It must also be recalled that the same 12mer, but modified only with Phn or cholesterol unit at its 3' terminus, has been demonstrated to be active against the HIV target[55].

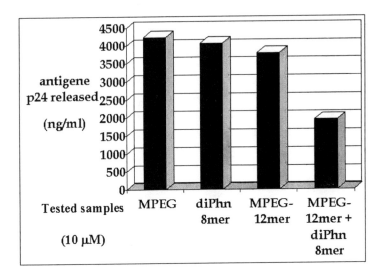

Figure 9. Antisense activity of the linear MPEG-conjugated anti-HIV 12mer in tandem with the 8mer sequence, and of the related compounds.

As shown in Figure 9, the antisense activity, expressed as inhibition of the production of the p24 antigene, indicates that the linear MPEG-conjugated 12mer, together with the Phn-modified 8mer, was active against the HIV infection. The branched MPEG-conjugate was not active in the same conditions, as well as the diPhn-8mer alone. The conjugated oligonucleotides, when tested without the tandem sequence, were fully inactive, as the free, native sequence. Of course, if a non-complementary 8mer is utilised in place of the right molecule, the observed activity will disappear.

4.   CONCLUSION

As a conclusion, it is clearly demonstrated that the presence of a high-molecular weight polyethylene glycol chain improves the activity of the conjugated oligonucleotide molecule, but the structure of the polymeric unit could interfere on this activity, as demonstrated by the effect of the branching MPEG. The stabilisation against the enzymatic degradation improves the activity of the antisense oligonucleotide, but the likely wider space occupied by the branched polymer could impede the correct interaction with the tandem 8mer. The effect of the position of the MPEG conjugation will be the subject of a further investigation.

Recently[56], a search for new polymers to substitute the polyethylene glycol as a successful conjugating moiety has been performed. Two polymers, polyvinylpirrolidone (PVP) and poly (N-acriloylmorpholine) (PacM), both bearing a single –OH functionality at one extreme, have been investigated. From the synthetic point of view, the PVP appears unable to support a liquid-phase process, owing to its unsuitable solubility, and solubilizing properties. On the other side, PAcM allows a convenient synthesis of an oligonucleotide chain, but with inferior results, mainly during the final purification procedures. The activity studies of the same 12mer previously tested on MPEG, but conjugated to PAcM, are in progress.

An additional evolution in this area will evaluate the effect of a large MPEG chain, when conjugated to a triplex-forming oligonucleotide.

A future, more difficult task will address the synthesis, and the activity studies, of mixed MPEG-conjugated biopolymers, as peptides and oligonucleotides, in search for new more specific and effective drugs.

On this regard, it is worth recalling a sentence of Thomas Carlyle, a famous Scottish essayist, who said: "The block of granite, which was an obstacle in the path of the weak, becomes a steppingstone in the path of the strong".

ACKNOWLEDGMENTS

This work was undertaken with the collaboration of Drs. B. Burcovich, A.M. De Franco, R. Rossin and S. Zaramella, all from the University of Padova (Italy) as far as all the synthetic and analytic aspects.

Prof. V. Zarytova and her colleagues from the Institute of Bioorganic Chemistry of Novosibirsk (Russia), together with the group of Prof. A. Pokrowsky and O. Pliasunova, from the Institute of Molecular Virology of Novosibirsk (Russia), have performed all the biological studies described.

The friendly collaboration, and the continuous, stimulating discussions, of Prof. F.M. Veronese from the Dept. of Pharmaceutical Sciences of the University of Padova (Italy) are greatly acknowledged.

The overall work has been supported by grants from MURST and CNR-Italy.

REFERENCES

1. Zamecnik P.C., Stephenson M.L., 1978, *Proc. Natl. Acad. Sci.* USA, 75: 280.
2. Takayama K.M., Inouye M., 1990, *Crit. Rev. Biochem.*, 25: 155.
3. Crooke S.T., 1996, *Med. Res. Rev.*, 16: 319.
4. Stein C.A., Cohen J.S., 1989, in *Oligonucleotide-Antisense Inhibition of Gene Expression*, Cohen J.S. Ed. Mac Millan Press, London, p. 97.
5. Miller, P.S., 1991, *Biotechnology*, 9: 358.
6. Froehler B., Ng P., Matteucci M., 1988, *Nucleic Acids Res.*, 16: 4831.
7. Sanghvy Y.S., Cook P.D., 1994, in *Carbohydrate Modifications in Antisense Research*, Sanghvi Y.S. and Cook P.D. Eds., American Chemical Society, Washington D.C., p. 1.
8. Imbach J-L., Rayner B., Morvan F., 1989, *Nucleosides Nucleotides*, 8: 627.
9. Seela F., Thomas H., 1995, *Helv. Chim. Acta*, 78: 94.
10. Gryaznov S.M., Schultz R.G., 1994, *Tetrahedron Lett.*, 35: 2489-2492
11. Goodchild J., 1990, *Bioconjugate Chem.*, 1: 165.
12. Hélène C., Thuong N.T., 1989, *Genome*, 31: 413.
13. Loskhov S.G., Podyminogin M.A., Sergeev D.S., Silnikov V.N., Kukiavin I.V., Shishkin G.U., Zarytova V.P., 1992, *Bioconjugate Chem.*, 3: 414.
14. Krieg A.M., Tonkison J., Matson S., Zhao Q., Saxon M., Zhang L.M., Bhanja U., Yakubov L., Stein C.A., 1993, *Proc. Natl. Acad. Sci.* USA, 90: 1048.
15. Eritja R., Pons A., Escarceller M., Giralt E., Albericio F., 1991, *Tetrahedron*. 47, 4113.
16. Bonfils E., Mendes C., Roche A.C., Monsigny M., Midoux P., 1992, *Bioconjugate Chem.*, 3: 277.
17. Calabretta B., Skorski T., Zon G., 1992, *Cancer Biol.*, 3: 391.

18. Wang S., Lee P.J., Cauchon G., Gorenstein D., Low PS., 1995, *Proc. NatL. Acad. Sci. USA*, 92: 3318.
19. Leonetti J-P., Degols EI., Lebleu B., 1990, *Bioconjugate Chem.*, 1: 149.
20. Bielinska A, Kukowska-Laballe J-F., Johnson J., Tomalia D.A., Baker jr J.R., 1996, *Nucleic Acids Res.,* 24: 2176.
21. Abdallah B, Hassan A, Benoist C., Gouil J., Behri. P., Demeneix, P.A., 1996, *Human Gene Therapy*, 7: 1947.
22. Edelman E. P, Simions M, Sirois M ., Rosenberg R.D. 1995, *Circ Res.*, 76: 176.
23. Cleek R.L., Rege A A , Denner L.D, Eskin S.G. , Mikos A G., 1997: *J Biomed. Mater. Res.*, 35: 525.
24. Buahin K.G., Judy K.D., Domb A.J., Hartke M., Maniar M., Coivin O.M., Brem A., 1992, *Polym. Adv. Technol.*, 3, 311.
25. Wang E., Kristensen J., Ruffner D. E., 1998, *Bioconjugate Chem.*, 9: 749.
26. Polson A., . 1997, *Prep. Biochem*, 7: 129.
27. Lentz B., 1994, *Chem. Phys. Lipids* 73: 91.
28. Pollock A., Witesides G.M, 1977, *J. Am. Chem. Soc.*, 98: 289.
29. Zalipsky S., 1995, *Adv. Drug Delivery Res.* 15: 157.
30. Poly(ethylene Glycol). *Chemistry and Biologial Applications*. J.M. Harris and S. Zalipsky Eds., ACS Symposium Series 680, American Chemical Society, Washington, D.C.(USA), 1997.
31. Mutter M., Bayer E. in *The Peptides*, Gross E. and Meienhofer J. Eds., 1979, Academic Press, New York, vol.2, pp.285.
32. Katze N.V., 1993, Adv. Drug Delivery Rev. 10: 91.
33. Harris J.M., 1985, *J. Macromol. Sci-Rev., Macromol. Chem. Phys.* C25, 325.
34. Zalipsky S., 1995, *Bioconjugate Chem* 6: 150.
35. Protocols for oligonucleotide conjugates. Agrawal S. ed., 1994, *Methods in Molecular Biology* vol.26, Humana Press, Totowa, N.J (USA),.
36. Durand M., Chevrie K., Chassignol M., Thuong N.T., Maurizot J.C., 1990, *Nucleic Acids Res.* 18: 6353.
37. Rumney S., Kool E.T., 1995, *J. Am. Chem. Soc.* 117: 5635.
38. Durand M., Peloille S., Thuong N.T., Maurizot J.C., 1992, *Biochemistry* 31: 9197.
39. Giovannangeli C., Thuong N.T., Hélène C. , 1993, *Proc. Natl. Acad. Sci.* USA 90: 10013.
40. Gao H., Chidambaram M., Chen B.C., PelhamD.E., Patel R., Yang M., Zhou L. Kook A., Cohen J.S., 1994, *Bioconjugate Chem* 5: 445.
41. Pyshanya L., Pyshnyi D.V., Ivanova E.M., Zarytova V.F., Bonora G.M., Scalfi-Happ C., Seliger H., 1998, *Nucleosides Nucleotides* 17: 1289.
42. Thomson J.B., Tuschl T., Eckstein F., 1993, *Nucleic Acids Res*, 21: 5600.
43. Jaeschke A., Fuerste J.P., Cech D., Erdmann V., 1993, *Tetrahedron Lett.* 34: 301.
44. Manoharan M., Tivel K.L., Andrade L.K., Mohan V., Condon T.P., Bennett C.F., Cook P.D., 1995, *Nucleosides Nucleotides* 14: 969.
45. Kawaguchi T., Askawa H., Tashiro Y., Juni K., Sueishi T., 1995, *Biol. Pharm. Bull.* 18: 174.
46. Jones, D.S., Hachmann J.P., Osgood S.A., Hayag M.S., Barstad P.A., Iverson G.M., Coutts S.M., 1994, *Bioconjugate Chem.* 5: 390..
47. Efimov V.A., Kalinkina A.L., Chakhmakhcheva O.G., 1993, *Nucleic Acids Res.*, 21: 5337.
48. Bonora G.M., 1995, *Applied. Biochem. Biotechnol,*. 54: 3.
49. Bonora G.M., Ivanova, E., Zarytova, V., Burcovich B., Veronese F.M., 1997, *Bioconjugate Chem.*, 8: 793.
50. Vsinarchuk F.P., Konevetz D.A., Plyasunova O.A., Pokrowsky A.G., Valssov V.V., 1993, *Biochimie*, 75: 49.

51. Bonora G.M., Zaramella S., Veronese F.M., 1998, *Biol. Proced. Online*, 1(1) at www.science.uwaterloo.ca/bpo

52. Burcovich B., Veronese F.M., Zarytova V., Bonora G.M., 1998, *Nucleosides Nucleotides* 17: 1567.

53. Walder R.J, Walder D.A., 1988, *Proc. Natl. Acad. Sci.* USA, 85: 5011.

54. Lokhov S.G., Podyminogin M.A., Sergeev D.S., Silnikov V.N., Kutyavin G.V., Shishkin V.P., Zarytova V.P., 1992, *Bioconjugate Chem.* 3: 414.

55. Pokrovsky A.G., Pliasunova O.A., unpublished results

56. Bonora G.M., De Franco A.M., Rossin R., Veronese F.M., Ivanova E., Pyshnyi D.V., Vorobjev P.E., Zarytova V., 2000, *Nucleosides, Nucleotides and Nucleic Acids*, 19, 8/9 (in press)

PHOTODYNAMIC ANTISENSE REGULATION OF HUMAN CERVICAL CARCINOMA CELL

[1]Akira Murakami, [1]Asako Yamayoshi, [1]Reiko Iwase, [2]Jun-ichi Nishida,
[1]Tetsuji Yamaoka, and [2]Norio Wake
[1]*Department of Polymer Science and Engineering, Kyoto Institute of Technology, Matsugasaki, Kyoto 606-8585, Japan, [2]Department of Reproduction Physiology and Endocrinology, Medical Institute of Bioregulation, Kyushu University, 4546, Tsurumihara, Oita 874-0838, Japan*

1.   INTRODUCTION

Oligonucleotides have been used to inhibit protein biosyntheses by suppressing the gene expression in a sequence specific manner [1,2,3]. A large number of reports have demonstrated the potency of the antisense strategy, and some antisense drugs are now under clinical trials[4]. Among various reports concerning with antisense strategy, many have presented clear evidences that the antisense mechanism actually functioned, and that, at the same time, non-antisense mechanisms also largely participated in the regulation of the cell growth. The clarification of the mechanism may be crucial for the future development of antisense strategy. To realize the purpose, functionalization of antisense molecules is among the possibilities[5,6,7,8]. We therefore adopted a photo-cross-linking reagent, 4,5',8-trimethylpsoralen, to functionalize antisense molecules. Psoralen derivatives have an ability to cross-link covalently with pyrimidine bases, favorably thymine and uracil, upon UVA-irradiation (320nm-380nm)[9]. Thus psoralen derivatives can inactivate gene expression *via* cross-linking. Psoralen-conjugated oligo(nucleoside methylphosphonate)s and oligonucleotides were found to cross-link with single stranded nucleic acids upon UVA-irradiation in a sequence specific manner[10,11,12]. In this review, we demonstrate the applicability of the photodynamic antisense regulation (PDAR) of cancer

Biomedical Polymers and Polymer Therapeutics
Edited by Chiellini *et al.*, Kluwer Academic/Plenum Publishers, New York, 2001   389

cells using a psoralen-conjugated antisense oligo(nucleoside phosphoro-thioate)[13].

2. EXPERIMENTAL

2.1 Materials

Human cervical carcinoma cell lines, C4II (HPV18 genome positive) and SiHa (HPV16 genome positive), were used. 4'-[[N-(aminoethyl)amino]-methyl]-4,5',8-trimethylpsoralen (aeAMT)[10] was introduced to 5'-hydroxyl group of S-Oligo on CPG support. The sequences of S-Oligos were as follows; PS-P-As: 5'-dAGCGCGCCATAGTAT, PS-P-scr: 5'-dACGTAG-CTCGATCAG. The psoralen conjugated S-Oligos were purified by reversed phase HPLC with acetonitrile gradient in triethylammonium acetate buffer (TEAA, pH7.0).

2.2 Photo-Irradiation of PS-S-Oligo

An equimolar mixture of PS-P-As (20µM) and its complementary oligo-ribonucleotide (rAUACUAUGGCGCGCUUUGAG, 20µM) in phosphate buffered saline (40µl, 0.135M NaCl, pH7.0) was UVA-irradiated on a transilluminator (365nm, 1.1mJ/cm^2•sec). The reaction solution was analyzed by HPLC. Cells in culture medium (100µl) were plated at the density of 2×10^4 cells / ml into a 96-well dish. After 24h, the cells were added by Ps-S-Oligos and were incubated for 2h. The dish was then transferred onto a transilluminator and cells were irradiated for 2min at room temperature. The cells were cultured for 72h and the growth was evaluated by a modified MTT assay (Cell Counting Kit. Dojindo, Japan). Experiments were performed in triplicate.

2.3 mRNA Isolation and RT-PCR

Total cellular mRNA from PS-S-Oligo treated cells was obtained by using mRNA Purification Kit (Quick Prep Micro, Amersham Pharmacia Biotech, USA) according to the manufacturer's protocol. The mRNA was stored at -80°C after treatment with DNaseI (Takara Shuzo Co., Ltd., Japan) for 2h at 37°C.

Total cellular mRNA (0.2µg) was subjected to the RT-PCR quantitation protocol. Oligo $(dT)_{16}$ primer was used for the reverse transcription (RT). RT reaction was done at 37°C for 1h using M-MLV reverse transcriptase

(Toyobo Co., Ltd., Japan). Primers for PCR were as follows; primer I (sense, from 24nt to 45nt of HPV18-E6*-mRNA) 5'-AAAAAGGGAGT-GACCGAAAACG, primer II (sense, from 126nt to 146nt of HPV18-E6*-mRNA) 5'-CGACCCTACAAGCTACCTGAT, and primer III (antisense, from 356nt to 376nt of HPV18-E6*-mRNA) 5'-TCTGCGTCGTTG-GAGTCTTTC. As an internal control, human glyceraldehyde-3-phosphate dehydrogenase (GAPDH) mRNA[14] was chosen and the primers were as follows; 5'-TGAAGGTCGGAGTCAACGGATTTGGT-3' and 5'-CAT-GTGGGCCATGAGGCTCACCAC-3'. The PCR was performed under the following condition: denaturation at 94°C for 6min followed by PCR (31 cycles; denaturation at 94°C for 1 min, primer-annealing at 58°C for 1 min, primer-extension at 74°C for 1 min).

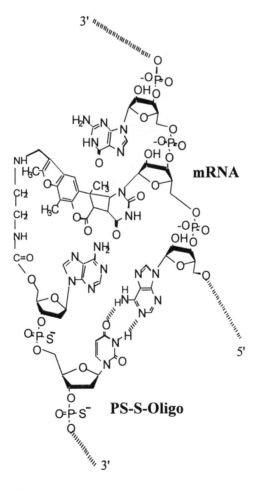

Figure 1. Schematic drawing of the photo-cross-linking between PS-S-Oligo and the target mRNA. A cyclobutane bridge is formed between the pyrone ring of the psoralen and the uracil residue of the target mRNA.

The RT-PCR transcripts were separated on 0.7% agarose-0.4% Synergel (Diversifield Biotech, Japan)(100V, 30min) and visualized by staining with ethidium bromide.

3.   RESULTS AND DISCUSSION

The photo-cross-linking reagent, 4,5',8-trimethylpsoralen was conjugated to oligonucleotides so that the antisense oligonucleotide could covalently bind to target mRNA upon UVA-irradiation[10]. Psoralen-conjugated oligo(nucleoside phosphorothioate) (PS-S-Oligo) might cross-link with the target mRNA. A cyclobutane bridge is supposed to be formed between the pyrone ring of the psoralen and the uracil residue of the target mRNA as shown in Figure 1. The sequences of PS-S-Oligo have to be designed to facilitate the interaction. As the target site of mRNA, we chose the initiation codon region, where many papers have successfully adopted.

The ability of PS-S-Oligo to form a hybrid was estimated by UV-melting profiles. The melting temperature (Tm) of the hybrid between PS-P-As and an oligoribonucleotide (20mer, sense-RNA) which consists of a fragmentary sequence of HPV18-E6*-mRNA was 45°C, whereas PS-P-scr did not show appreciable Tm. These results indicate that the binding PS-P-As to RNA was sequence-specific.

The cross-linking reaction of PS-P-As was evaluated by reversed phase HPLC. The equimolar mixture of PS-P-As and sense-RNA was irradiated with a UVA transilluminator (365nm). Figure 2 shows the time course of the production of photoproducts[13]. A new peak (arrow) increased as the irradiation time increased. These results indicate that Ps-S-Oligo has an ability to cross-link efficiently with sense-RNA.

Prior to cell treatments by Ps-S-Oligo, the growth susceptibility of C4II and SiHa cells against UVA-irradiation (365nm) was evaluated. The irradiation time in this study was decided to be 2min to avoid the cell damage.

S-Oligos, which do not contain psoralen did not show any inhibitory effect on the growth of C4II and SiHa up to 5μM. Also, Ps-S-Oligo did not show any inhibitory effects without UVA-irradiation. These results indicate that psoralen, which can intercalate between nucleobases and can stabilize the hybrid, did not enhance the inhibitory effect on C4II cell growth without UVA-irradiation. On the other hand, UVA-irradiation did drastically affect the cellular proliferation in the presence of PS-P-As in a concentration-dependent manner.

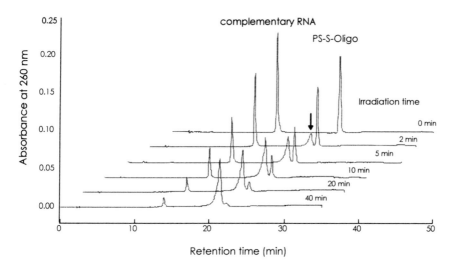

Figure 2. Photo-cross-linking reaction of psoralen-conjugated oligo(nucleoside phosphoro-thioate) (PS-P-As) with its complementary oligoribonucleotide (sense-RNA). [PS-P-As] = [sense-RNA] = 20μM. UVA-irradiation was carried out in phosphate buffered saline (0.135M NaCl, pH7.0) at 365nm with 1.1mJ/cm^2•sec.

As shown in Figure 3A, PS-P-As showed the remarkable inhibitory effect on the proliferation of C4II cells in a sequence dependent manner (IC$_{50}$ = 300nM) and the PS-P-As treated cells were almost completely deformed. Contrarily, PS-P-scr did not inhibit the cellular proliferation even with UVA irradiation. The difference in Tm among Ps-S-Oligos seems to be reflected in the inhibitory efficacy, suggesting that the antiproliferation effects were sequence-dependent. On the other hand, Ps-S-Oligo scarcely inhibited the proliferation of SiHa cells which don't contain HPV18 genome but HPV16 one (Fig.3B). These results indicate that the antiproliferation effects of Ps-S-Oligo were cell-line dependent, namely dependent on the presence of target messages. In summary, the specificity of Ps-S-Oligo revealed in this study was classified as follows; (a) sequence, (d) UVA-irradiation, (c) concentration, and (d) cell-line.

The mechanistic investigation was focused on two aspects. The one is the cellular localization of PS-S-Oligo. The second is the quantitation of mRNA. The fluorescence microscopic image using fluorescein-labeled PS-P-As showed that it was mainly accumulated in the proximity of the nuclei but scarcely inside the nuclei (data not shown).

RT-PCR studies were done to evaluate the amount of the intact mRNA in the Ps-S-Oligo treated C4II cells. Total mRNAs isolated from cells after Ps-S-Oligo treatment with UVA-irradiation were transcribed by reverse transcriptase in the presence of oligo(dT)$_{16}$ primer, followed by PCR. Two sets of PCR primers were designed. The one amplifies the region where PS-

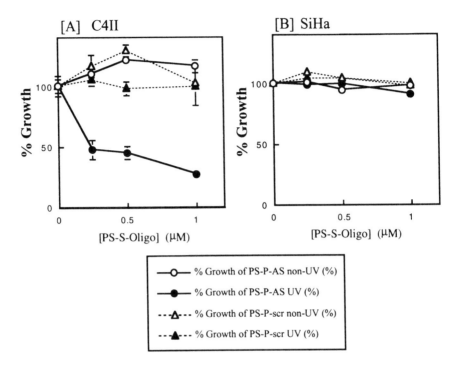

Figure 3. Effect of psoralen-conjugated oligo(nucleoside phosphorothioate)s (PS-S-Oligo) on the proliferation of human cervical carcinoma cell, C4II. and SiHa. [A] C4II cell line.: PS-P-As without UVA-irradiation (←), PS-P-scr without UVA-irradiation (Δ), PS-P-As with UVA-irradiation (●), PS-P-scr with UVA-irradiation (▲). [B] SiHa cell line. PS-P-As without UVA-irradiation (←), PS-P-scr without UVA-irradiation (Δ), PS-P-As with UVA-irradiation (●), PS-P-scr with UVA-irradiation (▲). UVA-irradiation: 365nm, 1.1mJ/cm²•sec. Cell growth was evaluated by modified MTT assay.

P-As might affect the transcription by photo-cross-linking (from 24nt to 376nt), giving 353bp transcript. The other does the region where PS-P-As might not affect the amplification (from 126nt to 376nt), giving 251bp transcript. Compared with untreated C4II cells, the distinguished decrease of 353bp transcript in PS-P-As system was observed, whereas the amount of 251bp transcript was not appreciably changed, indicating that the intact HPV18-E6*-mRNA scarcely remained. In PS-P-scr system, the amount of both transcripts, 353bp and 251bp, did not change. These results showed that PS-P-As might cross-link in the region where Primer set I-III could amplify, but not in the region where Primer set II-III could amplify. Table 1 shows the amount of the target mRNA quantitated with the RT-PCR protocol. Taking those results into consideration, it is concluded that the antiproliferation effect of PS-P-As on C4II cells was in an antisense manner, and that the UVA-irradiation could effectively enhance the effect.

Furthermore, as HPV18-E6* protein may be one of the crucial proteins for cervical carcinoma cells, the antisense effect might be attributed to the effective suppression of the protein synthesis. The detailed clarification of such inhibitory effects is now under investigation.

The concept presented in this review, photodynamic antisense regulation (PDAR), may provide useful information to design idealistic antisense molecules. Also it may be applicable to diseases whose gene can be UVA-irradiated.

Table 1. Amount of mRNA of PS-S-Oligo treated C4II cells

	Source of mRNA		
	untreated cell (C4II)	PS-P-As/UVA treated cell	PS-P-scr/UVA treated cell
[transcript (353bp)] [transcript (GAPDH)]	1.3 ± 0.4	0.1 ± 0.0	1.0 ± 0.2
[transcript (251bp)] [transcript (GAPDH)]	2.7 ± 0.4	3.1 ± 0.2	3.2 ± 0.2

ACKNOWLEDGMENTS

The authors are indebted to Emeritus Professor Junzo Sunamoto of Kyoto University for his generous consideration. This study was partly supported by the Grant-in-Aid for Scientific Research on Priority Area (A) of The Ministry of Education, Science, Sports and Culture of Japan (10153235, A.M., T.Y., R.I.), Terumo Life Science Foundation (A.M.), and Human Science Foundation (A.M.).

REFERENCES

1. Miller, P.S., Braiterman, L.T., and Ts'o, P.O.P., 1977, *Biochemistry* 16: *1988-1996.*
2. Zamecnik, P. C., Stephenson, M.L., 1978, *Proc.Natl.Acad. Sci.USA* 75: *280-284.*
3. Stephenson, M.L., Zamecnik, P.C., 1978, *Proc.Natl.Acad. Sci.USA.* 75, *285-288.*
4. Wickstrom, E., (ed) 1998, *Clinical Trials of Genetic Therapy with Antisense DNA and DNA Vector.* Marcel Dekker Inc. NY.
5. Isaacs, S.T., Shen, C.J., Hearst, J.E., Rapoport, H., 1977, *Biochemistry* 16: *1058-1064.*
6. Letsinger, R.L., Zhang, G., Sun, D.K., Ikeuchi, T., and Sarin, P.S., 1989, *Proc. Natl. Acad. Sci. USA,* 86: *6553-6556.*
7. Le Doan, T., Praseuth, D., Perrouault, L., Chassignol, M., Thuong, N.T., and Hélène, C., 1990, *Bioconjugate Chem.* 1: *108-113.*
8. Giovannangeli, C., Thuong, N.T., and Hélène, C., 1993, *Proc. Natl. Acad. Sci. USA,* 90: *10013-10017.*
9. Cimino, G.D., Gamper, H.B., Isaacs, S., and Hearst, J.E., 1985, *Ann. Rev. Biochem.* 54: *1151-1193.*

10. Lee, B.L., Murakami, A., Blake, K.R., Lin, S-B., Miller, P.S. 1988, *Biochemistry* 27: *3197-3203*.

11. Kean, J.M., Murakami, A., Blake, K.R., Cushman, C.D., Miller, P.S. 1988, *Biochemistry* 27: *9113-9121*.

12. Pieles U., English, U. 1989, *Nucleic Acids Res.* 17: *285-299*.

13. Murakami, A., Yamayoshi, A., Iwase, R., Nishida, J., Yamaoka, T., and Wake, N., *Europ. J. Pharm. Sci.*, in press.

14. Gima, T., Kato, H., Honda, T., Imamura,T., Sasazuki, T., Wake, N., 1994, *Int. J. Cancer,* 5: *480-485*

IN VITRO GENE DELIVERY BY USING SUPRAMOLECULAR SYSTEMS

Toshinori Sato*, Hiroyuku Akino, Hirotaka Nishi, Tetsuya Ishizuka, Tsuyoshi Ishii, and Yoshio Okahata

*Department of Biomolecular Engineering, Tokyo Institute of Technology, 4259 Nagatsuta, Midori-ku, Yokohama 226-8501, Japan, *present address; Department of Applied Chemistry, Keio University, Hiyoshi, Kouhoku-ku, Yokohama 223-8522, Japan*

1.   INTRODUCTION

Recently, a lot of new supramolecular systems have been developed to introduce a foreign DNA into animal cells. Cationic lipids[1-7] and cationic polymers[8] have been employed as non-viral gene transfer agents. These cationic gene transfer reagents form supramolecular systems with anionic DNA through electrostatic interaction. These cationic DNA complexes obtained were uptaken by cells because cell surface is negatively charged. Transfection efficiencies of these complexes have been investigated in vivo and in vitro. However, there are several problems to be overcome before practical use. Those involve insufficient transfection efficiency, lack of cell specificity, strong cytotoxicity, and inhibition by serum. In our previous paper,[9,10] we developed DNA complexes with cationic "lipoglutamates" such as α, γ-dibutyl glutamate and α, γ-dihexyl glutamate. We investigated the structure and cell uptake of the DNA/lipoglutamate complex. Next, we newly synthesized a cationic "lipoglutamide (PEG-lipid)" having tetra-ethylene glycol tails, N,N'-bis(18-hydroxy-7,10,13,16-tetraoxa-octadecyl)-L-glutamide hydrochloride (2EO4C6N$^+$, Fig.1)[5]. Formation and characterization of the DNA/2EO4C6N$^+$ complex were investigated by elemental analysis, melting temperature, and light-scattering measurement. Compaction of DNA by binding with the cationic 2EO4C6N$^+$ was confirmed

Biomedical Polymers and Polymer Therapeutics
Edited by Chiellini *et al.*, Kluwer Academic/Plenum Publishers, New York, 2001   397

by multi-angle light scattering method. Aggregation of the DNA/2EO4C6N$^+$ complex was significantly depressed compared with the DNA complex without tetraethylene glycol. The DNA complexes showed low cytotoxicity and efficient internalization into tumor cells compared with the native DNA. Besides 2EO4C6N$^+$, we synthesized other lipoglutamates such as *N,N'*-bis(22-hydroxy-11,14,17,20-tetraoxadocosanyl)-L-glutamide hydrochloride, having longer alkyl spacer chain (2EO4C10N$^+$, Fig.1) and *N,N'*-bis(22-methoxy-11,14,17,20-tetraoxadocosanyl)-L-glutamide hydrochloride, having a methoxy end group of ethylene glycol chains (2MeEO4C10N$^+$, Fig.1). We examined physicochemical features and cellular interaction of the DNA/PEG-lipid complexes.

$n = 6, R = H : 2EO_4C_6N^+$

$n = 10, R = H : 2EO_4C_{10}N^+$

$n = 10, R = CH_3 : 2MeEO_4C_{10}N^+$

Lipoglutamide

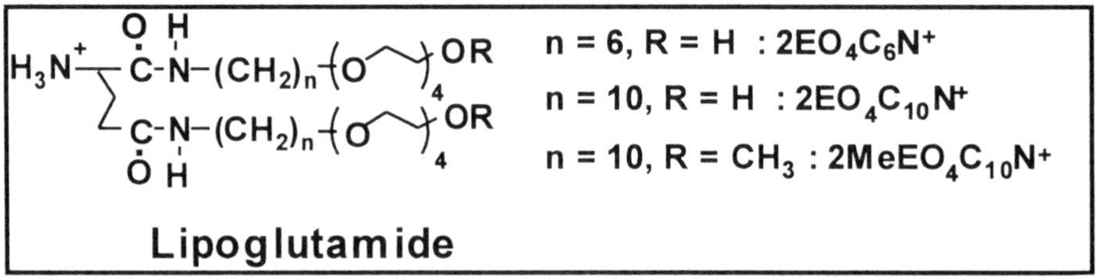

Figure 1. Chemical structures of lipoglutamide

2. THE DNA/PEG-LIPID COMPLEX

FITC-labelled DNA was employed to determine the cell uptake efficiency of DNA complexes by flow cytofluorometer. Fluorescence histograms of Hela cells incubated with DNA complexes at 37 °C for 6 h showed significant increase of fluorescence intensity compared with naked DNA. The DNA/2EO4C10N$^+$ complex interacted with cells more efficiently than the naked DNA and other DNA/PEG-lipid complexes (Fig.2)[5]. The DNA/2EO4C10N$^+$ complex having longer alkyl chain showed higher cellular uptake than the DNA/2EO4C6N$^+$ complex did (Fig.2). Conjugation with methoxy-terminated ethylene glycol resulted in the decrease of cellular uptake efficiency.

The influences of temperature on cellular uptake of the naked DNA and the DNA/2EO4C10N$^+$ complex were shown in Figure 3[11]. FITC-dextran was employed as a control which is known as endocytosis marker. Uptake efficiency of FITC-dextran and naked DNA at 4 °C were quite low, and largely increased at higher temperatures but saturated at 20 °C. On the other hand, uptake efficiency of the DNA/2EO4C10N$^+$ complex increases gradually with increasing temperatures. Linear correlation between cell uptake and incubation temperature has been observed in the case of fusion between liposome and plasma membrane of protoplast[12]. Furthermore, it is expected that passive diffusion would exhibit a gradual increase[13,14]. It is considered that the DNA/2EO4C10N+ complex would be likely to internalized into cells by the mechanisms different from endocytosis.

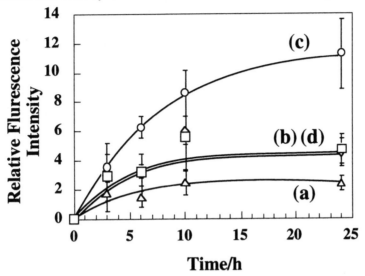

Figure 2. Time courses of cellular uptake on cellular uptake of (a) free DNA, (b) DNA/2EO4C6N$^+$ complex, (c) DNA/2EO4C10N$^+$ complex, and (d) DNA/2MeEO4C10N$^+$ complex. Incubation was carried out at 37 °C. [Cell] = 2 x 10^5 cells ml^{-1}. [DNA] = 20 μg ml^{-1}.[5]

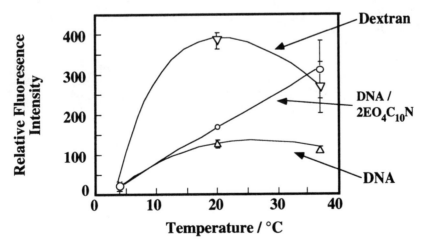

Figure 3. Temperature dependence for uptake of Dextran, free DNA, and DNA/2EO4C10N$^+$ complex into Hela cells[11]

Subcellular distribution of the DNA/2EO4C10N+ complex was visualized by a confocal fluorescence microscope[5].

The DNA complexes distributed almost in the cytoplasmic region, and only a small amount of fluorescence was observed in the nucleus. This characteristic distribution was observed up to 24 hours, suggesting that the DNA complexes localized in cytoplasmic region without digesting by intracellular enzyme. The DNA/2EO4C10N$^+$ complex showed both diffused fluorescence and punctuated fluorescence. The fluorescence image was different from subcellular distribution of endocytic marker (FITC-dextran). A vesicular distribution was not clearly observed. However, we could not confirm whether the DNA complex were released from lysosome or existed in lysosome. The cells incubated with free FITC-DNA did not show any clear fluorescence image because of low fluorescence intensity

2.1   The oligonucleotide/PEG-LIPID complex

Furthermore, we employed FITC-labelled thiooligonucleotide NT58 (S-oligo, 5'GCGGGTGAGCCCC3') and observed the subcellular distribution of the NT58 complexes[15]. The NT58/2EO4C10N$^+$ complex also showed both diffused fluorescence and punctuated fluorescence in cytoplasmic region as well as salmon sperm DNA/2EO4C10N$^+$ complex did. The punctuated fluorescence image of FITC-dextran (endocytosis marker).

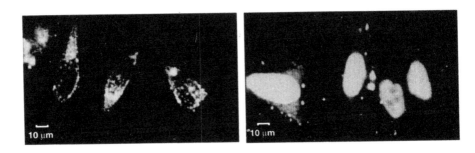

Figure 4. Subcellular distribution of the NT58/2EO4C10N$^+$ and NT58/lipofectin complexes observed by laser scanning fluorescence microscope.

The diffused fluorescence was considered to be due to NT58 released from endosome. For the NT58/lipofectin complex, NT58 was accumulated in nucleus. The similar distribution was confirmed by electron microscope (unpublished results). Hybridization assay suggested that the S-oligo/2EO4C10N$^+$ complex could bind target oligonucleotide. For this reason, even if the NT58/2EO4C10N$^+$ complex existed in cytoplasmic region, the NT58 complex may be able to bind target mRNA.

3.   THE DNA/GLYCOLIPID COMPLEX

In order to obtain cell-specific DNA complex, we synthesized new glycolipids as shown in Figure 5[16]. Each of glycolipids has saccharide end as a recognition site and primary amine end as a DNA-binding site. Formation of the DNA/glycolipid complexes were confirmed by light scattering method. Compaction of DNA by binding with the glycolipids was also observed.

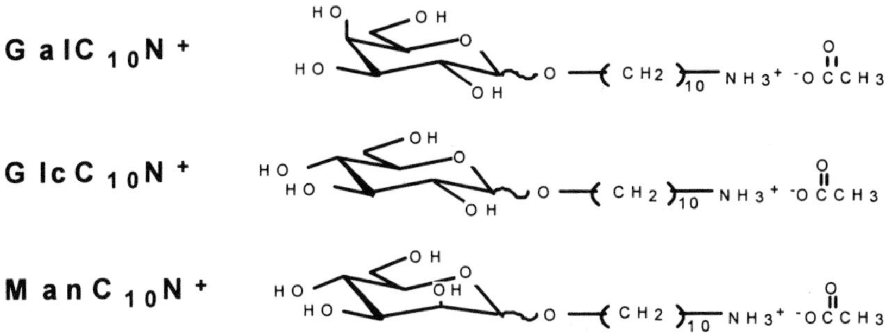

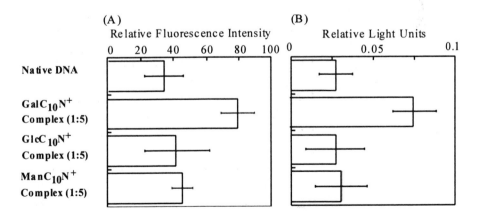

Figure 5. The structures of glycolipids synthesized in this study

When hepatocyte HepG2 cell was employed as a target cell, the DNA complex with GalC10N$^+$ having galactose end showed higher cell uptake and transfection efficiency of luciferase plasmid than the DNA complexes with glycolipids having glucose and mannose ends (Fig.6).

Figure 6. Effect of glycoside structures on the cell uptake (A) and transfection efficiency (B) of DNA/glycolipid complexes (1:5). Bars represent standard error (n = 2-3). ([DNA] = 20 μg/ml, 24 h). Cell line is HepG2. (reference 16)

For the DNA/GalC10N$^+$ complex, the effect of the stoicheometry (anion of DNA: cation of glycolipid ratio) on the cell uptake and transfection efficiency was significantly observed. The DNA/GalC10N$^+$ (1:2 and 1:5) complexes showed higher transfection efficiency than the DNA/GalC10N$^+$ (1:1) complex did, while the three DNA complexes were uptaken into cells at the same efficiency. These results suggest that the cell uptake of the DNA/GalC10N$^+$ complexes was mediated by sugar recognition, but the release of DNA from endosome was mediated by cationic charge of the DNA complex. The both processes are considered to be important for the efficient expression of the DNA/glycolipid complexes.

4.   THE DNA/CHITOSAN COMPLEX

Chitosan is a linear cationic polysaccharide composed of β1→ 4 linked glucosamine partly containing *N*-acetylglucosamine. Chitosan is known as biodegradable polymer[17,18]. Thus, chitosan microsphere is employed as drug delivery system[19-22]. We prepared the DNA/chitosan and DNA/ polygalactosamine (pGalN, α1→4 linked galactosamine) complexes, and investigated the interaction of the DNA complexes with tumor cells and white blood cells[23]. Especially, the DNA/chitosan complex was uptaken by tumor cells, but not by white blood cells. As shown in the recent papers[24,25], in vitro and in vivo gene delivery mediated by chitosan has been successful. However, optimum condition for the transfection efficiency of DNA/chitosan complex has not been clear. We investigated effects of the experimental conditions such as pH of medium, serum, and molecular mass of chitosan on the transfection efficiency of plasmid/chitosan complexes[26]. The transfection efficiency was largely dependent on pH of culture medium. The transfection efficiency at pH 6.9 was higher than that at pH 7.6. Chitosans of 15 kDa and 52 kDa largely promoted luciferase activity (Fig.7). Heptamer (1.3 kDa) did not show any expression. Transfection efficiency mediated by chitosan of >100 kDa was less than that by chitosan of 15 kDa and 52 kDa. Furthermore, Chitosan polymers (15 kDa and 52 kDa) were found to promote luciferase activities superior to lipofectin. On the other hand, the pGL3/pGalN complex did not show any gene expression. One of the practical problems for in vivo gene delivery mediated by cationic liposomes is that gene expression is inhibited by serum. However, the presence of serum resulted in enhancement of the gene transfer efficiency mediated by chitosan. At the serum content of 20%, gene expression by the pGL3/chitosan complex was promoted about 2-3 times higher than that without serum. Lipofectin-associated gene expression was markedly inhibited by serum.

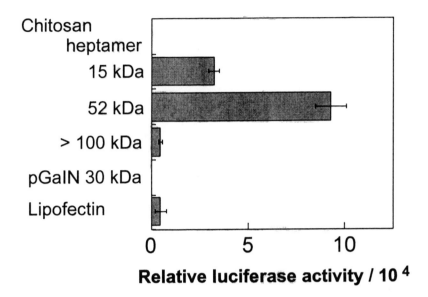

Figure 7. Effect of various gene transfer reagents on the luciferase activity of A549 cells in DMEM (pH 6.9) containing 10% FBS. The ratios of pGL3:chitosan and pGL3:pGalN were 1:5. The weight ratio of plasmid:lipofectin was 1:2. Concentration of pGL3 was 10 μg/ml.

5.   CONCLUSION

We have developed several gene delivery systems.   Low molecular weight transfection reagents such as lipoglutamates, lipoglutamides (PEG-lipids) and glycolipids are expected to compact DNA molecule and to control intracellular trafficking of DNA after endocytosis. Chitosan polymer is useful DNA carrier to enhance the expression of gene or antisence DNA. Based on this basic research in vitro, we are going to apply these transfection reagents to gene therapy.

ACKNOWLEDGMENTS

This work was supported in part by Terumo Life Science Foundation.

REFERENCES

1. Hodgson, C.P. and Solaiman, F., 1996, *Nature Biotech.*, 14: 339-342.
2. Budker, V., Gurevich, V., Hagstrom, J.E., Bortzov, F. and Wolff, J.A., 1996, *Nature Biotech.*, 14: 760-764.

3. Liu, Y., Mounkes, L.C., Liggitt, H.D., Brown, C.S., Solodin, I., Heath, T.D. and Debs, R.J., 1997, *Nature Biotech.*, 15, 167-173.

4. Templeton, N S., Lasic, D.D., Frederik, P.M., Strey, H.H., Roberts, D.D. and Pavlakis, G.N., 1997, *Nature Biotech.*, 15: 647-652.

5. Sato, T., Akino, H., Okahata, Y., Shoji, Y. and Shimada, J., 1997, *J. Biomaterial Science, Polym. Ed.*, 9: 31-42.

6. Sato, T., Kawakami, T., Shirakawa, N. and Okahata, Y., 1995, *Bull. Chem. Soc. Jpn.*, 68: 2709-2715.

7. Sato, T., Kawakami, T., Shirakawa, N., Akino, H. and Okahata, Y., 1996, Biomedical Functions and Biotechnology of Natural and Artificial Polymers, ATL Press, IL., pp. 135-144.

8. Goldman, C.K., Soroceanu, L., Smith, N., Gillespie, G.Y., Shaw, W., Burgess, S., Bilbao, G. and Curiel, D.T., 1997, *Nature Biotech.*, 15: 462-466.

9. Sato, T., Akino, H., and Okahata, Y., 1995, *Chem. Lett.*, 755-756.

10. Sato, T., Akino, H., and Okahata, Y., 1996, *Bull. Chem. Soc. Jpn.*, 69: 2335-2340.

11. Sato, T. Nishi, H., Okahata, Y., Shoji, Y., and Shimada, J., 1997, *Nucleic Acid Symp. Ser.*, 37: 303-304.

12. Arvinte, T., and Steponkus, P. L., 1988, *Biochemistry*, 27: 5671.

13. Shoji, Y., Akhtar, S., Periasamy, A., Herman, B., and Juliano,R. L., 1991, *Nucleic Acids Res.*, 19: 5543.

14. R. B. Genis, 1989, *Biochemistry*, Springer Verlag, NY.

15. Sato, T. Nishi, Akino, H., H., Okahata, Y., Shoji, Y., and Shimada, J., 1998, *Drug Delivery System*, 13: 359-364.

16. Sato, T., Ishizuka,, T., and Okahata, Y., 1998, *Kobunshi Ronbunnshuu*, 55: 217-224.

17. Hirano, S., Seino, H., Akiyama, Y. and Nonaka, I., 1988, *Polym. Eng. Sci.*, 59: 897-901.

18. Nordtveit, R.J., Vaarum, K.M. and Smidsroed, O., 1996, *Carbohydr. Polymers*, 29: 163-167.

19. Aspden, T.J., Illum, L. and Skaugrud, O., 1996, *Eur. J. Pharm.*, 4: 23-31.

20. Henriksen, I., Sminstad, G. and Karslen, J., 1994, *Int. J. Pharm.*, 101: 227-236.

21. Imai, T., Shiraishi, S., Saito, H. and Otagiri, M., 1991, *Int. J. Pharm.*, 67: 11-20.

22. Kristl, J., Smid-Korbac, J., Struc, E., Schara, M. and Rupprecht, H., 1993, *Int. J. Pharm.*, 99: 13-19.

23. Sato, T., Shirakawa, N., Nishi, H. and Okahata, Y., 1996, *Chem. Lett.*, 725-726.

24. Venkatesh, S. and Smith, T.J., 1997, *Pharm. Dev. Tech.*, 2: 417-418.

25. Roy, K., Mao, H-Q, Huang, S-K. and Leong, K.W., 1999, *Nature Med.*, 5: 387-391.

26. Sato, T., Ishii, T., Okahata, Y., 1999, *Proceed. Int'l Symp. Control. Rel. .Bioact. Mater.*,26: 803-804.

RIBOZYME TECHNOLOGY 1: HAMMERHEAD RIBOZYMES AS TOOLS FOR THE SUPPRESSION OF GENE EXPRESSION

[1]Hiroaki Kawasaki and [1,2]Kazunari Taira
[1]National Institute for Advanced Interdisciplinary Research, AIST, MITI, Tsukuba Science City 305-8562, Japan, [2]Department of Chemistry and Biotechnology, Graduate School of Engineering, University of Tokyo, Hongo, Tokyo 113-8656, Japan

1.   INTRODUCTION

A hammerhead ribozyme is one of the smallest catalytic RNA molecules[1-3]. Because of its small size and potential as an antiviral agent, numerous mechanistic studies[4-13] and studies directed towards application *in vivo* have been performed[6,14-19]. Many successful experiments, aimed at the use of ribozymes for suppression of gene expression in different organisms, have been reported[20-30]. However, the efficacy of ribozymes *in vitro* is not necessarily correlated with functional activity *in vivo*. Some of the reasons for this ineffectiveness *in vivo* are as follows. (i) Cellular proteins may inhibit the binding of the ribozyme to its target RNA or may disrupt the active conformation of the ribozyme. (ii) The intracellular concentration of the metal ions that are essential for ribozyme-mediated cleavage might not be sufficient for functional activity. (iii) Ribozymes are easily attacked by RNases. (iv) Transcribed ribozymes remain inactive because of improper folding and/or inability to colocalize with their target. Fortunately, we are now starting to understand the parameters that determine ribozyme activity *in vivo* [26,31-36]. Studies *in vivo* have suggested that the following factors are important for the effective ribozyme-mediated inactivation of genes: a high level of ribozyme expression[29]; the intracellular stability of the ribozyme[6, 37]; colocalization of the ribozyme and its target RNA in the same cellular compartment[31,34,38-40]; and the cleavage activity of the transcribed ribozyme[41]. Recently, it was shown that these various features depend on the expression system that is used [31,34].

The RNA polymerase II (pol II) system, which is employed for transcription of mRNAs, and the polymerase III (pol III) system, employed for transcription of small RNAs, such as tRNA and snRNAs, have been used as ribozyme-expression systems[19]. Transcripts driven by the pol II promoter have extra sequences at the 3' and 5' ends (for example, an untranslated region, a cap structure, and a polyA tail), in addition to the coding region. These extra sequences are essential for stability *in vivo* and functional recognition as mRNA. A transcript containing a ribozyme sequence driven by the pol II promoter includes all those sequences, unless such sequences are trimmed after transcription[15,42-45]. As a result, in some cases, the site by which the ribozyme recognizes its target may be masked, for example, by a part of the coding sequence. By contrast, the pol III system is suitable for expression of short RNAs and only very short extra sequences are generated. In addition, the level of expression is at least one order of magnitude higher than that obtained with the pol II system[46]. Indeed, in our hands, pol III driven ribozymes[26, 34-36] but not pol II driven ribozymes[43] could be detected by Northern blotting analysis, demonstrating the higher transcription level of the former RNA. Thus, it was suggested that the pol III system might be very useful for expression of ribozymes[29,47]. However, in many cases, the expected effects of ribozymes could not be achieved in spite of the apparently desirable features of the pol III system[31,48, 49].

We also used the pol III system for expression of a ribozyme[34]. In our system, a human tRNA^{Val} promoter was used to control expression of a hammerhead ribozyme. The ribozyme sequence was connected to the tRNA^{Val} promoter sequence with a short intervening linker sequence. We found that the putative structural differences influenced the cleavage activity *in vitro,* as well as the stability of the transcripts in cultured cells. The steady-state concentration of the ribozyme differed over a 26-fold range[34]. It should be useful to identify the structural factors that determine stability *in vivo* in order to enhance the functional activity of such ribozymes. Moreover, tRNA^{Val}-ribozymes that we designed were exported to the cytoplasm[34,46]. Thus, the ribozymes and their target were present within the same cellular compartment.   Under these conditions, our cytoplasmic tRNA^{Val}-ribozymes had significant activities in cultured cells[34, 46]. In this chapter, we will discuss our evidence that tRNA^{Val}-ribozymes are very useful tools in molecular biology, including a specific example of the use of such ribozymes in the functional analysis of a transcriptional co-activator protein.

2. PARTICIPATION OF THE TRANSCRIPTIONAL CO-ACTIVATOR CBP IN SEVERAL SIGNALING PATHWAYS

The co-activators CBP (<u>C</u>REB <u>b</u>inding <u>p</u>rotein, where CREB stands for cAMP-response-element-binding protein) and p300, a related protein, are functionally conserved proteins that act in concert with other factors to regulate transcription[50-55]. Many different classes of transcription factors, such as nuclear receptors, certain bHLH (<u>b</u>asic <u>h</u>elix <u>l</u>oop <u>h</u>elix) proteins involved in cell differentiation, and activators that respond to extracellular signaling events, all recruit CBP and/or p300 to their cognate promoters[55-58]. Since CBP and p300 have an intrinsic histone acetyltransferase activity, they might facilitate transcription by acting directly on chromatin[59-60]. Both proteins are also intimately involved in growth-control pathways, as is evident from the finding that several viral oncoproteins, such as adenovirus E1A and the T antigen of SV40, depend for their growth-stimulatory functions on the formation of complexes with CBP and p300[55, 61].

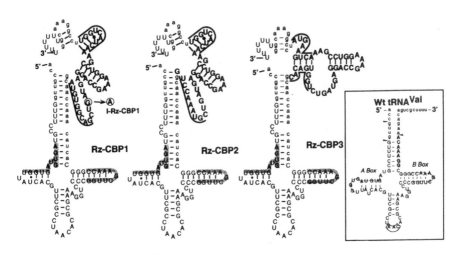

Figure 1. The secondary structures of tRNA^Val-ribozymes, as predicted by computer folding analysis. The sequence of the hammerhead ribozyme was ligated downstream of that of a seven-base-deleted tRNA^Val (capital letters) with linker sequences (lowercase letters). The sequences that correspond to the internal promoter of tRNA^Val, namely the A and B boxes, are indicated by shaded boxes. The recognition arms of ribozymes are indicated by underlining. The predicted secondary structure of human placental tRNA^Val is shown in the right panel. The tRNA is processed at three sites (arrowheads) to yield the mature tRNA^Val (capital letters). Rz-CBP1 through Rz-CBP3 are ribozymes targeted to CBP mRNA.

Upon exposure to retinoic acid (RA), embryonal carcinoma (EC) F9 cells can be induced to differentiate into three distinct extra-embryonic types of cell, namely, primitive, parietal and visceral endodermal cell, which resemble cells of the primitive extra-embryonic endoderm of the mouse embryo[62-64]. The RA-mediated differentiation of these cells results in dramatic changes in gene-expression programs. Concomitantly with the induction of the differentiated phenotype, RA can also trigger apoptosis (programmed cell death) in F9 cells[65, 66]. CBP and p300 interact with nuclear hormone receptors, such as RAR, RXR, TR and ER[56]. However, the functional significance of CBP during the RA-induced differentiation and apoptosis of F9 cells remains to be clarified. In an effort to shed some light on these issues, we decided to attempt to regulate the expression of the gene for CBP using tRNA^{Val}-driven hammerhead ribozymes and to examine the effects of such regulation in F9 cells.

The specific tRNA^{Val}-ribozymes used in the experiments described in this section are shown in Figure 1 (Rz-CBP1 through Rz-CBP3). In our pol III-driven ribozyme-expression cassette, in order to block 3'-end processing of the transcript without any negative effects on transcription, we first removed the last seven nucleotides from the mature tRNA^{Val} (Wt tRNA^{Val} in Fig.1; removed nucleotides are indicated by outlined capital letters in the box). The removed nucleotides were replaced by a linker (lowercase letters in Fig.1). The overall secondary structures were adjusted to maintain the cloverleaf tRNA-like configuration by the selection of an appropriate linker sequence. The selected target sequence might also be expected to influence the overall structure of the transcripts and, thus, to affect the accessibility of the recognition arms of the ribozyme (recognition arms are underlined). Naturally, it is important that the 5' and 3' substrate-recognition arms of the ribozyme be available to the substrate so that the ribozyme and substrate RNA together can form the stem structures that ensure subsequent cleavage of the substrate.

3.   CONSTRUCTION OF tRNA^{Val}-RIBOZYME EXPRESSION OF THE GENE FOR CBP

In general, for functional analysis, using ribozymes, of a gene whose role is unknown, it is necessary to choose appropriate target sites within the transcript for inactivation since some sites (*e.g.*, sites embedded within a stem structure) are inaccessible to a ribozyme. Three target sites that are not conserved in the mRNA for the related protein p300 were selected within CBP mRNA and the individual ribozymes targeted to these sites were designated Rz-CBP1 through Rz-CBP3 (Fig.1) in upstream to the downstream direction. Each of the ribozymes was transcribed under the control of the human tRNA^{Val} promoter, as described above. Each ribozyme had the same 24-nucleotide-long catalytic domain and each was equipped with nine nucleotides on both substrate-binding arms (recognition arms are underlined in Fig.1) that were targeted to relatively well-conserved sequences in CBP mRNA. The flexibility of the substrate-recognition arm of each ribozyme differed depending on the specific target site of each ribozyme. The inactive ribozyme (designated I-Rz) differed from the active ribozymes by a single G_5 to A mutation in the catalytic core [the numbering system follows the rule for hammerhead ribozymes[67]]. This single-nucleotide change would be expected to diminish cleavage activity while antisense effects, if any, should remain unaffected[9, 23].

The effects of tRNA^{Val}-ribozymes on the specific target mRNA, namely, the transcript of the CBP gene, were examined by measuring the level of CBP itself in F9 cells that had been transfected with the plasmids that encoded the various tRNA^{Val}- ribozymes.

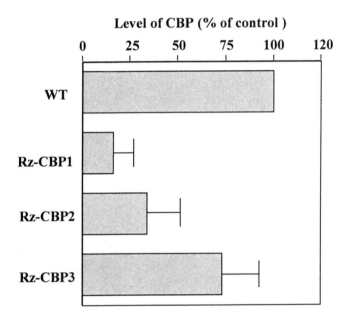

Figure 2. Effects of ribozyme-expression plasmids specific for cleavage of CBP mRNA on the levels of CBP in a transient-expression assay. Relative levels of CBP were compared by an amplified sandwich ELISA. Normalized levels of CBP in wild-type (WT) F9 cells were taken arbitrarily as 100%. All values are the averages of results from at least three experiments and the standard deviation for each value relative to the value for WT cells is indicated.

To determine the effectiveness of the ribozyme-mediated inhibition of expression of the CBP gene in a transient expression assay, we compared relative levels of CBP by an amplified sandwich ELISA. As shown in Figure 2, all the active ribozymes decreased the level of CBP *in vivo*. The results shown are the averages of results from three sets of experiments and are given as percentages relative to the control value of 100%. The extent of inhibition by the various ribozymes, expressed from the pol III promoter, ranged from 27% to 86% under the conditions of our studies. Stronger inhibition can, of course, be achieved by the introduction of a larger quantity of the ribozyme-coding plasmid.

4.   A ROLE FOR THE TRANSCRIPTIONAL CO-ACTIVATOR CBP DURING THE RA-INDUCED DIFFERENTIATION OF F9 CELLS

Our tRNA^{Val}-ribozymes, targeted to CBP mRNA, had high activity, which allows us to investigate the function of the transcriptional co-activator CBP in F9 cells.

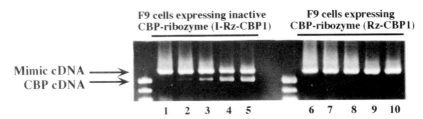

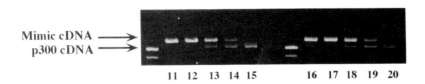

Figure 3A.   Effects of ribozyme vectors specific for cleavage of CBP mRNA on the levels of p300 and CBP mRNAs and on the levels of the proteins themselves. Analysis by competitive RT-PCR of mRNAs for p300 and CBP. The "mimic" DNA fragment specific for p300 or CBP was prepared as described in ref. 25. Products of PCR were subjected to electrophoresis on a 2% agarose gel. The concentrations of p300 or CBP mimic DNA were as follows: lanes 1, 6, 11 and 16, 1 attomole/μl; lanes 2, 7, 12 and 17, 1×10^{-1} attomole/μl; lanes 3, 8, 13 and 18, 1×10^{-2} attomole/μl; lanes 4, 9, 14 and 19, 1×10^{-3} attomole/μl; and lanes 5, 10, 15 and 20, 1×10^{-4} attomole/μl.

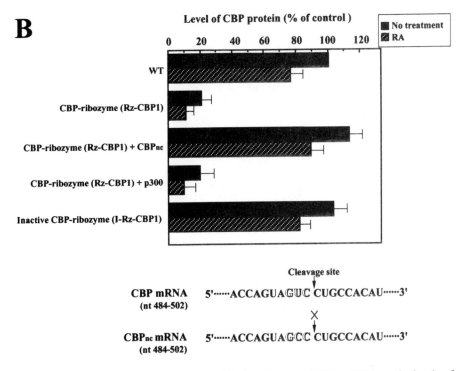

Figure 3B. Effects of ribozyme vectors specific for cleavage of CBP mRNA on the levels of p300 and CBP mRNAs and on the levels of the proteins themselves Relative levels of CBP were compared by an amplified sandwich ELISA. The normalized level of CBP in wild-type (WT) F9 cells was taken arbitrarily as 100%. CBPnc indicates an expression plasmid that encoded a mutant mRNA for CBP that was not cleaved (nc) by the CBP-ribozyme. All values are the averages of results from at least three experiments and the standard deviation for each value relative to the control is indicated. X indicates the absence of cleavage.

From among our active tRNA^{Val}-ribozymes (Fig.2), we chose the most active one, Rz-CBP1, to generate stable lines of F9 cells that expressed a tRNA^{Val}-ribozyme. The concentrations of CBP mRNA in these cell lines were measured by a competitive reverse transcription-polymerase chain reaction (RT-PCR; Fig.3A). The level of CBP mRNA fell dramatically from 1×10^{-2} attomole/µl to less than 1×10^{-4} attomole/µl in F9 cells that expressed Rz-CBP1 (in lanes 1-10 in Fig.3A, note that no CBP cDNA was detectable in cells that produced the active ribozyme). In the same cells, levels of p300 mRNA were unchanged (lanes 11-20 in Fig.3A), indicating that the Rz-CBP1 ribozyme had specifically cleaved the CBP mRNA without damaging the related p300 mRNA. An inactive ribozyme, I-Rz-CBP1, targeted to CBP mRNA had no effect on the levels of either p300 mRNA or CBP mRNA (lanes 1-5 and 11-15 in Fig.3A). These data indicate that the tRNA^{Val}-ribozyme acted, by catalyzing cleavage and not via an antisense effect, on CBP mRNA specifically despite the strong homology of p300 mRNA to CBP mRNA within conserved regions. Compared to levels of CBP mRNA in lines of F9 cells that expressed an inactive tRNA^{Val}-ribozyme, levels of CBP mRNA were about 100-fold lower in cells that expressed an active tRNA^{Val}-ribozyme. We next examined the levels of CBP itself. In F9 cells that expressed an active CBP-ribozyme, levels of CBP were at least 5-fold lower than levels in F9 cells that harbored an inactive CBP-ribozyme and than levels in WT F9 cells (Fig.3B).

In the experiments for which results are shown in Figure 3B, in order to rescue the expression of CBP, we also constructed a version of CBP mRNA that was not cleavable (nc) by the active CBP-ribozyme (CBP^{nc}; in which the triplet at the cleavage site was changed from the cleavable GUC to the uncleavable GCC, a mutation that does not change the translated amino acids). As expected, stable introduction of CBP^{nc} mRNA completely restored the production of CBP in F9 cells that expressed an active CBP-ribozyme.

We next asked whether F9 cells that lacked CBP could still respond to treatment with RA. When challenged with RA, F9 cells that synthesized an active CBP-ribozyme showed morphological signs of differentiation (Fig.4A; compare the RA-untreated F9 cells in the bottom panel with the RA-treated CBP-ribozyme-producing F9 cells in the top and middle panels). Upon treatment with RA, both WT F9 cells (data not shown) and F9 cells that synthesized the CBP-ribozyme down-regulated the expression of the cell-surface antigen SSEA-1. Expression of this antigen is characteristic of undifferentiated F9 cells (Fig.4A, right panels). Simultaneously, RA induced the synthesis in both WT and ribozyme-synthesizing cells of collagen IV (Fig.4B), which is specific for differentiated cells. These results indicate that CBP does not play a critical role in the RA-induced differentiation of F9 cells[25]. When a population of F9 cells is treated with RA, many cells differentiate but some undergo apoptosis[65, 66]. To explore the possible dependence of apoptosis on CBP, we evaluated levels of nicked DNA, using the TUNEL assay (TdT-mediated dUTP nick end labeling assay[68]), since nicked DNA is one of the hallmarks of apoptosis. F9 cells that synthesized the CBP-ribozyme were much less sensitive to the RA-induced nicking of DNA than were WT cells or those that synthesized the inactive ribozyme (Fig.5). Thus, it appeared that RA-induced apoptosis required the expression of intact CBP.

A

CBP-ribozyme (Rz-CBP1)

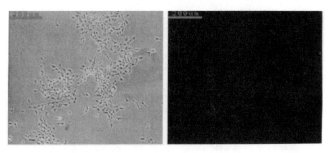

Inactive CBP-ribozyme (I-Rz-CBP1)

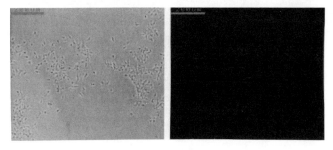

Undifferentiated WT F9 cells

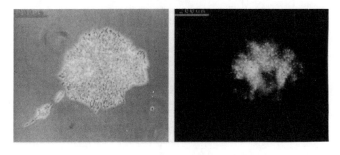

Figure 4A. Morphology of F9 cells that synthesized tRNA^{Val}-ribozymes and relativelevels of marker proteins that are specific for the undifferentiated and the differentiated state, respectively. Left panels show phase-contrast photomicrographs of F9 cells that synthesized either the active (upper) or the inactive (middle) CBP-ribozyme after treatment with retinoic acids (RA; 3×10^{-7} M) and undifferentiated wild type (WT) F9 cells (bottom). Right panels show immunostaining of F9 cells with antibodies specific for the surface antigen, SSEA-1.

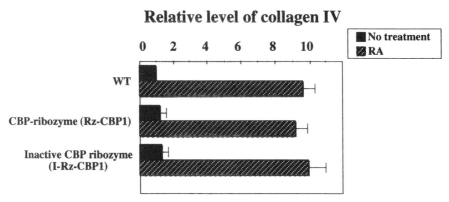

Figure 4B. Morphology of F9 cells that synthesized tRNA^{Val}-ribozymes and relativelevels of marker proteins that are specific for the undifferentiated and the differentiated state, respectively. Relative levels of collagen IV, a marker of differentiation, in WT F9 cells and in cells that expressed the indicated ribozyme after treatment with RA. Normalized levels of collagen IV were taken arbitrarily as 1.0. All values are the averages of results from at least three experiments and the standard deviation for each value relative to the control is indicated.

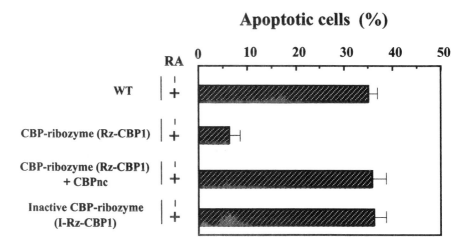

Figure 5. Percentages of apoptotic cells in cultures of wild-type (WT) F9 cells and F9 cells that synthesized a CBP-ribozyme in the presence and in the absence of retinoic acids (RA; 3 x 10^{-7} M). Apoptosis was inhibited in cells that synthesized the active CBP-ribozyme. Cell death was assessed by the TUNEL method[68].

Treatment of F9 cells with RA decreases the rate of proliferation of F9 cells[69]. We examined the consequences of decreasing levels of CBP on rates of cell proliferation by monitoring the incorporation of BrdU. In the absence of RA, the basal rate of proliferation of F9 cells that expressed the CBP-ribozyme was almost the same as that of WT F9 cells. To our surprise, exposure of cells to RA failed to reduce the rate of proliferation of F9 cells that expressed the CBP-ribozyme[25]. To gain some insight into the molecular mechanisms responsible for the absence of any change in the rate of proliferation, we examined the expression of inhibitors of cyclin-dependent kinases (cdk), such as p27Kip1 (Fig.6). The expression of p27Kip1 is induced during the differentiation of many types of cell, including F9 cells (WT; Fig.6), as well as after certain events that trigger cell-cycle arrest[70, 71]. Consistent with the failure of cells that expressed the active CBP-ribozyme to undergo apoptosis and to proliferate at a reduced rate, levels of p27Kip1 were not increased in these cells after treatment with RA. By contrast, expression of another inhibitor of cdk, p21Cip1, was normal in cells that expressed the active CBP-ribozyme[25]. These results indicate that the induced levels of p21Cip1 were insufficient to trigger cell-cycle arrest. It is significant that differential sensitivity of the expression of p21Cip1 and p27Kip1 to reduced levels of CBP itself was also observed in F9 cells that expressed the active CBP-ribozyme.

By specifically regulating the expression of CBP using tRNAVal-ribozymes, we demonstrated that CBP has important functions during RA-induced apoptosis and the cessation of cell proliferation. A schematic representation of the functional roles of CBP during RA-induced processes is shown in Figure 7.

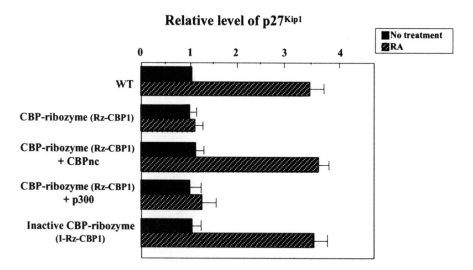

Relative level of p27Kip1

Figure 6. Levels of the cdk inhibitor p27Kip1 during the retinoic acid-induced (RA-induced) differentiation of wild-type (WT) and ribozyme-expressing F9 cells. Normalized levels of p27Kip1 in WT F9 cells were taken arbitrarily as 1.0. All values are the averages of results from three experiments and the standard deviation for each value relative to the control is indicated.

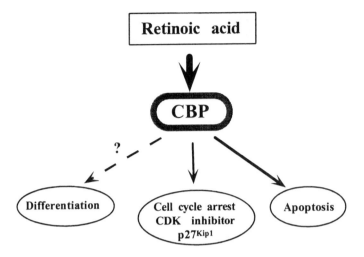

Figure 7. Model of retinoic acid-induced (RA-induced) differentiation. Normal expression of CBP is not required for the RA-induced differentiation of F9 cells. By contrast, CBP is required both for apoptosis and for expression of the cdk inhibitor p27^{Kip1}.

The observation that CBP was not necessary for the RA-induced differentiation of F9 cells was unexpected and suggests that CBP might not control the expression of target genes that are necessary for RA-induced differentiation. The same conclusion was reached in an analysis of mice that lacked CBP[72]. Moreover, the expression of the two inhibitors of cdk, p21^{Cip1} and p27^{Kip1}, was affected differently in F9 cells that synthesized the CBP-ribozyme. Induction of expression of p27^{Kip1} was dependent on normal levels of CBP. The fact that the level of p21^{Cip1} was normal in CBP-ribozyme-producing F9 cells indicates that p21^{Cip1} by itself is not sufficient to cause RA-induced apoptosis and complete cell-cycle arrest. Indeed, both the differentiation and apoptosis of cells were normal in mice that lacked p21^{Cip1}[73]. Apparently, induction of p27^{Kip1} might be responsible, at least in part, for these events. The results of earlier studies do not conflict with this conclusion [70, 71].

CBP is required for apoptosis and complete cell-cycle arrest. Since p53 is often involved in these two processes, the recent finding[74-76] that the functions of CBP and p300 are directly linked to the activity of p53 is very intriguing. In particular, CBP with a dominant negative mutation was shown to inhibit significantly both p53-mediated apoptosis and irradiation-induced arrest of the cell cycle[74]. Thus, the failure to activate p53 appropriately might explain, in part, the resistance of CBP-ribozyme-containing cells to apoptosis and complete cell-cycle arrest. These results demonstrate that, in the context of the regulation of cell growth and apoptosis, CBP plays an important role during RA-induced apoptosis and RA-induced arrest of the cell cycle. It is now necessary to dissect the molecular pathways by which CBP exerts its multiple functions in many divergent pathways.

5. CONCLUSION

The experiments described above provide an example of the utility of our ribozyme-expression system. In our functional analysis of the transcriptional co-activator CBP, we found that the CBP-ribozyme was very active in F9 cells. CBP exhibits strong homology to p300, in particular within the functional domains. However, despite the similarity between CBP and p300, the CBP-ribozyme cleaved CBP mRNA specifically, without affecting p300 mRNA, allowing us to elucidate the role of the transcriptional co-activator CBP in F9 cells. The advantages of ribozymes over conventional antisense RNA/DNA technology include the following: (i) inclusion in studies of a ribozyme that is rendered inactive by a point mutation allows discrimination of the effects of cleavage of the target RNA from other secondary effects, such as antisense effects; and (ii) as demonstrated in this section, it is usually possible to create an uncleavable version of the target mRNA, by introducing a point mutation into the target mRNA, that can be used to rescue cells from the effects of the ribozyme. The conclusion that we reached using tRNAVal-ribozymes[25] was identical to that reached in an analysis of transgenic mice[72]. Therefore, properly designed tRNAVal-ribozymes with high-level activity appear to be very useful tools in molecular biology, with potential utility in medicine.

REFERENCES

1. Guerrier-Takada, C., Gardiner, K., Marsh, T., Pace, N. and Altman, S. 1983, *Cell* 35: 849-857.
2. Haseloff, J. and Gerlach, W. L. 1988, *Nature* 334: 585-591.
3. Kruger, K., Grabowski, P. J., Zaug, A. J., Sands, J., Gottschling, D. E. and Cech, T. R. 1982, *Cell* 31: 147-157.
4. Dahm, S. C. and Uhlenbeck, O. C. 1991, *Biochemistry* 30: 9464-9469.
5. Dahm, S. C., Derrick, W. B. and Uhlenbeck, O. C. 1993, *Biochemistry* 32:13040-13045.
6. Eckstein, F. and Lilley, D. M. J. 1996, *Nucleic Acids and Molecular Biology* Vol. 10. Springer-Verlag, Berlin, Germany.
7. Lott, W. B., Pontius, B. W. and von Hippel, P. H. 1998, *Proc. Natl. Acad. Sci. USA* 95: 542-547.
8. Pontius, B. W., Lott, W. B. and von Hippel, P. H. 1997, *Proc. Natl. Acad. Sci. USA* 94: 2290-2294.
9. Ruffner, D. E., Stormo, G. D. and Uhlenbeck, O. C. 1990, *Biochemistry* 29: 10695-10702.
10. Zhou, D.-M., Zhang, L.-H., Kumar, P. K. R. and Taira, K. 1996, *J. Am. Chem. Soc.* 118: 8969-8970.

11. Zhou, D.-M., Zhang, L.-H. and Taira, K. 1997, *Proc. Natl. Acad. Sci. USA* 94: 14343-14348.
12. Zhou, D.-M., He, Q.-C., Zhou, J.-M. and Taira, K. 1998, *FEBS Lett.* 431: 154-160.
13. Zhou, D.-M. and Taira, K. 1998, *Chem. Rev.* 98: 991-1026.
14. Erickson, R. P. and Izant, J. 1992, *Gene Regulation; Biology of Antisense RNA and DNA*, Raven Press, New York, N. Y.
15. Murray J. A. H. 1992, *Antisense RNA and DNA*, Wiley-Liss, Inc, New York, N. Y.
16. Prislei, S., Buonomo, S. B. C., Michienzi, A. and I. Bozzoni. 1997, *RNA* 3: 677-687.
17. Rossi, J. J. 1995, *Trends Biotechnol.* 13: 301-306.
18. Scanlon, K. J. 1998, *Therapeutic applications of ribozymes; Methods in molecular medicine*, vol. 11. Humana Press, Towata, N. J.
19. Turner, P. C. 1997, *Methods in molecular biology, vol. 74. Ribozyme protocols*, Humana Press, Totowa, N. J.
20. Dropulic, B., Monika, N. and P. M. Pitha. 1996, *Proc. Natl. Acad. Sci. USA* 93: 11103-11108.
21. Ferbeyre, G., Bratty, J., Chen, H. and Cedergren, R. 1996, *J. Biol. Chem.* 271: 19318-19323.
22. Fujita, S., Koguma, T., Ohkawa, J., Mori, K., Kohda, T., Kise, H., Nishikawa, S., Iwakura, M. K. and Taira, K. 1997, *Escherichia coli. Proc. Natl. Acad. Sci. USA* 94: 391-396.
23. Inokuchi, Y., Yuyama, N., Hirashima, A., Nishikawa, S., Ohkawa, J. and Taira. K. 1994, *J. Biol. Chem.* 269: 11361-11366.
24. Kawasaki, H., Ohkawa, J., Tanishige, N., Yoshinari, K., Murata, T., Yokoyama, K. K. and Taira, K. 1996, *Nucleic Acids Res.* 24: 3010-3016.
25. Kawasaki, H., Ecker, R., Yao, T.-P., Taira, K., Chiu, R., Livingston, D. M. and Yokoyama, K. K. 1998, *Nature* 393: 284-289.
26. Kuwabara, T., Warashina, M., Tanabe, T., Tani, K., Asano, S. and Taira. K. 1998, *Mol. Cell* 2: 617-627.
27. Sarver, N., Cantin, E. M., Chang, P. S., Zaida, J. A., Ladne, P. A., Stepenes, D. A. and Rossi, J. J. 1990, *Science* 247: 1222-1225.
28. Yamada, O., Yu, M., Yee, J.-K., Kraus, G., Looney, D. and Wong-Staal, F. 1994, *Gene Ther.* 1: 38-45.
29. Yu, M., Ojwang, J. O., Yamada, O., Hampel, A.,Rappaport, J., Looney, D. and Wong-Staa, F. 1993, *Proc. Natl. Acad. Sci. USA* 90: 6340-6344.
30. Zhao, J. J. and Pick, L. 1993, *Nature* 365: 448-451.
31. Bertrand, E., Castanotto, D., Zhou, C., Carbonnelle, C., Lee, G. P., Chatterjee, S., Grange, T., Pictet, R., Kohn, D., Engelke, D. and Rossi, J. J. 1997, *RNA* 3: 75-88.
32. Gebhard, J. R., Perry, C. M., Mahadeviah, S. and Witton, J. L. 1997, *Antisense Nucleic Acid Drug Dev.* 7: 3-11.
33. Good, P. D., Krikos, A. J., Li, S. X. L., Lee, N. S., Giver, L., Ellington, A., Zaia, J. A., Rossi, J. J. and Engelke, D. R. 1997, *Gene Ther* 4: 45-54.
34. Koseki, S., Tanabe, T., Tani, K., Asano, S., Shioda, T., Nagai, Y., Shimada, T., Ohkawa, J. and Taira, K. 1999, *J. Virol.* 73: 1868-1877.
35. Kuwabara, T., Warashina, M. Orita, M., Koseki, S., Ohkawa, J. and Taira, K. 1998, *Nature Biotechnol.* 16: 961-965.
36. Kuwabara, T., Warashina, M., Nakayama, A., Ohkawa, J. and Taira, K. 1999, *Proc. Natle. Acad. Sci. USA* 96: 1886-1891.
37. Rossi, J. J. and Sarver., N. 1990, *Trends Biotechnol.* 8: 179-183.
38. Bertrand, E., Pictet, R. and Grange, T. 1994, *Nucleic Acids Res.* 22: 293-300.
39. Bertrand, E. and Rossi, J. J. 1996,. Eckstein, F. and D. M. J. Lilley (eds.), *Catalytic RNA: Nucleic Acids Mol. Biol.*, Vol. 10. Springer-Verlag, Berlin, p. 301-313

40. Sullenger, B. A., Lee, T. C., Smith, C. A. and Ungers, G. E. 1990, *Mol. Cell. Biol.* 10: 6512-6523.

41. Thompson, D. J., Ayers, F. D., Malmstrom, A. T., Ganousis, L. M., Chowrira, M. B., Couture, L. and Stinchcomb, T. D. 1995, *Nucleic Acids Res.* 23: 2259-2268.

42. Ohkawa, J., Yuyama, N., Takebe, Y., Nishikawa, S. and Taira, K. 1993, *Proc. Natl. Acad. Sci. USA* 90: 11302-11306.

43. Ohkawa, J., Yuyama, N., Koseki, S., Takebe, Y., Homman, M., Sczakiel, G. and Taira, K. 1998, *Gene Ther. Mol. Biol.* 2: 69-82.

44. Taira, K., Nakagawa, K., Nishikawa, S. and Furukawa, K. 1991, *Nucleic Acids Res.* 19:5152-5130.

45. Taira, K. and Nishikawa, S. 1992, Erickson, R. P. and J. G., Izant (eds.), *Gene Regulation: Biology of Antisense RNA and DNA.* Raven Press, New York, pp. 35-54

46. Cotten, M. and Birnstiel, M. 1989, *EMBO J.* 8: 3861-3866.

47. Perriman, R., Bruening, G., Dennis, E. S. and W. J. Peacock. 1995. *Proc. Natl. Acad. Sci. USA* 92: 6175-6179.

48. Ilves, H., Barske, C., Junker, U., Bohnlein, E. and Veres, G. 1996, *Gene* 171: 203-208.

49. Jennings, P. A. and Molloy, P. L. 1987, *EMBO J.* 6: 3043-3047.

50. Moran, E. 1993, *Curr. Opin. Genet. Dev.* 3: 63-70.

51. Eckner, R., Ewen, M. E., Newsome, D., Gerdes, M., DeCaprio, J. A., Lawrence, J. B. and Livingston, D. M. 1994, *Genes & Dev.* 8: 869-884.

52. Kwok, R. P. S., Lundblad, J. R., Chirivia, J. C., Richards, A., Bachinger, H. P., Brennan, R. G., Roberts, S. G. E., Green, M. R. and Goodman, R. H. 1994, *Nature* 370: 223-226.

53. Arany, Z., Newsome, D., Oldread, E., Livingston, D. M. and Eckner, R. 1995, *Nature* 374: 81-84.

54. Lundblad, J. R., Kwok, R. P. S., Laurence, M. E., Harter, M. L. and Goodman, G. H. 1995, *Nature* 374: 85-88.

55. Eckner, R., Ludlow, J. W., Lill, N. L., Oldread, E., Arany, Z., Modjtahedi, N., DeCaprio, J. A., Livingston, D. M. and Morgan, J. A. 1996, *Mol. Cell. Biol.* 16: 3454-3464.

56. Shikama, N., Lyon, J. and LaThangue, N. B. 1997, *Trends Cell Biol.* 7: 230-236.

57. Smith, C. L., Onate, S. A., Tsai, M-J. and O'Malley, B. W. 1996, *Proc. Natl. Acad. Sci. U.S.A.* 93: 8884-8888.

58. Kawasaki, H., Song, J., Eckner, R., Ugai, H., Chiu, R., Taira, K., Shi, Y., Jones, N. and Yokoyama, K. K. 1998, *Genes &Dev.* 12: 233-245.

59. Bannister, A. J. and Kouzarides, T. 1996, *Nature* 384: 641-643.

60. Ogryzko, V. V., Schiltz, R. L., Russanova, V., Howard, B. H. and Nakatani, Y. 1996, *Cell* 87: 953-959.

61. Avantaggiati, M. l., Carbone, M., Graessmann, A., Nakatani, Y., Howard, B. and Levine, A. S. 1996, *EMBO. J.* 15: 2236-2248.

62. Bernstine, E. G. and Ephrussi, B. 1976. M. I. Sherman, D. Solter (eds), *Teratomas and Differentiation.* New York, Academi,. pp. 271-87

63. Strickland, S. and Mahdavi, V. 1978, *Cell* 15: 393-403.

64. Strickland, S., Smith, K. K. and Marotti, K. R. 1980, *Cell* 21: 347-355.

65. Atencia, R., Garcia-Sanz, M., Unda, M. and Arechaga, J. 1994, *Exp. Cell Res.* 214: 663-667.

66. Clifford, J., Chiba, H., Sobieszczuk, D., Metzger, D. and Chambon, P. 1996, *EMBO J.* 15: 4142-4155.

67. Hertel, K. J., Pardi, A., Uhlenbeck, O. C., Koizumi, M., Ohtsuka, E., Uesugi, S., Cedergren, R., Eckstein, F., Gerlach, W. L., Hodgson, R. and Symons, R. H. 1992, *Nucleic Acids Res.* 20: 3252.

68. Gavrieli, Y., Sherman, Y. and Ben-Sasson, S. A. 1992, *J. Cell Biol.* 119: 493-501.

69. Dean, M., Levine, R. A. and Campisi, J. 1986, *Mol. Cell. Biol.* 6: 518-524.

70. Casaccia-Bonnefil, P., Tikoo, R., Kiyokawa, H., Friedrich, V. Jr., Chao, M. V. and Koff, A. 1997, *Genes Dev.* 11: 2335-2346.

71. Durand, B., Gao, F. B. and Raff, M. 1997, *EMBO J.* 16: 306-17.

72. Yao, T.-P., Oh, S. P., Fuchs, M., Zhou, N.-D., Ch'ng, L.-E., Newsome, D., Bronson, R. T., Li, E., Livingston, D. M. and Eckner, R. 1998, *Cell* 93: 361-372.

73. Deng, C., Zhang, P., Harper, J. W., Elledge, S. J. and Leder, P. 1995, *Cell* 82: 675-684.

74. Avantaggiati, M. L., Ogryzko, V., Gardner, K., Giordano, A., Levine, A. S. and Kelly, K. 1997, *Cell* 89: 1175-1184.

75. Gu. W., Shi, X-L. and Roeder, R. G. 1997, *Nature* 387:819-823. *Mol. Cell. Biol.* 16: 3454-3464.

76. Lill, N. L., Grossman, S. R., Ginsberg, D., DeCaprio, J. and Livingston, D. M. 1997,. *Nature* (London) 387: 823-827

RIBOZYME TECHNOLOGY 2: NOVEL ALLOSTERICALLY CONTROLLABLE RIBOZYMES, THE MAXIZYMES, WITH CONSIDERABLE POTENTIAL AS GENE-INACTIVATING AGENTS

[1,2]Tomoko Kuwabara, [1,2]Masaki Warashina, and [1,3]Kazunari Taira

[1]National Institute for Advanced Interdisciplinary Research, AIST, MITI, Tsukuba Science City 305-8562, Japan, [2]Institute of Applied Biochemistry, University of Tsukuba, Tennoudai 1-1-1, Tsukuba Science City 305-8572, Japan, [3]Department of Chemistry and Biotechnology, Graduate School of Engineering, University of Tokyo, Hongo, Tokyo 113-8656, Japan

1.   INTRODUCTION

The RNA cleavage reaction by ribozymes requires the presence of a divalent metal ion such as magnesium. Ribozymes are now recognized to be metalloenzymes[1,2,3]. The hammerhead ribozymes consists of an substrate-binding region (stem I and stem III) and a catalytic domain with a flanking stem-loop II section (Fig.1A). The catalytic domain is necessary for capturing catalytically indispensable Mg^{2+} ions. Various changes are possible in other regions of the molecule. In one case, for example, extra sequences were deleted from the stem/loop II region of the hammerhead ribozyme. Our ribozymes discussed below were developed as a result of studies to shorten the ribozyme. Discussions of the potential utility of the hammerhead ribozyme have been numerous since the discovery of this small ribozyme and many attempts have been made to create ribozymes with improved features. In using the ribozyme in medical applications, ease of design and economics dictate that smaller size is preferable. Therefore, the design and construction of short ribozymes, namely the minizymes, have

Biomedical Polymers and Polymer Therapeutics
Edited by Chiellini *et al.*, Kluwer Academic/Plenum Publishers, New York, 2001

been attempted by many investigators[4,5,6,7]. However, these attempts suffered from problems in that activity fell drastically compared to the wild type, a result that led to the suggestion that minizymes might not be suitable as gene-inactivating agents.

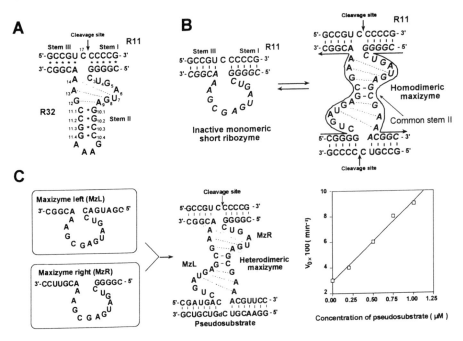

Figure 1. Secondary structures of (A) a wild-type hammerhead ribozyme (R32), (B) a maxizyme that is capable of forming a homodimer and (C) the heterodimeric maxizyme. The effects of the concentration of the uncleavable pseudosubstrate on the cleavage activity (V₀) of the heterodimeric maxizyme is also shown in Figure 1C.

2.   DISCOVERY OF MAXIZYMES THAT ACT AS A DIMERIC FORM OF SHORT RIBOZYMES

We tried to create variants of hammerhead ribozymes with deletions in the stem/loop II region and, fortunately, we found that some shortened forms of hammerhead ribozymes had high cleavage activity that was similar to that of the wild-type parental hammerhead ribozyme (R32; Fig.1A)[8]. Moreover, the active species appeared to form dimeric structures with a common stem II (Fig.1B). In the active short ribozymes, the linker sequences that replaced the stem/loop II region were palindromic so that two short ribozymes were capable of forming a dimeric structure with a common stem II. In order to distinguish monomeric forms of conventional minizymes that have extremely low activity from our novel dimers with high-level activity, we chose the name "maxizymes" for the very active short ribozymes that were capable of forming dimers. The activity of the homodimeric maxizymes (a dimer with two identical binding sequences) depended on Mg^{2+} ions and, in addition, interactions with the substrates also stabilized the dimeric structures.

We next synthesized heterodimeric maxizymes, in which one maxizyme left (MzL) and a one maxizyme right (MzR) were the monomers that together formed one heterodimeric maxizyme, as shown in Figure 1C. Such a heterodimeric maxizyme can recognize two independent RNA sequences as substrates. Moreover, we found that the cleavage rate of one of the substrates by the heterodimeric maxizyme increased linearly with increases in the concentration of the other substrate, a clear demonstration that the formation of the dimer was essential for cleavage of the substrate and, moreover, that the dimeric structure could be stabilized in the presence of the substrates (Fig.1).

3. DETECTION OF THE FORMATION OF MAXIZYME DIMERS BY HIGH-RESOLUTION NMR SPECTROSCOPY

Although the results of kinetic analysis supported the proposed dimerization of maxizymes (Fig.1C), we tried to obtain more direct evidence for the existence of homodimeric maxizymes by NMR spectroscopy[9]. In solution, in the absence of its substrate, the parental hammerhead ribozyme (R32, Fig.1A and 2A top) consisted of not only the expected GAAA loop II, stem II, and the non-Watson-Crick three-base-paired duplex (consisting of U_7:A_{14}, G_8:A_{13}, and A_9:G_{12}) but also of an originally unexpected four-base-paired duplex (consisting of $G_{2.1}$:$C_{15.5}$, C_3:$G_{15.4}$, U_4:$G_{15.3}$, and G_5:$C_{15.2}$) that included a wobble G:U base pair (Fig.2A, top left)[10]. In this configuration, the recognition arms of R32 were unavailable (the substrate-recognition regions formed intramolecular base pairs). NMR spectra of the maxizyme in the absence of its substrate were recorded under conditions identical to those used for the parental ribozyme[10]. As indicated by dotted lines in the spectrum, recorded at 5 °C (Fig.2A, right), we recognized all the resonances of imino protons that corresponded to protons that were expected to be in a similar environment to imuno protons in R32. The only imino proton in the maxizyme that was expected to be in a different environment from that in R32 was the imino proton of $G_{10.1}$. This proton should be sandwiched between A_9 and $C_{11.1}$ in the maxizyme while it is sandwiched between A_9 and $G_{10.2}$ in R32. The presence of the G_{12} signal strongly supported the proposed homodimeric structure for the maxizyme in the absence of its substrate since, if the maxizyme had existed in a monomeric form, it is unlikely that the resonance of the G_{12} imino proton would have had an identical chemical shift to that of the G_{12} imino proton in R32. Taken together with our kinetic results, these observations suggested that the major population of maxizymes was in a dimeric form not only in the absence of the substrate but also in the presence of the substrate.

Since the signals due to imino protons of the maxizyme were broader than those of R32 and since other signals were also detected, it appeared that the homodimeric maxizyme existed in the presence of other contaminating conformers. The NMR spectra of imino protons changed with changes in temperature (Fig.2B and 2C).

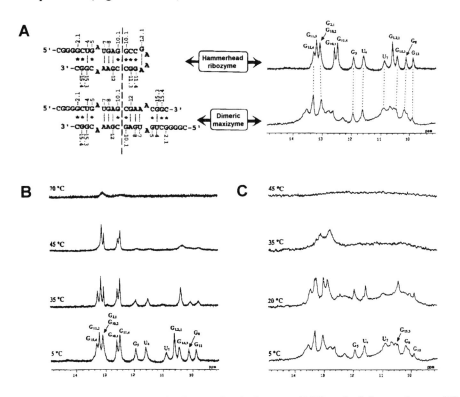

Figure 2. The NMR spectra of a hammerhead ribozyme (R32) and of the maxizyme. (A) Apparent secondary structures of R32 and the homodimeric maxizyme are shown on the left, and the NMR spectra of imino proton of R32 and a homodimeric maxizyme [in 0.1 M NaCl, 10 mM sodium phosphate buffer (pH 7.0) at 5 °C] are shown on the right . Effects of temperature on the NMR spectra of the imino protons of R32 (B) and the homodimeric maxizyme (C) are also shown.

The resonance of an imino proton is observed only if the lifetime of the imino proton is long compared to the rate of exchange with the solvent. As the temperature is increased, base pairs begin to denature and imino protons become better able to exchange with protons in the solvent. This phenomenon was observed in the NMR spectra as the broadening and eventual disappearance of the signals due to imino protons. It should be noted that the duplex of the homodimeric maxizyme melted at a lower temperature (Fig.2C) than the melting of R32 (Fig.2B). The lower melting temperature of the dimeric maxizyme was expected because of its intermolecular base-pairs, in contrast to the intramolecular base-pairs in R32 (Fig.2B).

4.   CONSTRUCTION OF tRNA^{Val}-EMBEDDED MAXIZYMES AND DETECTION OF THE FORMATION OF DIMERIC STRUCTURES BY tRNA^{Val}-EMBEDDED MAXIZYMES

For the application of maxizymes to gene therapy, they must be expressed constitutively *in vivo* under the control of a strong promoter. As described in the previous chapter, we succeeded in establishing an effective ribozyme-expression system that was based on the pol III promoter. High-level expression under control of the pol III promoter would be advantageous for maxizymes and enhance the likelihood of their dimerization. Therefore, we chose the expression system with the promoter of a human gene for tRNA^{Val}, which we had used successfully in the suppression of target genes by ribozymes[11,12].

We embedded the maxizymes similarly in the 3' portion of the tRNA^{Val} sequence. In the case of the tRNA^{Val}-embedded maxizymes, we worried initially about the possibility that the tRNA^{Val} portion might cause severe steric hindrance that would inhibit dimerization, with resultant production of monomers with extremely low activity. To our surprise, the tRNA^{Val}-embedded maxizymes still had significant cleavage activity *in vitro*[9]. We postulated that the tRNA^{Val}-embedded maxizymes might exist in an equilibrium between the monomeric and dimeric forms. Since the structural disturbance caused by the tRNA^{Val} portion appeared less significant in tRNA^{Val}-embedded maxizymes than in the tRNA^{Val}-embedded parental hammerhead ribozymes, despite the fact that the former could function properly only as dimers, we performed modeling studies, using the coordinates of the crystal structure of a hammerhead ribozyme[13] and those[14] of yeast tRNA^{Phe}, to confirm that the formation of dimeric structures by tRNA^{Val}-embedded maxizymes was theoretically feasible (Fig.3A). The resultant model structure appeared feasible and steric hindrance by the two tRNA moieties seemed not to be a problem[9].

More direct evidence for the formation of dimers by the tRNA^{Val}-embedded homodimeric maxizyme (tRNA^{Val}-Mz; Fig.3B) was provided by gel-shift analysis in the absence of the substrate[9]. As controls, we also analyzed tRNA^{Val} transcripts that contained a nonsense sequence between the tRNA^{Val} portion and the terminator sequence, as well as transcripts of the gene for the tRNA^{Val}-embedded parental hammerhead ribozyme (tRNA^{Val}-R32). Shifted bands (dimers) were observed only in the case of the tRNA^{Val}-Mz (Fig.3B).

A

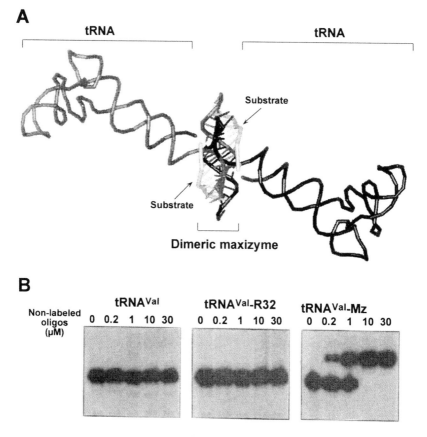

B

Figure 3. The dimeric form of a tRNA^{Val}-embedded maxizyme. (A) A model of the tRNA-embedded dimeric maxizyme (tRNA^{Val}-Mz). In this model, the internal loop or linker is ignored and is assumed to be an A-form duplex. (B) Detection of the dimeric form of tRNA^{Val}-Mz. Gel-shift analysis revealed that the dimerization occurred only in the case of tRNA^{Val}-Mz. 5'-^{32}P-labeled tRNA^{Val}-ribozyme (2 nM) was incubated with increasing amounts of the respective non-radiolabeled RNA in 25 mM MgCl$_2$ and 50 mM Tris-HCl (pH 8.0) at 37 °C for 20 minutes. The reaction products were separated on a non-denaturing gel.

The intensity of the band of the dimeric form increased with increasing concentrations of the non-radiolabeled tRNA^{Val}-Mz, supporting the proposed intermolecular dimerization process. By contrast, tRNA^{Val}-R32 and tRNA^{Val} transcripts remained in the monomeric form even at concentrations as high as 30 μM.

5. THE INTRACELLULAR ACTIVITIES OF tRNA^{VAL}-DIMERIC MAXIZYMES; SIMULTANEOUS CLEAVAGE OF HIV-1 TAT mRNA AT TWO INDEPENDENT SITES BY HETERODIMERIC MAXIZYMES

The high activity of the tRNA^{Val}-embedded maxizymes that had been transcribed *in vitro* prompted us to examine whether similar transcripts might also be active within cells. We designed heterodimeric maxizymes that should be able to cleave HIV-1 Tat mRNA at two independent sites simultaneously (Fig.4A). We confirmed *in vitro* that such heterodimeric maxizymes, with their two independent catalytic cores, were able to cleave HIV-1 Tat mRNA at two independent sites simultaneously[15]. We embedded the heterodimeric maxizymes in the 3' portion of the tRNA^{Val} sequence similarly as described above. In order to examine the intracellular activities of tRNA^{Val}-heterodimeric maxizymes, the following assay was performed. We used LTR-Luc HeLa cells which endogenously encoded a chimeric gene that consisted of the long terminal repeat (LTR) of HIV-1 and a gene for luciferase (Fig.4B)[16]. The LTR of HIV-1 contains regulatory elements that include a TAR region. The HIV-1 regulatory protein, Tat, binds to TAR and the binding of Tat stimulates transcription substantially. Therefore, luciferase activity originating from the chimeric LTR-Luc gene increases in response to increases in the concentration of Tat[16]. Measurements of luciferase activity allowed us to monitor the effects of tRNA^{Val}-embedded ribozymes on the Tat-mediated transcription of the chimeric LTR-Luc gene. After transient expression of both Tat and tRNA^{Val}-embedded ribozymes by co-transfection of cells with a Tat expression vector (pCD-SRα tat) and one of our tRNA^{Val}-embedded ribozyme-expression vectors (Fig.4B), we estimated the intracellular activity of each tRNA^{Val}-ribozyme by measuring the luciferase activity.

The luciferase activity recorded when we used only the Tat-expression vector (pCD-SRa tat) was taken as 100%. Ribozyme- and Tat-expression vectors were used at a molar ratio of 2:1 for co-transfection of LTR-Luc HeLa cells. The results shown in Figure 4C are the averages of results from five sets of experiments. As shown in Figure 4C, the tRNA^{Val}-dimeric maxizyme was extremely effective (D-L/R; >90% inhibition), despite the fact that the SRa promoter, which controlled the transcription of the target

Tat mRNA, is 10- to 30-fold more active than the SV40 early promoter regardless of the species and origin of cells[17]. tRNA^{Val}-hammerhead ribozymes were also effective, albeit to a lesser extent (>60% inhibition).

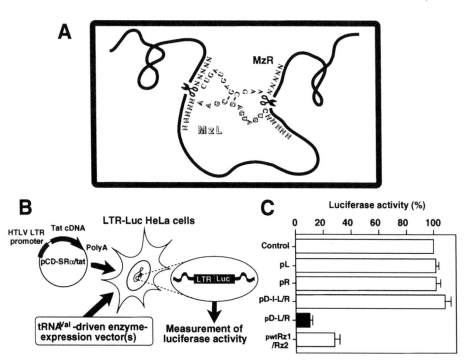

Figure 4. The heterodimeric maxizyme that can cleave an mRNA at two sites simultaneously. (A) Schematic representation of the heterodimeric maxizyme that cleave an mRNA at two sites simultaneously. (B) Assay system for measurements of activities of tRNA^{Val}-ribozymes in LTR-Luc HeLa cells and (C) the effects of tRNA^{Val}-ribozymes on the Tat-mediated transcription of the chimeric LTR-Luc gene. Luciferase activity was normalized by reference to the efficiency of transfection, which was determined by monitoring activity of a co-transfected gene for β-galactosidase.

It is important to note that, since each separate tRNA^{Val}-maxizyme (L, R) did not, by itself, have any inhibitory effects, the activities of the tRNA^{Val}-embedded maxizymes must have originated from the formation of active heterodimers in the mammalian cells. Moreover, since the inactive tRNA^{Val}-driven maxizymes (D-I-L/R) that had been created by a single G$_5$ to A$_5$ mutation within the catalytic core did not have any inhibitory effects, it is clear that the intracellular activities of the tRNA^{Val}-embedded ribozymes originated from their cleavage activities in cultured cells and not from any antisense effects[18].

The studies described above demonstrated that our heterodimeric maxizyme was more active than the parental hammerhead ribozymes *in vivo*.

The novel tRNA^{Val}-driven heterodimeric maxizymes, which can cleave one substrate at two independent sites simultaneously, can be designed very easily and, thus, they should be very useful tools *in vivo*. Nevertheless, it remains true that two independent tRNA^{Val}-driven parental ribozymes can also cleave such a substrate at the two different sites, albeit with lower efficiency[18]. In the following section, we shall describe tRNA^{Val}-driven heterodimeric maxizymes that specifically cleave a chimeric mRNA with high selectivity while conventional ribozymes failed to do so.

6. CHRONIC MYELOGENOUS LEUKEMIA (CML) AND THE POTENTIAL FOR RIBOZYME THERAPY

In some cytogenetic abnormalities, such as certain leukemias, chimeric fusion mRNAs connecting strange exons result from reciprocal chromosomal translocations and cause abnormalities. For the design of ribozymes that can disrupt such chimeric RNAs, it is necessary to target the junction sequence. Otherwise, normal mRNAs that share part of the chimeric RNA sequence will also be cleaved by the ribozyme, with resultant damage to the host cells.

Chronic myelogenous leukemia (CML) is a clonal myeloproliferative disorder of hematopoietic stem cells that is associated with the Philadelphia chromosome[19]. The reciprocal chromosomal translocation t(9; 22) (q34; q11) can be classified into two types: K28 translocations and L6 translocations (Fig.5A). These translocations result in the formation of the *BCR-ABL* fusion genes that encode two types of mRNA: b3a2 (consisting of *BCR* exon 3 and *ABL* exon 2) and b2a2 (consisting of *BCR* exon 2 and *ABL* exon 2)[20]. Both of these mRNAs are translated into a protein of 210 kDa (p210$^{BCR-ABL}$) which is unique to the malignant phenotype[21]. In the case of the b2a2 sequence, which results from reciprocal chromosomal translocations, there are no triplet sequences that are potentially cleavable by hammerhead ribozymes within two or three nucleotides of the junction in question. GUC triplets are generally the most susceptible to cleavage by a hammerhead ribozyme, and one such triplet is located 45 nucleotides from the junction. If this GUC triplet were cleaved by a ribozyme, normal *ABL* mRNA that shares part of the sequence of the abnormal *BCR-ABL* mRNA would also be cleaved by the ribozyme, with resultant damage to host cells. In designing ribozymes that might cleave b2a2 mRNA, we must be sure to avoid cleavage of normal *ABL* mRNA.

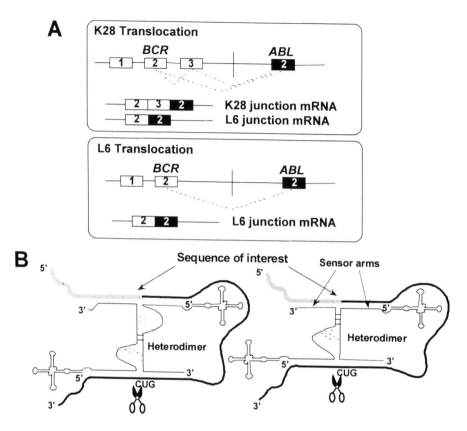

Figure 5. Types of translocation and the possible sites of cleavage within normal *ABL* mRNA and abnormal *BCR-ABL* fusion mRNAs by conventional ribozymes (A) and the design of the novel "maxizyme" (B).

7.   SPECIFIC DESIGN OF AN ALLOSTERICALLY CONTROLLABLE MAXIZYME

In extending our studies of dimeric maxizymes, we tried to create an allosterically controllable ribozyme, based on the heterodimeric maxizyme, that would have cleavage activity only when the target sequence of interest was present. Such a ribozyme should be controlled such that it is active only in the presence of an abnormal RNA sequence, for example, the junction sequence of a chimeric mRNA. We first thought about using one of the two substrate-binding regions of the heterodimeric maxizyme as sensor arms. Then, since one domain of the maxizyme was to be used solely for recognition of the target sequence of interest, we deleted its catalytic core completely to generate an even smaller monomeric unit (Fig.5B, right). Our goal was to control the activity of maxizymes allosterically by introducing sensor arms so that, only in the presence of the correct target sequence of interest, would the maxizyme be able to create a cavity for the capture of catalytically indispensable Mg^{2+} ions. Ribozymes are metalloenzymes[1-3], and we wanted to ensure that the active structure of the maxizyme with a Mg^{2+}-binding pocket would be formed only in the presence of the sequence of interest.

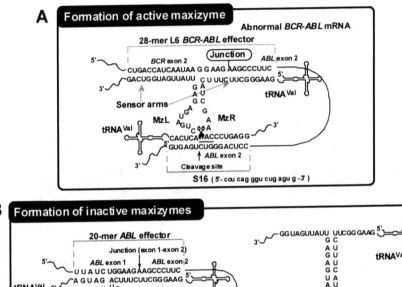

Figure 6. Secondary structures of the active (A) and inactive maxizymes (B).

In order to achieve high substrate-specificity, we wanted our maxizyme to adopt an active conformation exclusively in the presence of the abnormal *BCR-ABL* junction (Fig.6A), with the maxizyme remaining in an inactive conformation in the presence of normal *ABL* mRNA and in the absence of the abnormal *BCR-ABL* junction (Fig.6B). This phenomenon would resemble the changes in conformation of allosteric proteinaceous enzymes that occur in response to their effector molecules. The name *"maxizyme"* was initially chosen as an abbreviation of <u>m</u>inimized, <u>a</u>ctive, <u>x</u>-shaped (heterodimeric), and <u>i</u>ntelligent (allosterically controllable) ribo<u>zyme</u>[21]. Later, the term of maxizyme was used simply to distinguish monomeric forms of conventional minizymes that have extremely low activity from the novel dimeric short ribozyme with high-level activity.

8.   SPECIFICITY OF THE CLEAVAGE OF THE CHIMERIC BCR-ABL SUBSTRATE IN VITRO

In order to prove *in vitro* that changes in the conformation of our heterodimeric maxizyme depended on the presence or absence of the abnormal b2a2 mRNA, we prepared a short 16-nucleotide (nt) *BCR-ABL* substrate (S16) that corresponded to the target (cleavage) site indicated by capital letters in the upper panel of Figure 6B (within exon 2 of *ABL*).

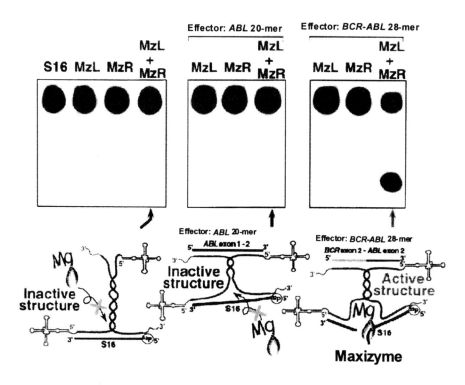

Figure 7. Allosteric control of the activity of the maxizyme *in vitro*. The specificity of maxizyme-mediated cleavage was examined by incubating tRNA^{Val}-driven component(s) with the 5'-^{32}P-labeled S16 in the presence and in the absence of an allosteric effector molecule, namely, either a normal *ABL* effector or a *BCR-ABL* effector. MzL and/or MzR were incubated at 0.1 μM with 2 nM 5'-^{32}P-labeled S16. When applicable, the concentration of the effector, 20-mer *ABL* or 28-mer *BCR-ABL*, was 1 μM. The catalytic action of Mg^{2+} ions is indicated by "Mg scissors".

Specificity was examined by incubating the maxizyme, which had been transcribed *in vitro*, with the 5'-^{32}P-labeled S16 in the presence and in the absence of either a 20-nt normal *ABL* effector molecule or a 28-nt *BCR-ABL* effector molecule (Fig.7). In this case, the 28-nt *BCR-ABL* effector molecule corresponding to the junction sequence in b2a2 mRNA should act *in trans* and it should be recognized, for annealing, by the sensor arms of MzL and MzR and should serve to direct the formation of the active dimer. The other recognition arms of the maxizyme should recognize the cleavage triplet in the S16 and specific cleavage should occur (Fig.7, on the right). Indeed, no products of cleavage of substrate S16 were detected in the absence of the *BCR-ABL* effector or in the presence of the normal *ABL* effector, demonstrating the expected high substrate-specificity of the maxizyme. The results shown in Figure 7 proved that the maxizyme was subject to complete allosteric control *in vitro*, in accord with the predicted conformational changes (Fig.6) that should occur in response to the effector molecule (the *BCR-ABL* junction) that was added *in trans*. Furthermore, the results confirmed that the tRNA^{Val} portion of the maxizyme did not interfere with the allosteric control.

9. THE ACTIVITY AND SPECIFICITY OF THE MAXIZYME AGAINST AN ENDOGENOUS *BCR-ABL* CELLULAR TARGET

We next examined the activity of the maxizyme against an endogenous *BCR-ABL* (b2a2 mRNA) target in cells. We established a line of murine cells, BaF3/p210$^{BCR-ABL}$, that expressed human b2a2 mRNA constitutively. Although the parental BaF3 cell line is an interleukin-3-dependent (IL-3-dependent) hematopoietic cell line, our transformed BaF3/p210$^{BCR-ABL}$ cells were IL-3-independent because of the tyrosine kinase activity of p210$^{BCR-ABL}$ and, thus, the latter transformed cells were able to grow in the absence of IL-3. However, if the expression of p210$^{BCR-ABL}$ were to be inhibited, BaF3/p210$^{BCR-ABL}$ cells should become IL-3-dependent and, in the absence of IL-3, they should undergo apoptosis. In the presence of IL-3, we generated BaF3/p210$^{BCR-ABL}$ cells that expressed constitutively a tRNA^{Val}, wtRz or maxizyme[21]. The transduced cells were cultured for a further 60 h and then IL-3 was eliminated from the medium for assays of subsequent apoptosis. In order to examine the specificity of the maxizyme, we also used H9 cells, which originated from human T cells and expressed normal *ABL* mRNA, as control cells.

The viability of cells was assessed in terms of their ability to exclude trypan blue dye. As shown in the left panel of Figure 8A, BaF3/p210^{BCR-}

ABL cells that expressed the maxizyme (pV-MzL/R) died rapidly, whereas the control-transfected BaF3/p210*BCR-ABL* (pV) cells were still alive 10 days after withdrawal of IL-3. Moreover, the maxizyme did not kill any H9 cells that expressed normal *ABL* mRNA (Fig.8A, middle), a result that demonstrates the high specificity of the maxizyme for its target, the chimeric *BCR-ABL* gene. By contrast, in the presence of the conventional hammerhead ribozyme wtRz, apoptosis was induced in both BaF3/p210*BCR-ABL* and H9 cells (Fig.8A, left and middle). The maxizyme also killed many more BV173 cells, derived from a leukemic patient with a Philadelphia chromosome, than did the wild-type ribozyme or the parental vector (Fig.8A, right).

10. DIRECT EVIDENCE FOR THE ENHANCED ACTIVATION OF CASPASE-3 BY THE MAXIZYME

Transduction of the apoptotic signal and execution of apoptosis require the coordinated actions of several aspartate-specific cysteine proteases, known as caspases. An inverse relationship between the *BCR-ABL*-mediated inhibition of apoptosis and the activation of procaspase-3 was recently established by Dubrez, *et al*[22]. Therefore, we investigated whether the maxizyme- (or ribozyme-) mediated apoptotic pathway might involve the activation of procaspase-3 in leukemic cells. We asked whether the specific depletion of p210*BCR-ABL* protein by the maxizyme might lead to the cleavage of inactive procaspase-3 to yield active caspase-3, with resultant apoptosis in BaF3/p210*BCR-ABL* cells. Immunoblotting analysis using the antibody which recognizes both the 32-kDa inactive precursor of caspase-3 (procaspase-3) and the processed, active protease, caspase-3, enabled us to follow the maturation process. In order to examine the specificity of the maxizyme, we performed a similar study using H9 cells. The basal level of procaspase-3 was almost the same in BaF3/p210*BCR-ABL* cells and H9 cells (Fig.8B). In maxizyme-transduced BaF3/p210*BCR-ABL* cells, the level of procaspase-3 decreased and the level of the p17 active subunit of caspase-3 increased. In stable maxizyme-transduced H9 cells, the level of procaspase-3 remained unchanged.

By contrast, expression of the wild-type ribozyme was associated with the processing of procaspase-3 in both BaF3/p210*BCR-ABL* cells and H9 cells. The extent of the conversion of procaspase-3 to caspase-3 in stable maxizyme-transduced BaF3/p210*BCR-ABL* cells was higher than that in wtRz-transduced BaF3/p210*BCR-ABL* cells.

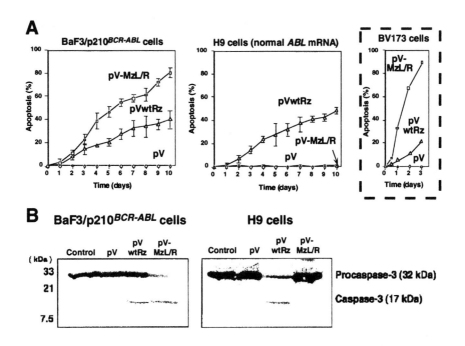

Figure 8. Efficiency of cleavage by the maxizyme of the endogenous *BCR-ABL* mRNA target. (A) Measurements of viability of tRNA^{Val}-enzyme-transduced BaF3/p210$^{BCR-ABL}$ cells and H9 cells. The viability of BV173 cells, which were derived from a patient with a Philadelphia chromosome and which were transiently expressed tRNA^{Val}-enzymes, is also shown. (B) Cleavage of inactive procaspase-3 to yield active caspase-3 upon specific depletion of p210$^{BCR-ABL}$ by the maxizyme. Immunoblotting analysis was performed using the antibody which recognizes both the 32-kDa precursor to caspase-3 (procaspase-3) and caspase-3 itself.

11. CONCLUSION

There have been many prior attempts using various approaches to construct artificial allosteric enzymes, but there have been no examples reported that can be used *in vivo*. The experimental results described above represent the first successful construction of an allosteric enzyme that functions at the tissue culture level. The maxizyme is new in that it imparts a sensor function to the short ribozymes that function as a dimer. By using this sensor function, it has become possible to specifically cleave abnormal chimeric mRNA only, without affecting the normal mRNA which should not be cleaved. This could not be achieved with the previously used ribozymes. There is more potent inhibition of expression of abnormal mRNA by the maxizyme than the previously used ribozymes, as has been demonstrated in cells derived from patients. Currently, the only known the effective treatment for leukemia is bone marrow transplantation. However, bone marrow transplantation cannot be performed on every patient, and the number of available donors is also insufficient. Thus, it is desirable to develop therapy that is generally applicable and safe. Since maxizymes are highly active and can be easily extended to clinical applications, they may find wide application in the gene therapy of chronic myelogenous leukemia in the future.

REFERENCES

1. Dahm, S. C., Derrick, W. B., and Uhlenbeck, O. C., 1993, *Biochemistry,* 32: 13040-13045.
2. Steitz, T. A., and Steitz, J. A., 1993, *Proc. Natl. Acad. Sci. USA,* 90: 6498-6502.
3. Zhou, D.-M., and Taira, K., 1998, *Chem. Rev.,* 98: 991-1026.
4. Long, D. M., and Uhlenbeck, O. C., 1994, *Proc. Natl. Acad. Sci. USA,* 91: 6977-6981.
5. Tuschl, T., and Eckstein, F., 1993, *Proc. Natl. Acad. Sci. USA,* 90: 6991-6994.
6. McCall, M. J., Hendry, P., and Jennings, P. A., 1992, *Proc. Natl. Acad. Sci. USA,* 89: 5710-5714.
7. Fu, D. J., Benseer, F., and McLaughlin, L. W., 1994, *J. Am. Chem. Soc.,* 116: 4591-4598.
8. Amontov, S. V., and Taira, K, 1996, *J. Am. Chem. Soc.,* 118: 1624-1628.
9. Kuwabara, T., Warashina, M., Orita, M., Koseki, S., Ohkawa, J., and Taira, K, 1998, *Nature Biotechnology,* 16: 961-965.
10. Orita, M., Vinayak, R., Andrus, A., Warashina, M., Chiba, A., Kaniwa, H., Nishikawa, F., Nishikawa, S., and Taira, K., 1996, *J. Biol. Chem.,* 271: 9447-9454.
11. Koseki, S., Tanabe, T., Tani, K., Asano, S., Shioda, T., Nagai, Y., Shimada, T., Ohkawa, J., and Taira, K., 1999, *J. Viol,.* 73: 1868-1877.
12. Kawasaki, H., Ohkawa, J., Eckner, R., Yao, T. P., Taira, K., Chiu, R., Livingston D. M., and Yokoyama, K. K., 1998, *Nature,* 393: 284-289.
13. Pley, H. W., Flaherty, K. M., and McKay, D. B., 1994, *Nature,* 372: 68-74.
14. Sussman, J. L., Holbrook, S. R., Warrant, R. W., Church, G. M., and Kim, S. H., 1978, *J. Mol. Biol.,* 123: 607-630.
15. Kuwabara, T., Amontov, S. V., Warashina, M., Ohkawa, J., and Taira, K, 1996, *Nucleic Acids Res.,* 24: 2302-2310.
16. Koseki, S., Ohkawa, J., Yamamoto, R., Takebe, Y., and Taira, K., 1998, *J. Control., Release,* 53: 159-173.
17. Takebe, Y., Seiki, M., Fujisawa, J., Hoy, P., Yokota, K., Arai, K., Yoshida, M., and Arai, N., 1988, *Mol. Cell. Biol,.* 8: 466-472.
18. Kuwabara, T., Warashina, M., Nakayama, A., Ohkawa, J., and Taira, K., 1999, *Proc. Natl. Acad. Sci. USA,* 96: 1886-1891.
19. Nowell P. C., and Hungerford, D. A., 1960, *Science,* 132: 1497-1499.
20. Rowley, J. D. 1973, *Nature,* 243: 290-293.
21. Kuwabara, T., Warashina, M., Tanabe, T., Tani, K., Asano, S., and Taira, K., 1998, *Mol. Cell,* 2: 617-627.
22. Dubrez, L., Eymin, B., Sordet, O., Droin, N., Turhan, A. G., and Solary, E., 1998, *Blood,* 7: 2415-2422.

INDEX

Index